# Socially Responsible AI Deployment in Public Systems

by

ant

**Table of Contents**

# Chapter 1: The Dawn of AI in Public Service

The public sector has long been on a journey of technological evolution, a path marked by a consistent, though often uneven, quest to harness innovation for the betterment of its citizens. From the clunky, room-sized mainframes of the mid-20th century that first automated basic record-keeping, to the widespread adoption of personal computers that brought digital capabilities to individual desks, and the subsequent explosion of the internet and mobile technologies that revolutionized communication and service delivery, each era has brought new tools and new possibilities. Artificial Intelligence, in this historical context, is not a disruptive anomaly, but rather the latest, and arguably most profound, chapter in this ongoing narrative of public service transformation. Understanding this trajectory is crucial for framing AI's current impact and its future potential, allowing us to learn from past successes and failures as we navigate this new frontier.

Early forays into computing within government agencies were often driven by the need for efficiency and the sheer volume of administrative tasks. Imagine the immense effort involved in managing census data, tax records, or social security benefits using purely manual systems. The introduction of early computers, while rudimentary by today's standards, represented a significant leap forward. These systems, though expensive and requiring specialized expertise, could process and store information at speeds previously unimaginable. They allowed for the automation of repetitive tasks, reducing human error and freeing up personnel for more complex duties. However, these early adoptions were also characterized by steep learning curves, high implementation costs, and a certain degree of skepticism from those accustomed to traditional methods. The lessons learned here were foundational: technology is a tool, and its effectiveness is deeply intertwined with the training, adaptation, and buy-in of the people who use it.

The advent of the personal computer and, subsequently, the widespread accessibility of the internet marked a more democratizing phase of technological adoption. Suddenly, digital tools were no longer confined to specialized data centers. Government departments could implement networked systems, enabling better internal communication and data sharing. The internet opened up new avenues for citizen interaction, initially through static websites providing information, and later through early online forms and portals for specific services. This era brought about a renewed focus on user experience, albeit often through a lens of functional utility rather than deep engagement. The challenges encountered included issues of digital divide - ensuring that access to these new technologies was equitable across all segments of the population - and the need for robust cybersecurity measures as more sensitive data moved online. The lesson here was potent: technological advancement must be paired with conscious efforts to ensure inclusivity and security. Without them, innovation risks exacerbating existing inequalities.

The mobile revolution further accelerated this trend, making digital access ubiquitous for a significant portion of the population. Public services began to explore mobile applications for citizen engagement, such as transit information apps, emergency alert systems, and platforms for reporting local issues. This shift highlighted the growing expectation that public services should be accessible anytime, anywhere, and on any device. The focus moved towards more dynamic, personalized interactions, and the potential for real-time data collection and analysis became more apparent. Yet, this period also underscored the challenges of maintaining privacy in an increasingly connected world, the complexities of managing a diverse range of devices and operating systems, and the critical need for accessible design that caters to users with varying abilities and digital literacy levels. The insight gained was clear: as technology becomes more embedded in daily life, its integration into public services must be thoughtful, secure, and universally accessible.

Each of these historical phases has contributed valuable lessons that are directly relevant to the current adoption of Artificial Intelligence in the public sector. The initial resistance to automation in earlier computing eras mirrors some of the concerns about job displacement and the perceived dehumanization of services that are voiced today regarding AI. The struggle to bridge the digital divide with internet and mobile technologies provides a stark warning about the potential for AI to create new forms of exclusion if not implemented with equity at its core. The growing emphasis on user experience and citizen engagement in the digital age has paved the way for a more sophisticated understanding of how AI can be designed to be truly citizen-centric.

Furthermore, past technological transformations have invariably involved significant investments, requiring careful planning, procurement, and implementation strategies. Public sector organizations have developed expertise, albeit imperfect, in managing large-scale IT projects, dealing with vendors, and navigating regulatory landscapes. The lessons learned from these past endeavors – the importance of clear requirements, robust project management, ongoing training, and effective governance – are directly transferable to the deployment of AI systems. For instance, the challenges faced in ensuring interoperability between different government systems in the past provide a valuable precedent for understanding the complexities of integrating AI solutions into existing public service infrastructure. Similarly, the efforts made to secure government data against cyber threats in the internet age lay the groundwork for the even more critical security and privacy considerations that AI demands.

AI, however, represents a qualitative leap. Unlike previous technologies that primarily automated existing processes or improved communication channels, AI has the potential to augment decision-making, predict outcomes, and even generate novel content. This shift from automation to augmentation and prediction introduces a new set of opportunities and, critically, a new set of ethical and societal challenges. The long history of technology adoption in public services has shown that innovation is rarely linear or universally beneficial without deliberate effort. It has demonstrated that the most successful implementations are those that are not driven solely by technological possibility, but by a clear understanding of citizen needs, societal values, and a commitment to equitable outcomes.

The narrative of public technology adoption is one of continuous learning. Early mainframe implementations taught us about the importance of infrastructure and specialized skills. The PC revolution highlighted the need for widespread access and user-friendly interfaces. The internet and mobile era emphasized connectivity, real-time interaction, and the growing awareness of privacy. Each stage has refined our understanding of what it means to leverage technology for public good. AI, as the latest evolution, builds upon this legacy, but demands an even greater degree of foresight, ethical deliberation, and citizen involvement. It is not merely about implementing a new tool; it is about reimagining the very fabric of public service delivery, with the potential for profound positive impact, but also with significant risks if not approached with careful consideration of the lessons from our technological past. This historical perspective is therefore not just a prologue, but an essential lens through which to view the promises and perils of AI in the public sphere. It underscores that progress in public service technology has always been a journey of adaptation, learning, and, most importantly, a persistent effort to ensure that technology serves, rather than dictates, the public interest.

The preceding discussion has underscored the public sector's long and evolving relationship with technology, tracing a path from the earliest computing machines to the pervasive connectivity of the mobile age. Each phase has imparted crucial lessons regarding implementation, accessibility, and the fundamental need for technology to serve the public interest. Now, as we stand at the precipice of the Artificial Intelligence era, it is imperative to establish a clear and accessible understanding of what AI truly signifies within this distinct public service context. This is not about engaging with abstract theoretical constructs, but rather about grasping the tangible capabilities and potential implications of AI for the citizens and the systems that serve them.

At its core, Artificial Intelligence, or AI, refers to the development of computer systems capable of performing tasks that typically require human intelligence. These tasks can range from simple recognition and decision-making to complex problem-solving and creative generation. However, when we speak of AI in the public sector, we are not necessarily envisioning sentient machines or the science fiction portrayals that often dominate public discourse. Instead, we are referring to a suite of technologies designed to analyze vast amounts of data, identify patterns, make predictions, and automate or augment various governmental functions. The ultimate goal is to enhance efficiency, improve service delivery, and foster more responsive and evidence-based policymaking, all while upholding the core values of fairness, transparency, and accountability that are foundational to public service.

One of the most transformative aspects of AI relevant to public services is

**Machine Learning (ML)**. In essence, ML is a subset of AI that allows systems to learn from data without being explicitly programmed. Imagine a system designed to identify potential fraud in social benefit applications. Instead of a human programmer writing thousands of rules to detect suspicious activity, an ML system can be fed historical data of both legitimate and fraudulent claims. By analyzing this data, the system learns to recognize the subtle patterns and correlations that indicate fraud. This learning process is iterative; the more data the system processes, the more accurate and sophisticated its predictions become. In healthcare, ML algorithms can analyze medical images to detect early signs of diseases like cancer, or they can predict patient readmission rates, allowing hospitals to allocate resources more effectively and provide proactive care. For housing authorities, ML could analyze demographic trends and housing market data to predict areas of potential housing shortages, enabling proactive planning and intervention. This ability to learn and adapt makes ML a powerful tool for tackling complex, data-intensive challenges within the public sector.

Another critical AI capability is

**Natural Language Processing (NLP)**. NLP enables computers to understand, interpret, and generate human language. Think about the sheer volume of communication that public sector organizations engage in daily, from citizen inquiries received via email or phone, to policy documents, legislative texts, and public feedback forms. NLP can be used to power intelligent chatbots on government websites, capable of understanding citizen questions asked in plain language and providing immediate, accurate answers. This not only improves citizen access to information but also frees up human staff from repetitive queries, allowing them to focus on more complex cases. Furthermore, NLP can be employed to analyze large volumes of text data, such as public comments on proposed regulations, to identify prevailing sentiments, common concerns, and emerging themes. This can provide policymakers with a more nuanced and comprehensive understanding of public opinion than traditional survey methods alone. In the realm of justice, NLP can assist in sifting through vast quantities of legal documents to identify relevant case law or evidence, accelerating legal processes.

**Predictive Analytics** is another key component of AI that holds immense promise for public sector applications. This involves using historical data, statistical algorithms, and machine learning techniques to make predictions about future events or trends. For transportation departments, predictive analytics can forecast traffic congestion patterns, enabling dynamic traffic light adjustments to optimize flow and reduce travel times. It can also predict infrastructure failures, such as potholes or bridge weaknesses, allowing for timely maintenance and preventing more costly repairs or safety hazards. In public safety, predictive analytics can help forecast areas with a higher likelihood of certain types of crime based on historical data and socio-economic factors, enabling more efficient deployment of law enforcement resources. It's crucial to emphasize that predictive analytics is not about fortune-telling; it is about leveraging data to make informed estimations, which can then guide proactive decision-making and resource allocation.

Beyond these core technologies, AI encompasses a broader range of capabilities.
**Computer Vision**, for instance, allows machines to "see" and interpret images and videos. This can be applied in various ways, from monitoring public infrastructure for damage or unauthorized activity, to analyzing satellite imagery for environmental monitoring and disaster response. In urban planning, computer vision can assist in analyzing street-level imagery to assess the condition of sidewalks, identify missing street signs, or even map green spaces. **Robotic Process Automation (RPA)**,

while not always considered "intelligent" in the same vein as ML or NLP, often works in conjunction with AI. RPA uses software robots to automate routine, rule-based digital tasks that humans typically perform, such as data entry, form processing, or account reconciliation. When combined with AI capabilities like decision-making or pattern recognition, RPA can automate more complex workflows, further enhancing operational efficiency.

It is important to distinguish between different levels of AI. At a basic level, we have

**Narrow AI** (also known as Weak AI), which is designed and trained for a specific task. Most AI applications currently deployed in the public sector fall into this category. For example, an AI system designed to process tax returns is a narrow AI; it cannot suddenly start diagnosing medical conditions. The AI systems that recommend relevant public services to citizens based on their profile, or that flag potential anomalies in financial transactions, are all examples of narrow AI. This is in contrast to **General AI** (or Strong AI), which would possess human-like cognitive abilities and be able to understand, learn, and apply its intelligence to any intellectual task that a human being can. General AI remains a theoretical concept and is not currently a practical consideration for public sector implementation. Our focus must remain on the practical, deployable applications of narrow AI.

Furthermore, understanding AI in the public sector requires acknowledging the role of

**Data**. AI systems are fundamentally data-driven. Their performance, accuracy, and fairness are directly dependent on the quality, quantity, and representativeness of the data they are trained on and operate with. This means that the successful adoption of AI is inextricably linked to robust data governance, data management, and data privacy practices. Public sector organizations collect and manage vast amounts of sensitive personal data, and any AI initiative must prioritize the secure and ethical handling of this information. The insights derived from AI are only as good as the data fueling them, and critically, biased data can lead to biased outcomes, a topic we will explore in greater depth.

When considering practical applications, imagine an AI-powered system for a city's public works department. This system could ingest data from various sources: sensor data from smart bins indicating fill levels, historical waste collection routes, traffic patterns, and weather forecasts. Using machine learning, it could then optimize waste collection routes dynamically, ensuring that trucks are dispatched only when bins are nearing capacity and are routed along the most efficient paths, considering current traffic conditions. This would not only reduce operational costs and fuel consumption but also improve the cleanliness and liveability of the city.

In healthcare, AI can assist in public health surveillance by analyzing anonymized search queries, social media trends, and news reports to detect early signals of disease outbreaks, allowing public health officials to respond more rapidly and effectively. Similarly, AI can help identify individuals at high risk of developing chronic diseases based on their electronic health records and demographic information, enabling targeted preventative health campaigns and early interventions.

For educational institutions, AI can personalize learning experiences for students. By analyzing a student's learning pace, strengths, and weaknesses, AI-powered platforms can recommend tailored educational content, exercises, and support, ensuring that each student receives the most effective learning path. This can be particularly valuable in addressing learning disparities and ensuring that no student is left behind.

The application of AI extends to areas like disaster management. AI algorithms can analyze satellite imagery and sensor data to quickly assess the extent of damage after a natural disaster, identify areas requiring immediate assistance, and optimize the deployment of emergency response teams and resources. This can significantly improve the speed and effectiveness of relief efforts, saving lives and mitigating suffering.

Consider also the domain of citizen engagement and service access. AI-powered recommendation engines can help citizens navigate complex government services by suggesting the most relevant forms, benefits, or information based on their stated needs or profile. This can make interacting with government less daunting and more efficient, particularly for vulnerable populations who may struggle with bureaucratic processes. For instance, an AI assistant could guide a newly arrived immigrant through the necessary steps to access social services, healthcare, and housing support, streamlining a process that might otherwise be overwhelming.

The underlying principle across all these examples is the ability of AI to process information at a scale and speed far beyond human capacity, identify complex relationships, and derive actionable insights. It is about augmenting human capabilities, not replacing them entirely, in many instances. The "intelligence" in AI, in this context, is task-specific, pattern-recognizing, and data-driven. It represents a powerful new set of tools for public servants to better understand the challenges they face, to design more effective interventions, and to deliver services more efficiently and equitably. This foundational understanding of what AI is, and what it can practically do, is the essential first step in navigating its responsible adoption within the public sector. It is about demystifying the technology and focusing on its potential to create tangible improvements in the lives of citizens and the functioning of government.

The advent of Artificial Intelligence presents a compelling vision for the future of public service—one marked by enhanced efficiency, deeper insights, and ultimately, a more profound commitment to the public good. This transformative potential lies not in abstract notions of artificial sentience, but in the practical application of AI technologies to address complex societal challenges and to refine the very mechanisms through which governments serve their citizens. Across a spectrum of governmental functions, AI offers the promise of a more responsive, equitable, and effective public sector, poised to tackle issues that have long strained human capacity and traditional methods.

One of the most significant avenues through which AI can contribute to public good is by revolutionizing service delivery. Imagine a scenario where a citizen seeking assistance with a complex benefit application no longer faces a labyrinthine bureaucratic process. Instead, an AI-powered virtual assistant, leveraging natural language processing and access to relevant databases, could guide them through each step, answering questions in real-time, identifying missing documentation, and even pre-filling certain fields based on existing government records. This not only expedites access to essential support but also democratizes it, ensuring that individuals, regardless of their technical proficiency or familiarity with administrative procedures, can readily access the services they are entitled to. Early iterations of such systems are already appearing, demonstrating the capacity of AI to act as a more accessible and patient front-door to government services, thereby reducing frustration and improving overall citizen satisfaction.

Beyond direct service interaction, AI holds immense power in optimizing resource allocation for maximum societal benefit. Consider the challenge of public transportation. AI algorithms can analyze real-time traffic data, passenger demand patterns, and even event schedules to dynamically adjust bus and train routes, frequencies, and schedules. This means fewer empty buses running during off-peak hours and more frequent service during peak demand, leading to reduced operational costs for transit authorities, lower carbon emissions, and a more reliable and convenient experience for commuters. Similarly, in the realm of emergency services, AI can forecast demand for ambulances and fire services based on historical data, weather patterns, and even public event calendars. This foresight allows for proactive deployment of resources, ensuring that help arrives faster when and where it is needed most, a critical factor in saving lives and minimizing harm.

The capacity of AI to analyze vast datasets and uncover subtle patterns offers unprecedented opportunities for improving decision-making within public administration. For instance, in public health, AI can sift through anonymized electronic health records, environmental data, and even social media trends to identify emerging public health threats far earlier than traditional surveillance methods. This could enable public health agencies to implement preventative measures, deploy resources to at-risk communities, and contain outbreaks more effectively, ultimately saving lives and reducing the economic and social burden of disease. In urban planning, AI can process satellite imagery, census data, and infrastructure condition reports to predict areas most vulnerable to flooding, heatwaves, or other climate-related impacts. This predictive capability allows city planners to prioritize infrastructure investments, develop targeted resilience strategies, and protect vulnerable populations from environmental hazards.

Furthermore, AI offers a powerful lens through which to examine and promote equity within public services. Many existing systems, by their very nature, can perpetuate or even exacerbate historical inequalities. AI, when developed and deployed with a conscious effort to mitigate bias, can help identify and address these disparities. For example, AI algorithms can analyze patterns in loan applications, housing permits, or educational attainment data to pinpoint systemic biases that may be disadvantaging certain demographic groups. By identifying these patterns, policymakers can then take corrective action, whether it's revising eligibility criteria for social programs, allocating more resources to underserved neighborhoods, or developing targeted educational interventions. The key here is a commitment to ethical AI development, ensuring that the data used to train these systems is representative and that the algorithms themselves are designed to be fair and impartial.

The application of AI in environmental stewardship is another area brimming with promise for the public good. Governments are increasingly tasked with monitoring and protecting natural resources, combating climate change, and managing waste. AI can significantly enhance these efforts. Computer vision algorithms, for example, can analyze satellite and drone imagery to monitor deforestation, track illegal fishing activities, detect pollution sources, and assess the health of ecosystems with a precision and scale previously unattainable. In waste management, AI can optimize recycling processes by identifying and sorting different materials with greater accuracy, or it can predict waste generation patterns to improve collection efficiency and reduce landfill burden.

In the realm of public safety and justice, AI can contribute to a more just and efficient system, provided it is implemented with extreme caution and rigorous oversight. AI can assist law enforcement in analyzing vast amounts of evidence, such as security footage or communication records, to identify patterns and connections that might elude human investigators. Predictive analytics, when used responsibly and with a deep understanding of its limitations and potential for bias, could potentially help in allocating patrol resources to areas where crime is statistically more likely to occur, not to pre-emptively target individuals, but to enable a more proactive and visible presence where it is most needed. Crucially, any deployment of AI in this sensitive domain must be accompanied by robust safeguards to prevent profiling, ensure transparency in decision-making, and uphold fundamental rights.

The potential for AI to enhance citizen engagement is also noteworthy. AI-powered platforms can help governments solicit and analyze public feedback on proposed policies or services more effectively. Natural Language Processing can sift through thousands of public comments, identify recurring themes, gauge sentiment, and even categorize concerns, providing policymakers with a more comprehensive and nuanced understanding of public opinion than traditional methods. This can lead to more inclusive and better-informed policymaking, fostering greater trust and collaboration between citizens and their government. Furthermore, AI can be used to create more personalized and accessible government communication, tailoring information about services and civic processes to the specific needs and contexts of individual citizens.

Consider the operational efficiencies that AI can unlock within government agencies themselves. Tasks such as processing permits, managing invoices, or responding to routine inquiries can be automated using AI and Robotic Process Automation (RPA). This frees up public servants from repetitive, low-value work, allowing them to focus on more complex, strategic, and citizen-centric tasks that require human judgment, empathy, and creativity. This not only boosts productivity but can also lead to greater job satisfaction for government employees, as they are able to engage in more meaningful work.

The narrative of AI for public good is one of augmentation, not replacement. It is about empowering public servants with tools that amplify their capabilities, enabling them to serve citizens more effectively and equitably. The promise is to move from reactive problem-solving to proactive intervention, from generalized service delivery to personalized support, and from opaque decision-making to transparent, data-informed governance. This optimistic vision, however, is contingent on a deliberate and ethical approach to AI adoption. The successful realization of AI's potential for the public good hinges on a clear understanding of its limitations, a steadfast commitment to addressing bias, a robust framework for data governance and privacy, and an unwavering focus on ensuring that technology serves the fundamental values of public service: fairness, accountability, and the well-being of all citizens. The chapters that follow will explore these critical considerations in greater detail, building upon this foundation of AI's promising potential.

The integration of Artificial Intelligence into the fabric of public service is not merely a technological upgrade; it is a profound societal undertaking that necessitates a deep and unwavering ethical compass. As governments begin to harness the power of AI to transform service delivery, optimize resource allocation, and enhance decision-making, they concurrently assume a heightened level of responsibility. The unique nature of public service—its inherent mandate to serve and protect all citizens, to uphold democratic values, and to ensure equitable access to resources and opportunities—places AI deployment under an intense ethical spotlight. The decisions made today regarding the design, implementation, and governance of AI systems in the public sector will reverberate for generations, shaping the very relationship between citizens and their government. Therefore, the ethical imperative is not an optional add-on, nor a secondary concern to be addressed once technological hurdles are cleared. Instead, it must be recognized as the foundational bedrock upon which all AI initiatives in public service are built, guiding innovation toward outcomes that are not only efficient and effective but also just, equitable, and respectful of human dignity.

This ethical imperative stems from the fundamental principles that underpin legitimate governance. Public institutions are entrusted with significant power and resources, and their actions must be characterized by a commitment to fairness, impartiality, and the common good. When AI is introduced, especially in ways that directly interact with citizens or influence critical life decisions, these principles become even more salient. The potential for AI to automate decisions, predict behaviors, or allocate resources at scale means that any embedded biases, errors, or opaque logic can have disproportionately large and harmful consequences for individuals and communities. Imagine an AI system used to determine eligibility for housing assistance, to flag individuals for increased scrutiny in law enforcement, or to guide healthcare resource allocation. In each of these scenarios, the stakes are incredibly high, and the margin for error, particularly when it comes to fairness and equity, is exceedingly narrow.

At the heart of this ethical imperative lies the principle of fairness. In the context of AI in public service, fairness demands that all citizens be treated equitably, regardless of their background, socioeconomic status, race, gender, or any other protected characteristic. AI systems learn from data, and if that data reflects historical societal biases, the AI will invariably perpetuate and potentially amplify those biases. For example, if a historical dataset shows that certain neighborhoods have been disproportionately policed, an AI system trained on this data might flag those neighborhoods for increased surveillance, creating a feedback loop of over-policing that further disadvantages residents. Achieving fairness requires a proactive and diligent approach: meticulously auditing training data for bias, employing fairness-aware machine learning techniques, and rigorously testing AI outputs to ensure they do not systematically disadvantage any group. This is not a one-time fix but an ongoing process of vigilance and adaptation, recognizing that the definition of fairness itself can be complex and contested, requiring ongoing dialogue with affected communities to ensure AI aligns with societal values.

Transparency is another cornerstone of the ethical imperative. For citizens to trust the government and the technologies it employs, they must understand how decisions are being made, especially when those decisions affect their lives. This is particularly challenging with complex AI systems, such as deep learning models, which can operate as "black boxes," making it difficult to discern the precise reasoning behind a particular output. In public service, however, opacity is unacceptable. Where AI is used to inform or automate decisions impacting citizens, there must be a clear and understandable explanation of how the system works, what data it uses, and what factors influence its outcomes. This does not necessarily mean revealing proprietary algorithms in their entirety, but it does require providing meaningful insights into the logic, limitations, and potential risks of the system. For instance, when an AI denies a citizen a particular service or flags them for further review, the citizen should be able to receive a clear explanation of the reasons, and there must be a readily accessible mechanism for appeal and human review. This transparency builds accountability and empowers citizens to challenge decisions they believe are unfair or erroneous.

Accountability is inextricably linked to transparency and fairness. When an AI system makes a decision that leads to a negative outcome for a citizen—a wrongly denied benefit, a misidentified suspect, or an inequitable resource allocation—someone must be held responsible.

This responsibility cannot be abdicated to the machine. Establishing clear lines of accountability for AI systems in public service is crucial. This involves defining who is responsible for the development, deployment, monitoring, and oversight of these systems. It means ensuring that there are human beings in the loop who can override AI decisions, correct errors, and take responsibility for the overall performance of the system. Furthermore, accountability requires mechanisms for redress, allowing individuals who have been harmed by an AI system to seek recourse and compensation. Without robust accountability frameworks, the adoption of AI risks creating a diffusion of responsibility, where no single entity or individual can be held liable for the negative consequences, undermining public trust and the very notion of good governance.

The ethical imperative also demands a profound respect for human dignity. Public service is fundamentally about people, and the deployment of AI must always prioritize the well-being and inherent worth of every individual. This means ensuring that AI systems are designed to augment human capabilities and to enhance human experiences, rather than to dehumanize or degrade them. For example, while AI can automate routine tasks, it should not replace human interaction where empathy, compassion, and nuanced understanding are essential, such as in frontline social work or complex healthcare consultations. Moreover, the use of AI must never infringe upon fundamental human rights, including the right to privacy, the right to due process, and the right to be free from discrimination. The collection and use of citizen data for AI training and operation must be conducted with the utmost care, adhering to strict privacy regulations and ensuring that data is anonymized and secured to prevent misuse or breaches.

Furthermore, the ethical imperative requires a commitment to ongoing evaluation and adaptation. The impact of AI systems can evolve over time, and new ethical challenges may emerge as the technology matures and its applications expand. Therefore, public sector organizations must establish robust processes for continuous monitoring, auditing, and impact assessment of their AI deployments. This includes regularly reviewing system performance, assessing fairness metrics, gathering feedback from citizens and public servants, and being prepared to make necessary adjustments, recalibrations, or even decommissions of systems that prove to be detrimental or fall short of ethical standards. This iterative approach, grounded in a commitment to learning and improvement, is essential for ensuring that AI remains a tool for public good, aligned with societal values and serving the evolving needs of the populace.

The concept of the "digital divide" also takes on new ethical dimensions when AI is involved in public service. If AI-powered services become the primary mode of interaction with government, or if the benefits of AI-driven efficiencies are not equitably distributed, those who lack digital literacy, internet access, or the necessary technological infrastructure risk being further marginalized. The ethical obligation is to ensure that the adoption of AI does not create new barriers to accessing essential public services. This might involve maintaining traditional service channels, investing in digital literacy programs, or designing AI systems with universal design principles to be accessible to a wide range of users. The goal is to leverage AI to bridge existing divides, not to deepen them.

The imperative for ethical AI in public service also extends to the domain of public procurement and vendor relationships. Governments often rely on third-party vendors for AI solutions. It is crucial that public sector entities conduct thorough due diligence on these vendors, ensuring they adhere to ethical standards in AI development and data handling.

Contracts should include clauses that mandate transparency, accountability, and fairness in the AI systems provided. Public bodies must also resist the temptation to outsource ethical responsibility to vendors, maintaining ultimate oversight and accountability for the AI systems they deploy.

Finally, fostering a culture of ethical AI within public service is paramount. This involves comprehensive training for public servants on AI ethics, the responsible use of data, and the potential implications of AI technologies. It means encouraging open discussion about ethical dilemmas, establishing clear guidelines and codes of conduct, and empowering employees to raise concerns without fear of reprisal. An ethically-minded workforce is the most effective safeguard against the unintended negative consequences of AI, ensuring that technological advancement remains in service of public value and democratic principles. The ethical imperative, therefore, is not a static set of rules but a dynamic, evolving commitment that requires constant attention, critical reflection, and a deep understanding of the human-centric mission of public service.

This guide is meticulously crafted for public sector professionals—policymakers, IT leaders, service designers, program managers, and frontline staff—who are navigating the complex terrain of Artificial Intelligence adoption. Our aim is to demystify AI's capabilities and challenges within the unique context of public service, moving beyond abstract theoretical discussions to provide concrete, actionable strategies for responsible implementation. We recognize that public sector organizations operate under a distinct set of mandates and values, prioritizing equity, accountability, transparency, and the public good above all else. Therefore, this book focuses on how AI can be leveraged to *enhance* these core public service principles, rather than compromise them.

The scope of this guide encompasses the entire lifecycle of AI deployment in public service. We begin by establishing a foundational understanding of AI, its various forms, and its potential applications across different government functions, from optimizing infrastructure and public health responses to improving citizen engagement and administrative efficiency. However, our exploration quickly pivots to the critical ethical and practical considerations that must underpin any AI initiative. We delve into the nuances of data governance, addressing the paramount importance of data quality, privacy, and security. This includes examining how to curate and manage datasets responsibly, ensuring they are free from bias and representative of the diverse populations government serves.

A significant portion of this guide is dedicated to the principles of ethical AI implementation. This involves dissecting concepts such as fairness, accountability, and transparency in the context of public service AI. We will explore methodologies for detecting and mitigating bias in AI algorithms, ensuring that automated decision-making processes do not perpetuate or exacerbate existing societal inequities. Furthermore, we will provide practical frameworks for establishing robust accountability mechanisms, defining roles and responsibilities for AI systems, and ensuring that human oversight remains a critical component of any AI-driven process. The "black box" problem, a common challenge in AI, will be addressed with strategies for achieving meaningful transparency, enabling citizens and oversight bodies to understand how AI systems arrive at their conclusions, and facilitating avenues for redress when errors occur.

Our objectives extend to equipping readers with the practical skills and knowledge required to manage AI projects effectively and ethically. This includes guidance on developing clear procurement strategies for AI solutions, ensuring that vendor partnerships align with public sector values and ethical standards. We will explore best practices for pilot testing and iterative deployment, emphasizing the importance of continuous monitoring, evaluation, and adaptation of AI systems in response to real-world performance and evolving societal needs. This iterative approach is crucial for ensuring that AI remains a flexible and responsive tool that serves the public interest, rather than a rigid system that becomes obsolete or detrimental over time.

Moreover, this guide aims to foster a culture of ethical AI within public sector organizations. We understand that technological adoption is only as effective as the people who manage and utilize it. Therefore, we will offer insights into building capacity within government agencies, including strategies for training public servants on AI literacy, ethical considerations, and the responsible use of AI tools. We believe that empowering government employees with the knowledge and confidence to engage with AI critically is fundamental to its successful and ethical integration. This includes encouraging open dialogue about potential challenges and creating safe channels for employees to voice concerns or ethical dilemmas.

The scope of our coverage also extends to the critical issue of public trust. In an era where data-driven decision-making is becoming increasingly prevalent, maintaining and strengthening the trust citizens place in their government is paramount. This guide will explore how responsible AI implementation—characterized by fairness, transparency, and accountability—can serve as a catalyst for building and reinforcing this trust. Conversely, we will also address the potential risks to public trust that arise from poorly designed or ethically compromised AI systems, offering proactive strategies to mitigate these risks.

Furthermore, we are keenly aware of the diverse technological landscapes and resource constraints that public sector organizations face. This guide is designed to be adaptable, providing principles and frameworks that can be applied by agencies of all sizes and levels of technical maturity. We recognize that not all public sector entities will have the same capacity for AI development or deployment. Therefore, our advice emphasizes scalable solutions and practical approaches that can be tailored to specific organizational contexts, focusing on achieving meaningful impact with available resources. The objective is not to mandate cutting-edge, resource-intensive solutions, but rather to guide organizations toward the most appropriate and ethically sound AI applications for their particular needs and operational realities.

A core objective of this book is to demystify AI for public servants who may not have a technical background. We avoid overly technical jargon, opting instead for clear, accessible language that focuses on the implications and practical applications of AI technologies. The aim is to empower every reader with a sufficient understanding to engage in informed discussions, make sound decisions, and contribute to the responsible development and deployment of AI within their organizations. This includes explaining fundamental concepts like machine learning, natural language processing, and computer vision in ways that highlight their relevance to public service challenges and opportunities.

We also intend to address the evolving nature of AI and its governance. The field of AI is rapidly advancing, and the ethical and regulatory frameworks surrounding it are still developing. This guide is therefore framed not as a static set of rules, but as a dynamic resource that encourages continuous learning and adaptation. We will highlight emerging trends, potential future challenges, and the importance of staying abreast of best practices and evolving societal expectations regarding AI in government. The objective is to equip public sector professionals with the mindset and tools to navigate this evolving landscape proactively.

The scope of this guide includes examining the impact of AI on the public sector workforce. We will discuss how AI can automate certain tasks, augment human capabilities, and potentially create new roles, while also considering the implications for job displacement and the need for reskilling and upskilling initiatives. Our objective is to provide a balanced perspective on how AI can be integrated in a way that supports, rather than undermines, the public sector workforce, ensuring that human expertise and judgment remain central to public service delivery.

Ultimately, the overarching objective of this guide is to serve as a comprehensive resource that enables public sector organizations to harness the transformative potential of AI while upholding their fundamental commitment to serving the public interest. We aim to provide a roadmap that balances innovation with caution, efficiency with equity, and technological advancement with democratic values. By equipping public servants with the knowledge, tools, and ethical frameworks necessary for responsible AI implementation, we seek to ensure that AI becomes a powerful force for positive change in public service, leading to more effective, equitable, and trustworthy governance for all. This guide is intended to be a catalyst for thoughtful adoption, encouraging a measured, deliberate, and ethically grounded approach to integrating AI into the core functions of government.

## Chapter 2: Understanding Your Stakeholders

In embarking on the journey of integrating Artificial Intelligence (AI) into the fabric of public service, it is paramount to first recognize and meticulously map out the diverse constellation of individuals and groups who will be touched by, or have a stake in, these transformative deployments. This isn't a monolithic entity known as "the public"; rather, it is a vibrant, complex ecosystem with a multitude of interests, perspectives, and levels of engagement. A superficial understanding of stakeholders as simply "citizens" or "government" will invariably lead to blind spots, missed opportunities, and potentially detrimental outcomes. To truly foster inclusive development and responsible deployment, we must delve into the granular layers of this stakeholder landscape, acknowledging their varied interests and the potent influence they wield over the success or failure of any AI initiative within the public sector. This detailed comprehension is the bedrock upon which ethical, equitable, and effective AI implementation is built.

At the most fundamental level, we encounter the **citizenry**—the direct and indirect beneficiaries, and sometimes recipients, of public services. This group is far from homogenous. It encompasses individuals from all walks of life, each with unique needs, experiences, and levels of digital literacy. Consider, for instance, the elderly population who may rely on traditional service channels and possess limited familiarity with digital interfaces, or those with disabilities for whom AI-driven services must be designed with stringent accessibility standards in mind. There are also specific **vulnerable populations**—low-income households, recent immigrants, individuals with chronic health conditions, or those facing socioeconomic disadvantages—who are often disproportionately affected by the quality and accessibility of public services. AI systems, if not carefully designed, could exacerbate existing disparities for these groups, creating new barriers or reinforcing old ones. Conversely, AI holds immense potential to *improve* access and equity for these very populations if designed with their needs at the forefront. For example, AI-powered chatbots can offer 24/7 access to information in multiple languages, assisting those who cannot access help during standard business hours. Predictive analytics in public health could help identify at-risk communities for targeted interventions, preventing outbreaks before they become widespread. However, the data used to train such systems must be representative and free from historical biases that might unjustly flag certain communities as high-risk.

Beyond the broad category of individual citizens, we must also consider **community organizations and advocacy groups**. These entities often serve as crucial liaisons between specific populations and government agencies. They possess deep, nuanced understandings of the communities they represent, their challenges, and their aspirations. For example, a local tenants' rights group will have invaluable insights into the challenges faced by renters in accessing housing assistance or understanding complex housing regulations. A disability advocacy organization can highlight the critical need for AI systems to be developed with universal design principles, ensuring they are usable by individuals with a wide range of abilities. These groups can act as vital partners in co-designing AI solutions, providing essential feedback during pilot phases, and helping to disseminate information about new AI-powered services to their constituents. Their involvement is not merely consultative; it is instrumental in building trust and ensuring that AI deployments are perceived as legitimate and beneficial by the communities they are intended to serve. Engaging with them proactively can reveal potential pitfalls related to equity, accessibility, and cultural appropriateness that might otherwise be overlooked.

Within the government structure itself, the stakeholder landscape is equally intricate, extending far beyond a single department or agency. **Various government departments and agencies** are not just implementers but also key stakeholders with their own operational mandates, existing technological infrastructure, and established workflows. An AI system designed to streamline permit applications, for instance, will impact not only the planning department but also the building inspection unit, the finance department responsible for fee collection, and potentially the legal department if it involves interpreting regulations. Each of these departments may have different priorities, data silos, and levels of technical readiness. Their buy-in and cooperation are essential for the seamless integration of AI. Furthermore, **elected officials**—mayors, council members, legislators—are critical stakeholders. They are accountable to their constituents and are often responsible for approving budgets, setting policy priorities, and championing or opposing new technological initiatives. Their understanding of AI's potential and risks, and their perception of public sentiment towards it, will significantly influence the political feasibility of AI projects. Their support can unlock resources and provide the necessary political will, while their opposition can halt progress.

The
**public sector workforce** itself is a vital stakeholder group. This includes frontline staff who will directly interact with citizens using AI-enabled systems, as well as IT professionals, data scientists, program managers, and policy analysts who will be involved in the development, implementation, and oversight of these technologies. Frontline workers, in particular, possess invaluable practical knowledge about the day-to-day realities of service delivery and the nuances of citizen interactions. Their insights can identify practical challenges and opportunities for AI to genuinely improve service. Failure to involve them in the design process can lead to AI tools that are cumbersome, inefficient, or even counterproductive in practice. Simultaneously, investing in training and upskilling these employees is crucial. They need to understand how AI systems work, how to interpret their outputs, and how to intervene when necessary. This not only ensures the effective operation of AI but also addresses potential anxieties about job displacement and empowers

the workforce to leverage AI as a tool to enhance their own capabilities and job satisfaction.

The
**technology vendors and private sector partners** play a significant role in the public sector AI ecosystem. These are the companies that develop and supply AI solutions, platforms, and expertise. Their engagement is often necessary due to the specialized skills and resources required for AI development. However, their primary objective is typically profit, which may not always align perfectly with the public sector's mandate of serving the public good, ensuring equity, or maintaining stringent privacy standards. Therefore, the relationship with vendors must be managed carefully, with clear contractual agreements, robust oversight mechanisms, and a commitment to transparency. Public sector entities need to develop clear procurement strategies that prioritize ethical considerations, data ownership, and long-term value over short-term cost savings. Understanding the vendor's motivations, their track record in ethical AI, and their willingness to collaborate on co-design and ongoing support is crucial. This relationship is a partnership, but one where the public sector retains ultimate responsibility and control over the ethical application of the technology.

Crucially, we must not overlook the
**oversight bodies**, both internal and external, which play a vital role in ensuring accountability and adherence to ethical standards. This includes internal audit departments, ethics committees, and data protection officers within government agencies. Externally, it encompasses regulatory agencies, legislative oversight committees, and independent watchdog organizations. These bodies are tasked with ensuring that AI systems comply with existing laws and regulations, uphold democratic values, and protect citizens' rights. Their involvement is essential for building public trust, as they provide an independent

layer of assurance that AI is being used responsibly. Engaging with these oversight bodies early and often can help identify potential compliance issues, risk mitigation strategies, and ensure that AI deployments are designed with robust governance frameworks from the outset. Their role is not to impede progress, but to safeguard the public interest and maintain the integrity of public institutions.

Furthermore, in a networked governance landscape, other **intergovernmental bodies** and **research institutions** can also be considered stakeholders. Regional or national bodies might be involved in setting standards or sharing best practices for AI deployment across different jurisdictions. Universities and research institutions can provide valuable expertise, conduct independent evaluations, and contribute to the ethical discourse surrounding AI in public service. Collaborating with these entities can foster innovation, enhance the rigor of AI implementations, and ensure that public sector AI efforts are informed by the latest academic research and emerging best practices.

Understanding this multifaceted stakeholder environment requires more than just creating a list; it demands an ongoing process of engagement, dialogue, and active listening. Each stakeholder group brings a unique perspective, a set of expectations, and varying degrees of influence. For instance, a citizen group might prioritize accessibility and ease of use, while a finance department might focus on cost-effectiveness and ROI. An IT department might be concerned with technical integration and data security, while an elected official will be attuned to public perception and policy implications.

The effective identification and understanding of these diverse stakeholders lay the essential groundwork for developing AI strategies that are not only technologically sound but also socially responsible and politically viable. It is through this comprehensive mapping—

acknowledging the varied interests, potential influence, and unique contributions of each group—that we can begin to forge pathways for AI adoption that truly serve the public good, enhance trust, and promote equity in public service delivery. This detailed awareness moves us beyond a technology-centric approach towards a human-centric one, ensuring that AI is a tool that empowers and benefits all members of society, rather than a force that inadvertently marginalizes or disadvantages them. The complexity is not a deterrent, but an invitation to engage more thoughtfully, more inclusively, and ultimately, more effectively.

Understanding the intricate web of stakeholders is a crucial first step, but it is only the beginning. To effectively navigate this landscape and ensure that AI initiatives in the public sector are not only technically sound but also ethically grounded and genuinely beneficial, we must move beyond mere identification to a systematic process of **mapping influence and interest**. This analytical endeavor is what transforms a list of names and groups into a dynamic understanding of power, impact, and crucial dependencies. Without this mapping, even the best-intentioned AI projects can falter, encountering unexpected resistance, unintended consequences, or failing to achieve their desired positive outcomes.

The core of mapping influence and interest lies in discerning two fundamental dimensions for each stakeholder group: their **level of interest** in the AI initiative and their **degree of influence** over its success or failure. Interest refers to how much a stakeholder group cares about or will be affected by the AI project. Influence, on the other hand, speaks to their ability to shape its trajectory, whether through formal authority, public opinion, technical expertise, or resource control. By plotting these dimensions, we can create a strategic overview that informs engagement strategies and resource allocation.

One of the most widely recognized frameworks for this analysis is the **Power/Interest Grid**, often attributed to earlier strategic management literature but highly applicable to AI deployment in public service. This grid categorizes stakeholders into four quadrants based on their high or low levels of power (influence) and high or low levels of interest.

**High Power, High Interest (Key Players):** These are the stakeholders who have both significant influence and a strong interest in the AI project. They are the most critical to engage actively and collaboratively. For example, a department head responsible for budget allocation and operational oversight, who also sees the strategic benefits of an AI-powered efficiency tool, would fall into this category. Similarly, a powerful citizens' advocacy group representing a large, vocal segment of the population that will be directly impacted by a new AI-driven service would also be a key player. Their support can be instrumental, while their opposition can be a major impediment. Strategies here involve close management, ensuring their needs are met, and involving them deeply in decision-making processes. They should be considered partners in driving the project forward.

**High Power, Low Interest (Keep Satisfied):** These stakeholders possess significant influence but have a relatively low interest in the day-to-day details of the AI project. Their power stems from their position, authority, or control over resources, but they may not be intimately concerned with the specific technology or its immediate outcomes, as long as it doesn't disrupt their broader objectives or create significant problems. A senior executive or a legislative committee that approves broad policy directives but doesn't engage in the granular operational details might fit here. The goal is to keep them satisfied by addressing their overarching concerns, ensuring the project aligns with their strategic priorities, and avoiding actions that might pique their interest in a negative way. They need to be informed, but not necessarily consulted on every detail.

**Low Power, High Interest (Keep Informed):** This quadrant includes stakeholders who are highly interested in the AI project and will be significantly affected by it, but have limited direct influence. This often includes frontline staff who will use the AI system daily, or citizen groups representing those directly receiving the service. For instance, a community organization advocating for improved accessibility of public transit through an AI-powered routing system, but lacking formal decision-making power, would be in this category. Their interest is high because the AI will directly impact their lives or the lives of those they represent. While they may not have the power to unilaterally stop or approve the project, their engagement is crucial for understanding real-world impacts, identifying unintended consequences, and building buy-in at the grassroots level. Keeping them informed through regular updates, feedback mechanisms, and opportunities for input is vital for gathering valuable insights and fostering a sense of inclusion. Their collective voice, even if individually low in power, can influence public opinion and put pressure on higher-power stakeholders.

**Low Power, Low Interest (Monitor):** These stakeholders have both minimal influence and minimal interest in the AI project. They are unlikely to actively support or oppose the initiative, and their engagement is unlikely to be critical to its success. This might include a tangential government department with no direct involvement or a very small, unorganized segment of the public. The strategy here is simply to monitor them for any potential shifts in their interest or influence. Minimal effort is typically required, though it's prudent to remain aware of their existence in case circumstances change.

Applying the Power/Interest Grid requires careful assessment. "Power" can manifest in various forms: formal authority (e.g., budget control, regulatory power), expertise (e.g., technical knowledge, deep domain understanding), positional influence (e.g., leadership roles), or referent power (e.g., ability to mobilize public opinion or attract media attention). "Interest" can be gauged by the degree to which a stakeholder group's objectives, operations, or daily lives will be directly or indirectly impacted by the AI initiative. This is not a static assessment; stakeholder positions can evolve as a project progresses, so periodic re-evaluation is essential.

Beyond the Power/Interest Grid, other analytical lenses can enrich our understanding of stakeholder dynamics. The **Stakeholder Salience Model**, for instance, suggests that managers' attention and responses are directed towards stakeholders who possess one or more of three attributes: **power** (ability to impose their will), **legitimacy** (perceived validity of their claim), and **urgency** (need for immediate attention). Stakeholders with all three attributes (definitive stakeholders) demand the highest priority. Those with two attributes (expectant stakeholders) are also important, and those with only one (latent stakeholders) are less critical but still require monitoring.

For an AI project in the public sector, this means considering not only who can stop the project (power) but also who has a rightful claim to be heard (legitimacy) and whose concerns are time-sensitive or demand immediate action (urgency). A group with a legitimate claim, such as a community facing potential surveillance risks from an AI system, even if they lack overt power, should be prioritized due to the legitimacy of their concerns. Similarly, if an AI system is intended to address a critical public health crisis, the urgency of that crisis elevates the importance of all stakeholders involved in its management and impact.

Visual tools can be incredibly effective in synthesizing this information. Beyond simple grids, creating **stakeholder maps** can illustrate the relationships and interdependencies between different groups. These maps can depict direct lines of communication, influence pathways, and potential alliances or conflicts. For example, a map might show how a community organization (low power, high interest) can channel its concerns through a sympathetic elected official (high power, potentially medium interest) to influence a government department (high power, high interest). Understanding these indirect influence pathways is critical for effective strategy development.

Another useful approach is to analyze stakeholders based on their **role in the AI lifecycle**. Different groups will have varying levels of engagement and influence at different stages:

**Conception and Design:** At this nascent stage, identifying "visionary" stakeholders who can articulate the problem AI can solve, and "expert" stakeholders who understand the technical feasibility and ethical considerations, is paramount. This includes policy advisors, domain experts, and often, forward-thinking community leaders who can articulate future needs. Citizen groups may also be involved in defining desired outcomes and principles.

**Development and Testing:** During the technical build and pilot phases, the input of data scientists, IT professionals, vendors, and, crucially, end-users (including frontline staff and representatives of the affected citizenry) becomes essential. Their feedback ensures the AI is functional, accurate, and aligned with practical realities. Testing with diverse user groups helps uncover biases and accessibility issues.

**Deployment and Implementation:** This phase requires engagement with operational managers, frontline staff who will integrate the AI into their workflows, and citizens who will interact with the new service. Training and support for public sector employees become a major focus, as does clear communication and onboarding for citizens.

**Operation and Maintenance:** Ongoing engagement with IT support, system administrators, oversight bodies, and user feedback channels is necessary to ensure the AI continues to function effectively, ethically, and in line with evolving needs and regulations. This stage also involves monitoring for unintended consequences and adapting the system over time.

By overlaying the influence and interest analysis onto these lifecycle stages, we gain a more nuanced understanding of where and when specific stakeholders need to be engaged most intensely. For instance, a vendor might have high influence during development but their interest might wane after deployment unless contractual obligations remain. Conversely, a citizens' group's interest might be low during the technical design phase but spike dramatically during deployment and operation when the service becomes tangible.

Furthermore, it is vital to consider the **potential impact of the AI initiative on different stakeholder groups**. This involves not just identifying who is affected, but *how* they are affected. Are the impacts positive or negative? Are they direct or indirect? Are they immediate or long-term? Are they equitable or disproportionately distributed? For example, an AI system designed to optimize traffic flow might benefit commuters (positive, direct, immediate) but could negatively impact residents in newly created congestion zones (negative, direct, immediate) or lead to job losses for certain public transport workers (negative, indirect, medium-term). Analyzing these differential impacts helps to identify potential conflicts, equity concerns, and areas where mitigation or compensatory measures might be necessary. This depth of analysis moves beyond simply mapping power to understanding the ethical implications of that power and influence.

Consider the development of an AI-powered predictive policing system.

**High Power, High Interest:** Police leadership (formal authority, operational control), city council members (budgetary power, political accountability), and potentially, influential community leaders who have a vested interest in public safety and the methods used to achieve it.
**High Power, Low Interest:** State-level legislative bodies that set broader legal frameworks for policing but may not be involved in the specific technical implementation. They need to be satisfied that the system complies with state laws.
**Low Power, High Interest:** Community members, especially in historically over-policed neighborhoods, who will be directly affected by increased surveillance or targeted enforcement. Civil liberties organizations and advocacy groups focused on racial justice and privacy rights would also fall here, as they have a strong interest but often limited formal power to unilaterally halt the system's deployment.
**Low Power, Low Interest:** General taxpayers who are not directly engaged with the criminal justice system and may have only a passing awareness of the technology.

Mapping these groups would reveal that while police leadership holds significant power, the success and ethical legitimacy of the system would heavily depend on how the high-interest, low-power community groups and advocacy organizations perceive and react to it. Their ability to mobilize public opinion, raise legal challenges, or lobby sympathetic officials could turn them into significant, albeit indirect, influencers.

Another example could be an AI system for optimizing social service eligibility determination.

**High Power, High Interest:** The Director of Social Services (operational oversight, budget), the Lead IT Architect (technical feasibility, integration), and key legislative committees overseeing social welfare budgets.
**High Power, Low Interest:** The Mayor's office, provided the system improves efficiency and doesn't generate significant public backlash. Their interest is in overall city performance and public perception.
**Low Power, High Interest:** Social workers who will use the system daily, and critically, the diverse range of citizens applying for or receiving social services. This includes individuals with varying levels of digital literacy, language barriers, and specific accessibility needs (e.g., for disabled applicants, elderly applicants). Organizations representing these groups, such as senior centers or immigrant support services, would also have high interest.
**Low Power, Low Interest:** The municipal finance department, assuming the AI system is budgeted for and doesn't create accounting anomalies.

In this scenario, the high-interest, low-power groups (social workers and citizens) are absolutely critical. If the AI system is difficult to use, generates incorrect eligibility decisions, or is perceived as unfair, it can lead to significant service delivery failures, increased appeals, and erosion of public trust, thereby drawing the attention of higher-power stakeholders. Proactive engagement with these groups during design and pilot testing is paramount to ensure the AI serves its intended purpose equitably.

The process of mapping influence and interest is not a one-time exercise. It is an iterative and dynamic undertaking. As an AI project progresses, the influence and interest levels of various stakeholders can shift. New stakeholders might emerge, existing ones might become more or less engaged, and their priorities can change. Therefore, regular reassessment and updating of stakeholder maps are crucial to maintain an accurate understanding of the landscape and to adapt engagement strategies accordingly. This might involve periodic surveys, focus groups, informal check-ins, or formal reviews of stakeholder perceptions.

Ultimately, mapping influence and interest serves as a foundational tool for ethical AI implementation in the public sector. It moves beyond a simplistic view of "stakeholders" to a sophisticated understanding of power dynamics, dependencies, and the distribution of impact. By systematically analyzing who has power, who cares, and who is most affected, public sector organizations can proactively identify risks, foster collaboration, build trust, and design AI systems that are more likely to be successful, equitable, and truly serve the public good. This analytical rigor is indispensable for responsible innovation.

In the intricate process of understanding and engaging with stakeholders for public sector AI initiatives, a crucial and often overlooked step is the deliberate identification of communities that are historically marginalized or disproportionately vulnerable to systemic biases. These groups, frequently on the periphery of traditional power structures, may not self-identify as stakeholders in ways that are immediately apparent to project teams. Yet, they are precisely the populations most likely to experience the unintended, adverse consequences of AI systems if their unique needs, potential risks, and perspectives are not proactively sought out and integrated into the design and implementation phases. This subsection delves into the strategies and ethical imperative of ensuring these voices are not just heard, but are central to the AI development lifecycle.

Recognizing and reaching out to marginalized and vulnerable communities requires a fundamental shift from broad-stroke engagement to a deeply nuanced and targeted approach. It means moving beyond the assumption that all affected parties will naturally come forward or that standard communication channels will suffice. These communities may face barriers to participation, including digital divides, language differences, historical distrust of government institutions, lack of time or resources to engage, or the very invisibility that arises from being systematically overlooked. Therefore, identifying them is not a passive observation but an active, investigative endeavor.

One of the primary strategies for uncovering these communities is to engage with organizations and individuals who are already working with them. This involves identifying and partnering with community-based organizations (CBOs), non-governmental organizations (NGOs), faith-based groups, and local advocacy bodies that have established trust and deep understanding of the populations they serve. These entities act as vital bridges, providing insights into the specific challenges, concerns, and priorities of their constituents. For instance, an AI project aimed at improving public health access through a digital platform would need to engage with organizations that support the elderly, individuals with disabilities, low-income families, or specific ethnic or linguistic minorities, as these groups may face unique barriers to digital access or understand health information differently.

Furthermore, a thorough analysis of demographic data, disaggregated by race, ethnicity, income level, age, disability status, geographic location, and other relevant factors, can illuminate areas where specific populations are concentrated and may require special consideration. While data alone cannot reveal lived experiences, it can point towards communities that are statistically more likely to be vulnerable or to have faced historical discrimination, which can then inform more targeted qualitative outreach. Mapping these communities geographically and understanding their social service utilization patterns can also reveal potential touchpoints for engagement. For example, an AI system intended to manage social housing applications might disproportionately affect low-income families, single-parent households, or individuals experiencing homelessness, all of whom might be concentrated in specific neighborhoods or utilize particular support services.

When considering the potential risks that AI systems pose to these communities, it is imperative to think beyond technical failures and consider the amplification of existing societal inequalities. Algorithmic bias is a prime concern, where AI systems trained on data reflecting historical discrimination can perpetuate or even exacerbate unfair outcomes. This could manifest in various ways: an AI-powered hiring tool might unfairly screen out candidates from certain backgrounds; a facial recognition system used in public spaces could have higher error rates for individuals with darker skin tones, leading to wrongful identification; or an AI system for allocating public resources might inadvertently deprioritize neighborhoods with historically under-resourced populations. Identifying vulnerable communities involves understanding their specific experiences with systemic bias and anticipating how an AI system might interact with or amplify these existing vulnerabilities.

For example, in the context of justice system AI, communities that have historically been subject to over-policing and discriminatory practices are inherently vulnerable. An AI tool designed to predict crime hotspots or assess recidivism risk, if not rigorously audited for bias, could disproportionately target these same communities, leading to increased surveillance, arrests, and further entrenchment of systemic inequalities. Engaging with representatives from these communities, and not just relying on official crime statistics (which themselves can be biased), is crucial to understanding their lived realities and concerns about such technologies.

Similarly, in healthcare, an AI system for patient diagnosis or treatment recommendations must be examined for its potential to disadvantage certain demographic groups. If clinical trial data used to train an AI model predominantly features one racial or gender group, its diagnostic accuracy for other groups may be compromised. Communities that have historically faced barriers to accessing quality healthcare, or those whose specific health conditions are less represented in medical research, are particularly vulnerable. Targeted engagement with these communities would involve understanding their health beliefs, preferred communication styles, and any historical distrust of the medical establishment, ensuring the AI system is designed to be inclusive and equitable in its health outcomes.

The process of engagement itself must be adapted to the specific needs and contexts of vulnerable and marginalized communities. This often means moving away from one-size-fits-all approaches and embracing flexibility, cultural sensitivity, and humility. Instead of simply hosting public forums at convenient times and locations for the majority, initiatives might need to conduct outreach during community events, hold meetings in trusted local spaces, offer translation services, provide childcare, or compensate participants for their time and expertise. For communities with lower digital literacy, providing access to technology and training during engagement sessions might be necessary. The language used in communications and during engagement must be clear, accessible, and free of jargon, respecting diverse linguistic backgrounds and educational levels.

Furthermore, genuine engagement requires establishing trust, which can be a significant challenge when dealing with communities that have experienced historical harm or broken promises from institutions. This means being transparent about the AI project's goals, limitations, and potential impacts, and demonstrating a genuine commitment to incorporating feedback. It also involves actively listening and validating concerns, even if they are difficult to hear or challenge existing project assumptions. Simply presenting a fait accompli or treating feedback as a mere formality will not foster the deep engagement necessary to uncover and address vulnerabilities.

A key aspect of proactive identification is to consider the intersectionality of identities. Vulnerability is rarely a single-issue phenomenon. For example, an elderly individual who also belongs to an ethnic minority and has a disability may face a confluence of challenges that are not adequately addressed by engaging solely with senior advocacy groups or ethnic organizations. Understanding how multiple forms of marginalization intersect is vital for designing truly inclusive AI systems. This requires a nuanced understanding of social dynamics and a willingness to engage with diverse perspectives that reflect these intersecting identities.

In practice, this might involve co-design workshops where members of vulnerable communities work alongside AI developers and policymakers to conceptualize, design, and test AI solutions. This participatory approach empowers these communities, ensuring their lived experiences and values are embedded in the technology from its inception. For instance, if an AI system is being developed to manage urban planning and resource allocation, including residents from low-income neighborhoods or communities historically affected by environmental injustice in the design process can help ensure that the AI does not perpetuate patterns of neglect or inequitable development. Their input might highlight critical local needs or potential negative externalities that would be missed by those with less direct experience.

Another crucial strategy is to leverage the expertise of social scientists, ethicists, and community liaisons who specialize in working with marginalized populations. These professionals can provide invaluable guidance on culturally appropriate engagement methods, help interpret complex social dynamics, and act as trusted intermediaries between the project team and the community. Their involvement can help navigate potential misunderstandings and ensure that engagement efforts are sensitive and effective.

The identification of vulnerable and marginalized communities is not a static process but an ongoing commitment. As an AI initiative evolves, new stakeholders may emerge, and the needs and concerns of existing ones may change. Regular reassessment and continued dialogue are therefore essential to ensure that the project remains aligned with the principles of equity and inclusion. This involves establishing feedback loops that are accessible and responsive throughout the AI lifecycle, from initial conception and design through to deployment and ongoing monitoring.

Ultimately, the ethical imperative to identify and actively engage with vulnerable and marginalized communities stems from the fundamental principle of justice and the responsibility of public sector organizations to serve all members of society equitably. Failing to do so not only risks creating or exacerbating harm but also undermines the legitimacy and effectiveness of AI in public service. By prioritizing the voices and needs of those most often overlooked, public sector AI initiatives can move towards a future where technology is a tool for empowerment and inclusion, rather than a mechanism for reinforcing existing inequalities. This deliberate, targeted, and sustained engagement is not merely a best practice; it is a moral and practical necessity for responsible AI deployment.

Establishing robust and effective stakeholder engagement is not a one-size-fits-all endeavor, particularly within the complex landscape of public sector AI initiatives. Once stakeholders have been identified, the critical next step is to implement a suite of techniques designed to foster genuine dialogue, build trust, and ensure that a diverse range of voices can contribute to the AI lifecycle. This involves a strategic blend of broad outreach and targeted interventions, recognizing that different stakeholders will require different methods of interaction. The overarching goal is to create a continuous feedback loop, allowing for adaptation and improvement as the AI system evolves.

One of the foundational techniques for broad stakeholder engagement is the organization of public forums or town hall meetings. These events serve as valuable platforms for disseminating information about an AI project, explaining its objectives, potential benefits, and anticipated impacts, as well as gathering initial feedback from a wide audience. To maximize their effectiveness, careful consideration must be given to accessibility. This includes selecting venues that are physically accessible to individuals with disabilities, choosing times that accommodate working individuals and families, and offering childcare services where feasible. Furthermore, providing information in multiple languages and utilizing clear, accessible language, free from overly technical jargon, is paramount to ensuring inclusivity. For public sector AI initiatives, these forums can help demystify complex technologies and foster a sense of shared ownership and understanding within the community. For instance, a city exploring an AI-powered traffic management system might host public forums to explain how the system will collect data, how it will optimize traffic flow, and crucially, how it will address potential concerns like increased surveillance or impact on pedestrian safety. The feedback gathered might inform the specific algorithms used or the transparency measures implemented.

Complementing public forums, the establishment of stakeholder advisory committees or working groups offers a more structured and sustained approach to engagement. These committees typically comprise representatives from various stakeholder groups, including community leaders, subject matter experts, civil society organizations, and affected individuals. By bringing together a diverse set of perspectives, advisory committees can provide in-depth feedback on specific aspects of the AI system, such as data privacy, ethical considerations, fairness, and user experience. The key to a successful advisory committee lies in its composition and mandate. It should be representative of the identified stakeholder landscape, with mechanisms in place to ensure that marginalized or less vocal groups are adequately represented. The committee's mandate should be clearly defined, outlining its role in providing recommendations, reviewing designs, and offering ongoing guidance. For example, a government agency developing an AI tool for personalized learning in public schools might establish an advisory committee that includes teachers, parents, students, educational technologists, and child psychologists. This group could collaboratively shape the AI's pedagogical approach, review privacy safeguards for student data, and ensure the system is equitable across different socioeconomic backgrounds. The committee's ongoing work can provide a critical check and balance throughout the development process.

Focus groups represent another powerful technique for delving deeper into specific issues and understanding nuanced perspectives. Unlike public forums, focus groups typically involve smaller, more homogeneous groups of individuals who share certain characteristics or are likely to be affected by the AI system in similar ways. Facilitated by skilled moderators, these sessions allow for in-depth discussions on particular themes, such as user experience, potential biases, or communication strategies. For example, if an AI system is being designed to automate aspects of social service applications, focus groups might be convened with different segments of the applicant population – single parents, individuals with disabilities, or recent immigrants – to understand their specific challenges, concerns about the automated process, and preferred methods of communication. This qualitative data can reveal critical insights that might be missed in broader consultations. The careful selection of participants and the design of discussion guides are essential for eliciting rich and relevant information.

For stakeholders whose participation might be hindered by geographical distance, time constraints, or other logistical barriers, individual consultations offer a more personalized and accessible form of engagement. This can involve one-on-one meetings, interviews, or even online consultations tailored to the individual's availability and preferences. This approach is particularly valuable for engaging high-level officials, key decision-makers, or individuals with specialized knowledge who may not be able to attend group sessions. It also provides an opportunity to directly address the unique concerns of specific individuals or organizations. For instance, when developing an AI system for environmental monitoring, individual consultations with representatives from affected industries, environmental advocacy groups, and local indigenous communities can help gather diverse perspectives on data interpretation, regulatory compliance, and potential impacts on natural resources. Such tailored interactions can foster a sense of being valued and heard, building stronger relationships and more robust input.

Beyond these established methods, innovative approaches are increasingly being employed to enhance stakeholder engagement in public sector AI. Digital platforms and online engagement tools can broaden reach and facilitate continuous communication. This might include dedicated project websites with information portals, feedback forms, and even participatory design tools where stakeholders can contribute ideas or comment on prototypes. Online forums and moderated social media groups can also foster ongoing dialogue, allowing for questions to be answered and discussions to continue outside of scheduled meetings. However, it is crucial to ensure these digital tools are accessible to all, considering the digital divide and providing alternative methods for those who lack reliable internet access or digital literacy. A municipality implementing an AI-powered waste management system, for example, could use an online platform to share real-time data on collection schedules, allow residents to report issues, and gather feedback on proposed changes to recycling programs. This can be supplemented with physical information kiosks in community centers for those without digital access.

Co-design and participatory design methodologies represent a more profound level of engagement, where stakeholders are actively involved in the creative and decision-making processes of AI development. Instead of merely providing feedback on pre-defined options, participants collaborate with the project team to envision, design, and refine the AI system. This can take the form of workshops, design sprints, or hackathons where diverse groups work together to solve specific challenges. For example, a public health agency developing an AI-driven disease surveillance tool could involve clinicians, public health officials, and community health workers in a co-design process to ensure the tool is practical, user-friendly, and ethically sound. This collaborative approach not only leads to more effective and user-centered solutions but also builds a stronger sense of ownership and buy-in among stakeholders. It acknowledges that those with lived experience often possess invaluable insights into how technology can best serve community needs.

Gamification and simulation exercises can also be employed to make engagement more interactive and engaging, particularly for complex systems or for younger demographics. By turning feedback into a game or allowing participants to interact with a simulated version of the AI system, their understanding and involvement can be enhanced. For instance, an AI system designed to manage emergency response resources might use a simulation where participants, acting as emergency managers, have to allocate resources under various scenarios, providing feedback on the AI's recommendations and their own decision-making processes. This experiential learning can reveal practical challenges and preferences that might not emerge through traditional discussions.

Crucially, the techniques chosen must be tailored to the specific context and the nature of the stakeholders involved. What works for a group of technical experts may not work for a community organization, and what is effective for a large urban population might differ from engagement strategies for a rural community. A flexible and adaptive approach is therefore essential. This involves not only selecting the right methods but also consistently evaluating their effectiveness and being prepared to pivot when necessary. Regular check-ins with stakeholders, opportunities for informal feedback, and a commitment to transparency about how feedback is being used are vital for building and maintaining trust. This might involve establishing a clear process for responding to feedback, even if the feedback cannot be incorporated directly, explaining the rationale behind decisions.

Furthermore, building trust is a continuous process that underpins all engagement techniques. For public sector AI initiatives, where systems can have significant societal impact, trust is not merely desirable but essential for legitimacy and public acceptance. This involves demonstrating authenticity, consistency, and a genuine commitment to serving the public interest. Transparency about data sources, algorithmic processes (to the extent possible without compromising intellectual property or security), and decision-making criteria is paramount. When unintended consequences arise, swift and honest communication, coupled with a commitment to remediation, is critical for rebuilding trust. For example, if an AI system used for resource allocation is found to be inadvertently disadvantaging a particular neighborhood, an immediate public acknowledgment of the issue, a clear plan for investigation and correction, and a commitment to ongoing dialogue with the affected community are vital steps in addressing the breach of trust.

The commitment to engagement should extend throughout the entire lifecycle of the AI system, from initial conceptualization and design through to deployment, monitoring, and eventual decommissioning. Early engagement helps shape the direction of the project and avoid costly missteps. Ongoing engagement during development allows for course correction and refinement based on evolving needs and feedback. Post-deployment monitoring and feedback loops are crucial for identifying and addressing any emergent issues or biases, ensuring the system continues to serve its intended purpose equitably. For instance, an AI system designed to optimize public transportation routes might require continuous feedback from riders and transit operators to adapt to changing commuter patterns, infrastructure developments, or unforeseen disruptions. This iterative process of engagement ensures that the AI remains a responsive and beneficial tool for the public.

In essence, effective stakeholder engagement for public sector AI is a dynamic and multifaceted process that requires a strategic selection and skillful application of various techniques. It demands a commitment to inclusivity, accessibility, transparency, and continuous dialogue. By embracing a range of methods, from broad public forums to intimate co-design sessions, and by prioritizing the building of trust and the incorporation of feedback, public sector organizations can foster stronger relationships with their stakeholders and ensure that AI systems are developed and deployed in a manner that is both effective and ethically sound, ultimately serving the public good.

Building trust is not merely a desirable outcome of stakeholder engagement; it is the bedrock upon which all successful public sector AI initiatives are built. Without it, even the most meticulously designed engagement processes can falter, leaving stakeholders feeling unheard, undervalued, and ultimately, resistant. Fostering genuine partnerships requires a deliberate and sustained effort, moving beyond transactional data collection to cultivate relationships characterized by mutual respect, transparency, and a shared commitment to the public good. This involves recognizing that community members, civil servants, academics, and industry professionals each bring invaluable knowledge and perspectives that are essential for the ethical and effective deployment of AI.

One of the most potent strategies for building trust is through the demonstration of active, empathetic listening. This means more than just hearing words; it involves striving to understand the underlying concerns, aspirations, and lived experiences of stakeholders. When engaging with community members, for instance, it is crucial to approach conversations with humility, acknowledging that they possess a deep understanding of their local context, the existing challenges, and the potential unintended consequences of technological interventions. This requires moving away from a top-down approach where information is disseminated and feedback is passively received, towards a collaborative dialogue where all voices are genuinely considered and respected. For example, when a municipality is considering an AI-powered system for predictive policing, actively listening to the concerns of communities that have historically been over-policed is paramount. This might involve setting up listening sessions in community centers, employing facilitators who are trusted members of those communities, and dedicating time to understand their historical experiences with law enforcement and technology. The insights gained from such sessions might inform crucial decisions about the system's design, the data it utilizes, and the oversight mechanisms required to prevent bias and abuse. This deep listening is not about yielding to every demand, but about ensuring that the perspectives of those most affected are at the forefront of decision-making, thereby building a foundation of trust and legitimacy.

Beyond listening, establishing genuine partnerships necessitates a commitment to reciprocity. This means that the relationship should be mutually beneficial, with all parties feeling that they are contributing and gaining something valuable. In the context of public sector AI, this reciprocity can manifest in several ways. For community stakeholders, it can mean seeing their input directly shape the development and deployment of AI systems, leading to solutions that are more responsive to their needs and values. It can also mean gaining greater agency and understanding of how technology impacts their lives. For public sector agencies, the benefit lies in acquiring richer, more contextualized information, developing solutions that are more effective and sustainable, and building public buy-in that is essential for successful implementation. For example, a government agency developing an AI tool to streamline access to social services could partner with community organizations that serve vulnerable populations. The agency could provide training and resources to these organizations on AI literacy, helping them to better understand and advocate for their constituents. In return, these organizations can provide critical feedback on the AI tool's usability, accessibility, and potential for bias, ensuring it serves all users equitably. This symbiotic relationship, where knowledge and capacity are shared, fosters a sense of shared ownership and strengthens the partnership.

Demonstrating respect for community knowledge is a critical component of building trust. This involves recognizing that expertise is not solely confined to technical professionals or academic researchers. Local knowledge, embodied experience, and practical wisdom are equally valuable, and often indispensable, for understanding the complex social implications of AI. This respect should be evident in how stakeholder input is solicited, processed, and acted upon. It means valuing the time and effort stakeholders dedicate to participating in consultations, and ensuring that their contributions are acknowledged and meaningfully incorporated. For instance, when designing an AI system for urban planning and resource allocation, local residents often possess an intimate understanding of neighborhood dynamics, informal economies, and the historical evolution of their communities that might not be captured in official data sets. Engaging with these residents through participatory workshops or citizen science initiatives, where their local knowledge is explicitly sought and integrated into the AI's parameters or decision-making frameworks, demonstrates genuine respect. This could involve co-designing data collection methods that capture local nuances or developing AI algorithms that are informed by community-defined priorities. By valuing and incorporating this diverse knowledge, public sector organizations can move beyond superficial consultation and build deeper, more trusting relationships.

Transparency is another cornerstone of trust-building, particularly when dealing with AI systems that can be complex and opaque. Stakeholders need to understand not only what an AI system is intended to do, but also how it works, what data it uses, and what its limitations are. This transparency should be tailored to different audiences, with clear and accessible explanations provided to the general public, while more detailed information may be shared with technical experts or oversight bodies. When discussing algorithms, for example, it is important to move beyond jargon and provide understandable explanations of the underlying logic, potential biases, and decision-making processes. For an AI system used in the justice system, such as one for risk assessment, complete transparency about the factors influencing its recommendations, the data sources, and the potential for racial or socioeconomic bias is not just ethical, but essential for public acceptance. This might involve making code publicly available (where feasible), publishing regular impact assessments, and establishing independent review boards. Moreover, transparency extends to acknowledging uncertainties and potential risks. No AI system is perfect, and openly discussing potential failure modes or unintended consequences, along with plans for mitigation, can foster greater trust than an unrealistic portrayal of infallibility. For example, a public health agency deploying an AI system for disease outbreak prediction should be transparent about the model's accuracy rates, the conditions under which it might err, and the protocols for verifying its predictions, thereby building confidence in its responsible use.

Building reciprocal relationships also involves actively working to dismantle power imbalances that often exist between public institutions and the communities they serve. This means creating inclusive engagement processes that actively seek out and amplify the voices of marginalized or underrepresented groups. This might involve going to where people are, rather than expecting them to come to established institutions. It could mean holding meetings in community spaces, offering stipends for participation to compensate for lost wages or childcare costs, and utilizing communication channels that are accessible and familiar to different communities. For instance, when engaging with elderly populations about an AI system that manages public services, it is crucial to consider their potential digital literacy challenges and preferences. This might involve using traditional mail, telephone outreach, or in-person consultations in senior centers, rather than relying solely on online platforms. Furthermore, it involves a commitment to long-term engagement, recognizing that trust is not built overnight but through consistent, reliable interaction. This means establishing clear channels for ongoing communication and feedback, and demonstrating a commitment to continuous improvement based on stakeholder input. A public transit authority implementing an AI-powered route optimization system, for example, should establish a permanent feedback mechanism for riders and transit workers, and regularly report on how their input has influenced system adjustments, thereby fostering a lasting partnership.

Moreover, the process of building trust must be iterative and adaptive. Public sector organizations should be prepared to reflect on their engagement strategies, assess what is working and what is not, and make adjustments accordingly. This requires a willingness to be self-critical and to solicit honest feedback on the engagement process itself. For instance, after a series of public consultations, an agency might realize that certain demographic groups were underrepresented. Instead of simply continuing with the same approach, they should actively seek to understand why these groups were disengaged and modify their outreach methods to be more inclusive in the future. This could involve partnering with specific community leaders or organizations that have deep connections with those groups. This ongoing commitment to learning and improvement signals a genuine dedication to partnership and reinforces the idea that stakeholder input is valued and acted upon.

Finally, successful partnerships are cemented by a shared vision and a commitment to common goals. While stakeholders may have diverse interests, identifying overarching objectives that align with the public good can unite disparate groups. For public sector AI, this shared goal is often the improvement of public services, the enhancement of community well-being, or the responsible stewardship of public resources. When stakeholders understand that the AI initiative is driven by these fundamental aspirations, and that their input is instrumental in achieving them, they are more likely to invest in the partnership. For example, an AI system designed to manage disaster relief efforts should have a clear and universally understood goal: to save lives and mitigate suffering. By framing the engagement process around this shared objective, and by involving stakeholders in defining what effective disaster response looks like, a stronger, more cohesive partnership can be forged. This alignment ensures that the technology serves a purpose that resonates with the values and priorities of the community, thereby building not just trust, but also a sense of collective responsibility and shared success.

# Chapter 3: Integrating Participatory Design Principles

Participatory design represents a fundamental shift in how we conceive, create, and deploy technology, moving beyond a model where experts unilaterally develop solutions for others. At its core, it is a philosophy and a practical approach that champions the active involvement of end-users, stakeholders, and communities directly impacted by a system in its entire lifecycle – from initial conceptualization and problem definition through to design, development, testing, and ongoing iteration. It's about democratizing the design process, recognizing that those who will live with, work with, or be affected by a technology possess invaluable insights that are often overlooked in traditional top-down development. This approach is rooted in the belief that co-creation leads to more effective, equitable, and desirable outcomes, fostering a sense of ownership and ensuring that technology serves genuine human needs and societal values.

The relevance of participatory design to Artificial Intelligence (AI) cannot be overstated. AI systems, by their very nature, are distinct from many other technologies. They are often characterized by complexity, opacity, and the capacity for autonomous decision-making, which can lead to profound and sometimes unforeseen consequences. Unlike a static piece of software, AI systems learn, adapt, and evolve, frequently based on vast datasets that can themselves embed societal biases. When these systems are deployed in public sector contexts – whether for optimizing resource allocation, informing judicial decisions, predicting public health trends, or enhancing law enforcement capabilities – their impact can ripple across entire communities, affecting fundamental rights, access to services, and the very fabric of public life. Without the direct input of those who will be most affected, AI systems risk perpetuating existing inequalities, creating new forms of discrimination, or simply failing to meet the needs they are intended to address because their underlying assumptions do not align with the lived realities of their users.

Consider an AI system designed to allocate public housing. A purely technocratic approach might focus on optimizing for efficiency metrics like minimizing vacancy rates or administrative costs. However, if the community members who are experiencing housing insecurity, who understand the complex web of social determinants of homelessness, and who are familiar with the specific challenges of different neighborhoods are not involved in the design process, the system could inadvertently create new barriers. For instance, it might prioritize certain types of documentation that are inaccessible to individuals without stable addresses or consistent internet access, or it might overlook the crucial role of informal community networks in supporting vulnerable populations. Participatory design, in this scenario, would involve bringing together potential recipients of housing, housing advocates, social workers, and city planners to collectively define what constitutes "fair" and "effective" allocation. This might lead to the development of AI parameters that account for familial support structures, proximity to essential services that are valued by the community, or even incorporate qualitative assessments alongside quantitative data, ensuring the system is not just efficient but also humane and equitable.

The shift inherent in participatory design is from designing *for* people to designing *with* people. This is a crucial distinction. Designing *for* people often implies a benevolent but detached stance, where developers, armed with their expertise, attempt to anticipate and fulfill the needs of an assumed user. It can lead to well-intentioned but ultimately flawed solutions because the designers lack the deep, contextual understanding that comes from lived experience. Designing *with* people, on the other hand, is an invitation to partnership. It acknowledges that expertise is distributed and that the most profound knowledge about the problems and potential solutions resides within the communities themselves. This collaborative ethos requires a commitment to humility, active listening, and a willingness to cede control to some extent, allowing the process to be shaped by the collective intelligence of the participants.

In the realm of AI, this collaborative approach is particularly vital due to the potential for unintended consequences. AI algorithms learn from data, and if that data reflects historical biases—whether racial, gender, socioeconomic, or otherwise—the AI will amplify those biases. For example, an AI system trained on historical loan application data might learn to associate certain demographic markers with a higher risk of default, not because of inherent traits, but because of systemic discrimination in past lending practices. Without the input of community members who understand these historical injustices and their ongoing effects, an AI designed to streamline loan approvals could end up reinforcing those same discriminatory patterns, shutting out deserving individuals and perpetuating economic inequality. A participatory design process, however, would engage individuals from affected communities to scrutinize the data, question the assumptions embedded in the algorithms, and help define what constitutes a "fair" assessment of creditworthiness, potentially leading to the development of more equitable lending models. This could involve identifying proxies for creditworthiness that are less susceptible to historical bias, or developing mechanisms for human review that are explicitly designed to counteract algorithmic predispositions.

The methods employed in participatory design are diverse and adaptable, often drawing from fields like human-computer interaction, community organizing, and action research. They include techniques such as co-design workshops, user journey mapping, focus groups, ethnographic studies, citizen juries, and scenario planning, all adapted to foster genuine dialogue and shared decision-making. What unites these methods is their emphasis on empowering participants, providing them with the tools and information necessary to contribute meaningfully, and ensuring that their contributions have a tangible impact on the final design. For instance, in a co-design workshop for an AI-powered public safety application, community members might be given simplified representations of algorithmic logic, or tasked with sketching out desired features and functionalities that address their specific safety concerns. They might also be involved in generating and labeling data that represents real-world scenarios, thereby embedding their understanding of nuance and context directly into the system's training.

Furthermore, participatory design is not a one-off event; it is an ongoing commitment that extends throughout the lifecycle of an AI system. It acknowledges that needs evolve, that systems can drift from their intended purpose, and that unforeseen challenges can emerge over time. Therefore, continuous engagement and feedback loops are essential. This might involve establishing community advisory boards for AI systems, creating accessible channels for reporting issues or suggesting improvements, and conducting regular impact assessments in collaboration with affected communities. When an AI system for traffic management is deployed, for example, ongoing dialogue with residents, local businesses, and transit operators can help identify new traffic patterns, unexpected congestion points, or accessibility issues that the original design did not anticipate. This continuous feedback allows for agile adjustments and ensures that the AI system remains responsive and beneficial to the community it serves.

The importance of participatory design for AI in the public sector is amplified by the ethical imperatives at play. Public sector organizations have a duty to serve all citizens equitably, to uphold democratic values, and to ensure accountability. When AI systems are developed and deployed without public involvement, they risk undermining these very principles. Opaque algorithms can make it difficult for citizens to understand why a particular decision was made, hindering their ability to seek redress or challenge unfair outcomes. AI systems that are not designed with diverse user needs in mind can exacerbate existing digital divides, further marginalizing those who are already underserved. Participatory design offers a powerful antidote to these risks. By bringing diverse voices into the design process, it helps to ensure that AI systems are transparent, accountable, and aligned with public values. It allows for the proactive identification and mitigation of ethical risks, such as bias, discrimination, and privacy violations, before they become entrenched in deployed systems. For instance, engaging civil liberties advocates and privacy experts alongside community members in the design of an AI surveillance system can help establish robust safeguards against misuse and ensure that the system's deployment is proportionate and respects fundamental rights.

In essence, participatory design for AI in the public sector is about more than just improving user experience; it is about building legitimate, equitable, and effective public services in an increasingly technologically driven world. It is an investment in democratic governance, ensuring that the technologies that shape our lives are built by and for the people they are intended to serve. It recognizes that in the complex and often sensitive domain of public AI, the most robust and ethical solutions emerge not from isolated expertise, but from collective wisdom and shared commitment. This approach fosters trust, promotes innovation that truly addresses societal challenges, and ultimately leads to AI systems that are not only technically sound but also socially responsible and democratically legitimate.

The transition from a traditional, expert-driven model of technology development to one that embraces the active participation of end-users and affected communities is profound. Within this evolution, a critical distinction emerges: the difference between mere consultation and genuine co-creation. While consultation might involve seeking opinions or gathering feedback, it often remains a unidirectional process, with developers holding the reins of decision-making. Co-creation, conversely, signifies a fundamental shift towards partnership, where stakeholders are not just informants but active collaborators, integral to the very fabric of shaping the AI system. This journey from consultation to co-creation is not merely semantic; it represents a qualitative leap in how we can ensure that public sector AI is not only functional and efficient but also deeply aligned with the needs, values, and lived realities of the people it serves.

Consultation, in its most basic form, can be seen as an initial step, a way to gauge public sentiment or identify potential pain points. It might involve surveys, public forums, or even one-on-one interviews where developers ask specific questions about desired features or potential problems. For example, when designing an AI-powered system to manage public transport routes, a city council might consult with residents about their commute times, preferred modes of transport, and general satisfaction with current services. This feedback can inform the initial requirements gathering, highlighting areas where an AI solution might offer improvements. However, in a purely consultative model, the developers retain the ultimate authority to interpret this feedback and translate it into design decisions. The insights gathered might be valuable, but they are filtered through the existing understanding and biases of the development team. There is a risk that critical nuances, deeply embedded contextual knowledge, or alternative problem framings that are not explicitly solicited or easily articulated might be overlooked. The "why" behind certain user preferences, the historical context of service provision, or the subtle social dynamics of how people interact with public spaces might not surface in a way that fundamentally alters the design direction.

Consider a hypothetical scenario where an AI system is being developed to assist in the allocation of social welfare benefits. A purely consultative approach might involve asking recipients about their current challenges and what information they think is important for an aid worker to know. The feedback might highlight the need for faster processing times or more accessible application forms. While these are valid concerns, a consultative process might stop there, leading to an AI that prioritizes speed and administrative efficiency. However, if the process stops at consultation, it might miss the opportunity to understand the complex, often informal, support networks that exist within communities, or the specific barriers faced by individuals with disabilities, those experiencing domestic violence, or recent immigrants who may lack established documentation. Without deeper engagement, the AI might be designed to optimize for quantifiable data points, potentially disadvantaging those whose needs are best understood through qualitative assessment or relational trust. The system might, for instance, require digital signatures or online verification that are inaccessible to many, or it might prioritize applications based on easily verifiable metrics that fail to capture the full complexity of an individual's circumstances. The risk is that the system, while appearing efficient on paper, could inadvertently exclude or penalize the very individuals it is intended to support, perpetuating existing inequalities rather than alleviating them.

Co-creation, on the other hand, elevates stakeholders from passive recipients of information to active architects of the solution. It is a collaborative journey where the line between developer and user blurs, and the process of discovery and problem-solving is shared. This involves bringing together a diverse group of individuals – including end-users, community representatives, domain experts, ethicists, and technologists – to work together from the earliest stages of conceptualization through to implementation and iteration. In a co-creative model for the social welfare benefits AI, instead of simply asking for feedback, recipients would be invited to workshops to help define what "fairness" and "equity" mean in the context of benefit allocation. They might participate in scenario-building exercises, identifying edge cases and potential unintended consequences of different algorithmic approaches. They could be involved in designing the user interface, ensuring it is intuitive and accessible for people with varying levels of digital literacy. Crucially, they would have a voice in defining the parameters and ethical guidelines of the AI itself, helping to identify what data is relevant, how it should be interpreted, and what safeguards are necessary to prevent bias and discrimination.

This deeper level of involvement allows for the uncovering of needs and solutions that might never emerge from a purely consultative process. When community members are actively involved, they bring to the table lived experiences, contextual knowledge, and an understanding of social dynamics that are invaluable. For instance, in developing an AI system to manage urban park maintenance, a consultative approach might involve asking residents about their preferred types of flowers or the frequency of lawn mowing. A co-creative approach, however, would involve park users, horticulturalists, local environmental groups, and city planners in sessions where they collaboratively map out the park's ecosystem, discuss accessibility needs for different user groups (e.g., parents with strollers, individuals with mobility issues, dog walkers), and consider the biodiversity impact of different maintenance strategies. This could lead to an AI system that not only optimizes for aesthetic appeal but also promotes ecological health, ensures inclusivity, and reflects the community's multifaceted relationship with its green spaces. The AI might be trained to recognize and prioritize native plant species, adjust watering schedules based on microclimates identified by local users, or even incorporate feedback mechanisms for reporting damaged play equipment or identifying areas needing improved lighting, directly informed by the collective insights of those who use and care for the park.

The progression from consultation to co-creation is often marked by a shift in power dynamics and a willingness to embrace emergent, rather than predetermined, outcomes. In a consultative model, the project plan and objectives are typically set by the developers or the commissioning agency. Feedback is then sought within these pre-defined boundaries. In co-creation, the objectives themselves can be collectively defined and refined. Participants are empowered to question the initial problem framing and to propose entirely new avenues for exploration. This requires a significant commitment to flexibility and a willingness to adapt. For example, if during a co-design workshop for an AI tool to support small business loan applications, community entrepreneurs begin to articulate systemic barriers to capital access that go beyond the scope of loan eligibility, a co-creative process would allow for these broader issues to be explored. Instead of simply tweaking the AI's credit scoring algorithm, the group might decide to advocate for policy changes or develop complementary resources that address these root causes, with the AI design then evolving to support these broader goals.

The methods employed in co-creation are inherently more interactive and hands-on than those used in consultation. They often involve participatory workshops, design sprints, hackathons focused on societal challenges, citizen assemblies, and collaborative prototyping. In these settings, participants are given the tools and the space to actively contribute to the design. This might involve sketching out user interfaces, mapping out desired data flows, using simplified visual programming tools to define algorithmic logic, or even contributing to the labeling of training data in a way that reflects community values and understanding. For instance, when developing an AI system to predict and manage crowd flows at public events, co-creation might involve participants using physical modeling kits or digital simulation tools to visualize different crowd dynamics and then collaboratively designing interventions. They could work together to define what constitutes "safe" or "comfortable" crowd density, identify potential bottlenecks from their perspectives as event-goers, and co-design the alerts and guidance systems that the AI would generate. This ensures that the system's algorithms are grounded in real-world experience and designed to prioritize safety and positive attendee experience, rather than solely focusing on abstract flow metrics.

Moreover, co-creation recognizes that the design process is not linear but iterative. It anticipates that as an AI system is developed and potentially piloted, new insights will emerge, and the system's performance and impact will need to be continuously evaluated and refined in partnership with the community. This moves beyond a single feedback loop to establish ongoing collaborative relationships. For a public sector AI system, this could translate into establishing community oversight boards, creating ongoing feedback channels that are actively monitored and responded to by diverse stakeholders, and conducting regular participatory impact assessments. For an AI system designed to optimize waste management routes, for instance, co-creation would extend beyond the initial design phase. It would involve ongoing dialogue with sanitation workers, residents, and environmental advocates to assess the system's effectiveness in different neighborhoods, identify any emerging issues (like uncollected recyclables due to misclassification by the AI), and collaboratively develop adjustments to the algorithms or operational procedures. This continuous loop of involvement ensures that the AI system remains a living, evolving tool that is responsive to the community's changing needs and circumstances.

The ethical implications of moving from consultation to co-creation are substantial. Consultation can, at best, lead to incremental improvements or a superficial sense of public involvement. At worst, it can be used as a tokenistic exercise, providing a veneer of legitimacy without genuinely incorporating user perspectives. Co-creation, by its very nature, has the potential to embed ethical considerations directly into the design process. When individuals from affected communities are active partners, they can raise critical questions about bias, fairness, transparency, and accountability from the outset. They can help identify how data might be misinterpreted, how algorithms could perpetuate discrimination, or how the system's outputs might disproportionately impact vulnerable groups. For example, in the development of an AI-powered tool for predictive policing, a purely consultative approach might ask communities about their safety concerns and what types of crime they experience. A co-creative approach, however, would bring together residents, civil rights advocates, legal experts, and law enforcement to collaboratively define what constitutes "predictive policing" in a way that respects privacy, avoids profiling, and is demonstrably equitable. Participants could help scrutinize the data sources, question the historical biases embedded within them, and design mechanisms for human oversight and appeal that are robust and accessible. This collaborative scrutiny is far more likely to produce an AI system that is not only effective in its stated goals but also ethically sound and aligned with democratic values.

The difference can be starkly illustrated by considering the development of an AI system for personalized learning in public schools. A consultative approach might involve surveying teachers about their preferred digital tools or asking parents about their child's learning challenges. This feedback might lead to an AI that adapts the pace of instruction or offers additional practice exercises. However, a co-creative approach would bring together students, teachers, educational psychologists, curriculum developers, and parents to collectively define what constitutes "personalized learning." Students themselves could articulate how they best learn, what motivates them, and what forms of digital interaction feel supportive rather than intrusive. Teachers could share their insights into pedagogical best practices and classroom dynamics. This collaborative process could lead to an AI that does more than just adjust the speed of content delivery; it might foster critical thinking, encourage collaborative learning, identify learning styles that are often overlooked by standardized assessments, and ensure that the system's recommendations are pedagogically sound and ethically aligned with educational goals. The students, as active co-creators, would have a direct hand in shaping the tools that influence their learning journey, fostering a sense of agency and ownership over their education.

Ultimately, the shift from consultation to co-creation represents a commitment to a more democratic and equitable approach to technology development. It acknowledges that the most innovative and effective solutions are often born from the intersection of diverse perspectives and lived experiences. In the public sector, where technologies are intended to serve the common good, this deeper engagement is not just a best practice; it is an ethical imperative. It moves beyond simply gathering data *about* people to actively building solutions *with* them, fostering trust, building legitimate public services, and ensuring that the AI systems that shape our future are truly representative of and beneficial to all. This embrace of co-creation is essential for building public trust in AI and for

ensuring that public sector technology serves as a force for positive social change, rather than a perpetuator of existing inequities. It is about recognizing that the collective intelligence of communities, when harnessed through genuine partnership, holds the key to unlocking the true potential of AI for the public good.
In moving from abstract principles to tangible implementation, the practice of co-design demands a robust toolkit. These are not merely theoretical constructs but practical, hands-on methods that empower diverse stakeholders to actively shape the development of AI systems. The goal is to translate lived experiences and community needs into concrete design elements, fostering a shared understanding and ownership of the technology being created. This section delves into specific tools and methodologies that facilitate this collaborative process, illustrating how they can be adapted and applied within the unique context of public sector AI development.

One foundational methodology that proves invaluable in co-design is **journey mapping**. Traditionally used to visualize a user's experience with a product or service, in the AI context, it expands to encompass the entire lifecycle of interaction with an AI-driven system. This involves bringing together a diverse group of participants – perhaps citizens who will interact with a public service AI, the frontline workers who will use it, and the developers or policymakers overseeing it – to collectively map out the steps an individual takes when engaging with the service. This isn't just about documenting a current process; it's about envisioning a future state influenced by AI. For instance, if a municipality is developing an AI system to streamline the process of applying for building permits, a journey map co-design session might begin with participants sketching out the current, often arduous, steps involved: researching requirements, gathering documents, filling out forms, submitting applications, waiting for approvals, and dealing with potential rejections.

As the co-design progresses, the map evolves. Participants, armed with an understanding of the AI's potential capabilities, would then collaboratively overlay potential AI interventions. Where does an AI chatbot offer initial guidance and answer common questions, reducing the need for in-person visits? At what point might an AI-powered document analysis tool pre-emptively identify missing or incorrect information, saving applicants time and frustration? Crucially, the journey mapping process would also highlight potential pain points and ethical considerations introduced by the AI. Participants might identify moments where an AI's decision-making process feels opaque, where the system offers insufficient recourse for errors, or where the reliance on digital interfaces creates barriers for those with limited digital literacy or access. The visual representation of these interactions, collaboratively built, becomes a powerful diagnostic tool. It illuminates not just functional pathways but also moments of potential friction, anxiety, or inequity, providing clear visual cues for where design interventions are most needed. For example, a community member might point to the "decision-making" phase on the journey map and express concern that an AI algorithm's approval or rejection criteria might be biased against certain neighborhoods, leading to a collaborative discussion about transparency and appeal mechanisms that are integrated into that specific touchpoint.

Complementing journey mapping is the development of **personas**. Personas are semi-fictional representations of key user groups, crafted based on research and user data, but in co-design, they are brought to life and refined through collaborative workshops. Instead of developers creating personas in isolation, co-design involves community members contributing to and validating these profiles. This ensures that the personas are not stereotypes but nuanced reflections of real people with distinct needs, motivations, and technological proficiencies. For an AI system aimed at improving access to local government services, personas might represent different demographic groups: a digitally savvy young professional seeking quick online access to information, an elderly resident who relies on phone calls and in-person interactions, or a small business owner navigating complex regulations.

In a co-design workshop, participants might be asked to embody these personas. They could be presented with a design challenge – for example, how an AI system should handle a citizen's inquiry about a newly introduced waste collection schedule – and then tasked with responding from the perspective of their assigned persona. This exercise helps the development team understand how different individuals might perceive and interact with the AI. A persona representing someone with low digital literacy might express confusion or frustration if an AI chatbot uses technical jargon or offers only online self-service options. Conversely, a persona representing a busy parent might prioritize speed and convenience, valuing an AI that can quickly provide information via SMS or a voice interface. Through this role-playing and discussion, the team gains empathy and a deeper understanding of the diverse needs that the AI must address, leading to design decisions that are more inclusive and user-centered. For instance, if the persona of an elderly resident struggles with the chatbot's initial greeting, this immediately flags a need to rethink the onboarding experience for that persona, potentially introducing a simpler, more direct question or an option to connect with a human agent earlier in the interaction.

**Scenario planning** offers another potent co-design methodology, particularly effective for exploring the potential futures and unintended consequences of AI systems. This technique involves creating plausible narratives about how AI systems might be used or misused in specific contexts. In a co-design setting, participants collaboratively develop these scenarios, drawing on their diverse experiences and foresight. For a public sector AI aimed at optimizing traffic flow or managing public spaces, scenarios could explore: "What happens if the AI misinterprets a weather forecast and reroutes all traffic into a residential area during a storm?" or "How might an AI-powered facial recognition system in a public park be used to identify protestors, and what are the implications for freedom of assembly?"

These collaboratively generated scenarios serve as powerful thought experiments. They push the development team beyond standard use cases to consider edge cases, ethical dilemmas, and emergent behaviors. Participants can then collectively brainstorm mitigation strategies, design safeguards, or even pivot the AI's intended functionality based on the insights gained. For example, if a scenario highlights the risk of an AI system for benefit allocation inadvertently flagging individuals with specific medical conditions as potentially fraudulent due to historical data biases, the co-design team could then work together to design specific checks and balances, such as requiring human review for all flagged cases or developing a system that accounts for legitimate reasons for certain patterns in data. This proactive identification and resolution of potential issues before the AI is even fully built is a hallmark of effective co-design. It transforms abstract ethical concerns into concrete design requirements that are embedded into the system from its inception.

**Prototyping**, in its various forms, is the tangible manifestation of co-design ideas. It moves beyond paper sketches to create interactive mock-ups that allow participants to "test drive" potential AI functionalities. For public sector AI, this can range from low-fidelity paper prototypes to more sophisticated interactive digital mock-ups. For instance, when co-designing an AI-powered scheduling system for public health appointments, a low-fidelity prototype might involve participants using sticky notes to represent appointment slots and role-playing the interaction with a receptionist (simulating the AI). This allows for quick iteration and feedback on the basic flow and logic.

As the design progresses, higher-fidelity prototypes can be developed. These might be interactive wireframes or even basic functional prototypes that simulate key AI interactions. Imagine co-designing an AI tool to assist residents in reporting local infrastructure issues, like potholes or broken streetlights. Participants could interact with a prototype chatbot that asks them to describe the problem, upload a photo, and confirm the location. The co-design team can observe how participants navigate these steps, where they encounter confusion, and what language they use to describe issues. This direct interaction with a tangible representation of the AI allows for invaluable feedback. It helps identify whether the AI's natural language processing capabilities are likely to understand user inputs, whether the confirmation steps are clear and reassuring, and whether the overall user experience feels efficient and empowering. Crucially, these prototypes can also be used to test abstract concepts like fairness and transparency. For example, if an AI is designed to prioritize repair requests based on certain criteria, the prototype can be used to present participants with scenarios where different types of requests are prioritized, allowing them to discuss whether the outcomes feel equitable and to suggest adjustments to the underlying logic or the way these priorities are communicated.

Another powerful tool is
**card sorting**, often used in information architecture, which can be adapted for AI co-design, particularly for understanding how users categorize information or desired functionalities. If developing an AI that helps citizens navigate complex legal information or public services, participants could be given cards, each representing a piece of information, a service, or a potential AI function. They are then asked to

group these cards in a way that makes sense to them. This reveals users' mental models and helps structure the AI's knowledge base or interface. For example, if an AI aims to provide information on various social services, card sorting could reveal whether users naturally group services by need (e.g., housing, employment, healthcare) or by eligibility criteria, informing how the AI organizes its responses.

Furthermore,
**impact mapping** is a strategic co-design tool that extends beyond the immediate user interface to consider the broader societal impact of the AI system. This involves collaboratively defining the desired outcomes and then working backward to identify the actors, deliverables, and activities needed to achieve them. For a public sector AI, this could involve mapping out how an AI-driven public safety system aims to reduce crime rates (the outcome), who needs to be involved (police, community leaders, residents - the actors), what specific features or reports the AI needs to generate (the deliverables), and how these are developed and deployed (the activities). This holistic view, developed collaboratively, ensures that the AI is not just functional but also contributes positively to broader civic goals and that potential negative impacts are considered and addressed from the outset. For instance, an impact map for an AI designed to improve public school resource allocation might identify "equitable access to educational resources" as a key outcome. This could lead to discussions about how the AI can be designed to identify and mitigate disparities in funding or teacher allocation across different schools, involving community members in defining what "equitable" means in their specific context and ensuring the AI's metrics and algorithms reflect these shared values.

The collaborative nature of these tools is paramount. They are not simply techniques to be applied
*to* a community but methods to be employed *with* a community. The facilitators of these co-design sessions play a crucial role in creating an

environment of trust, respect, and open communication. This involves active listening, ensuring all voices are heard, managing potential power imbalances, and translating technical concepts into accessible language. The output of these methodologies – journey maps, refined personas, detailed scenarios, interactive prototypes, and impact maps – all converge to provide a rich, nuanced, and community-informed foundation for the design and development of public sector AI systems. They bridge the gap between abstract ideals of participatory design and the concrete reality of building AI that is truly for, and with, the people it serves.

To further illustrate the application of these tools, consider a hypothetical project involving the development of an AI-powered system to assist in managing local park maintenance and improvements in a mid-sized city. This system would aim to optimize resource allocation, predict maintenance needs, and potentially facilitate community input on park enhancements.

**1. Journey Mapping for Park Management AI:**
The co-design team comprises city park officials, maintenance crews, local environmental advocacy groups, and representative park users (e.g., parents, dog walkers, elderly residents).
**Current State Mapping:** The team begins by mapping the current journey of a park user reporting an issue (e.g., a broken swing set) or requesting an improvement (e.g., more benches). This involves steps like identifying the problem, finding the correct department contact (often difficult), filling out a form or making a call, waiting for a response, and the eventual resolution (or lack thereof). They also map the journey of a maintenance crew member receiving a work order, assessing the issue, scheduling the repair, and completing the task.

**Future State Co-Design:** The team then envisions how an AI system could improve these journeys. They might propose an AI chatbot accessible via the city's website or a dedicated park app.
**User Journey:** A user reports a broken swing: the AI guides them through providing details and location via text or voice, potentially analyzes an uploaded photo to identify the specific part needed, and provides an estimated timeframe for repair. The user receives updates. Potential pain points identified: what if the AI misunderstands the issue? What if the user doesn't have a smartphone? This leads to designing a clear escalation path to a human operator and ensuring phone-based reporting is also integrated.
**Maintenance Crew Journey:** The AI aggregates reported issues, prioritizing them based on severity (e.g., safety hazards over aesthetic concerns) and location efficiency. It can predict routine maintenance needs (e.g., lawn mowing based on weather and grass growth models). This might lead to the AI suggesting optimized routes for maintenance vehicles. Potential pain points identified: will the AI's predictions be accurate? What about unforeseen issues the AI can't predict? This leads to incorporating a mechanism for crew members to override AI suggestions and provide feedback on the accuracy of predictions, feeding back into the AI's learning.
**Ethical Considerations:** The journey mapping highlights opportunities for the AI to solicit community feedback on proposed park improvements (e.g., "Which new play equipment would you prefer?"). It also raises concerns about data privacy if the AI tracks park usage patterns for optimization. The team collectively decides on anonymizing usage data and ensuring that community input requests are transparently presented.

**2. Persona Development for Park Management AI:**

Based on initial discussions and demographic data, personas are developed:

**"Eco-Warrior Emily"**: A passionate environmentalist, concerned about native plant preservation and sustainable practices. She wants the AI to prioritize ecological health in maintenance and suggest eco-friendly improvements. She is digitally literate and active on community forums.
**"Busy Parent Ben"**: Primarily concerned with safety and accessibility for his young children. He wants quick reporting of hazards and assurances that play areas are well-maintained and suitable for all abilities. He uses a smartphone for convenience.
**"Senior Citizen Sophia"**: Values quiet contemplation and easy access to amenities like clean restrooms and accessible pathways. She prefers clear, simple instructions and may not be comfortable with advanced digital interfaces. She often communicates via phone or in person.
**"Maintenance Manager Mike"**: Focused on efficiency, budget constraints, and team safety. He needs the AI to provide reliable data for scheduling, resource allocation, and predicting equipment needs. He is practical and needs to see tangible benefits to his team's workflow.

During a workshop, participants role-play these personas. When faced with a decision about whether to plant more flowers or restore a natural habitat, "Emily" advocates for habitat restoration, while "Ben" might prioritize colorful, child-friendly plants. "Sophia" might emphasize the need for accessible seating near any new planting areas. "Mike" would be concerned about the cost and labor involved in each option. This dialogue helps the team balance competing needs and understand how different users will perceive and react to AI-driven recommendations for park improvements. It might lead to a design where the AI presents multiple options with clear trade-offs, allowing human decision-makers (or even citizens through voting mechanisms) to choose.

**3. Scenario Planning for Park Management AI:**
The team brainstorms potential scenarios:

**Scenario: AI-Driven "Beautification Blitz"**: The AI, analyzing peak visitor times and social media sentiment, proposes a large-scale, rapid planting of non-native, visually appealing flowers across all parks.
Co-design Discussion: "Emily" raises concerns about disrupting local ecosystems and the water usage of non-native species. "Ben" might worry that the quick planting displaces areas where children can play. "Mike" points out the significant cost and labor for such a rapid deployment. This leads to designing an AI that balances aesthetic appeal with ecological sustainability and suggests phased, localized improvements rather than sweeping changes. It also highlights the need for human oversight on large-scale aesthetic decisions.
**Scenario: AI Prioritizes Play Area Safety Over Natural Habitats**: The AI, designed to minimize liability, flags a mature, partially overgrown natural area in a park as a "potential hazard" and recommends extensive clearing and paving.
Co-design Discussion: "Emily" strongly objects, arguing that natural areas provide essential habitats and are valuable for educational purposes. "Sophia" might find the natural area a peaceful retreat. "Ben" might see it as a missed opportunity for a new playground. "Mike" calculates the cost of clearing. This scenario prompts the team to refine the AI's definition of "hazard" and to incorporate ecological and recreational value into its prioritization algorithms, ensuring that the AI doesn't oversimplify complex park environments.
**Scenario: Digital Divide in Reporting**: The AI app is highly effective for reporting issues, but a significant portion of the community, like "Sophia," does not use smartphones.
Co-design Discussion: This leads to reinforcing the importance of phone-in reporting options and potentially exploring community outreach programs to ensure all residents can utilize the system. It might also prompt the development of a simpler, web-based interface with larger fonts and clearer navigation.

**4. Prototyping for Park Management AI:**
**Low-Fidelity Paper Prototype:** The team sketches out the screens of a park reporting app. Participants are given paper cutouts representing buttons and text fields. They "drag and drop" these elements to simulate reporting an issue. For example, they might draw a "Report Issue" button, then a field for "Description," followed by a "Location" selector, and an "Upload Photo" button. This quickly reveals if the sequence of steps is logical or if any essential fields are missing.
**Interactive Digital Prototype:** A more refined prototype is built using tools like Figma or Adobe XD, simulating the AI chatbot interaction. Users can type queries like "Report broken bench" or "Request more trash cans." The prototype responds with pre-written text and options. The team observes how users interact, where they get stuck, and what kind of responses they find most helpful. They might test different ways the AI could ask clarifying questions: "Is the bench in the main picnic area or near the playground?" They can also simulate scenarios where the AI suggests optimal locations for new amenities based on usage data (visualized on a park map interface). This allows for testing the clarity of the AI's suggestions and the user's ability to provide feedback on them. For instance, the prototype might show the AI suggesting a new water fountain location, and users can then provide feedback like, "That spot gets too much direct sun," or "There's already one nearby."

These tools, when used collaboratively, transform the development of public sector AI from a top-down technical exercise into a dynamic, iterative, and community-driven process. They ensure that the AI systems being built are not only technologically sound but also deeply aligned with the values, needs, and realities of the people they are intended to serve, fostering trust and leading to more equitable and effective public services.

The journey toward integrating participatory design principles into public sector AI development is undeniably transformative, promising more equitable, effective, and trusted technologies. However, this path is not without its formidable obstacles. Recognizing and proactively addressing these challenges is as crucial as mastering the co-design methodologies themselves. Without this foresight, even the most well-intentioned participatory efforts can falter, leading to frustration, disillusionment, and ultimately, AI systems that fail to serve the diverse needs of the public.

One of the most pervasive and insidious challenges is the inherent **power imbalance** that often exists between different stakeholders. In public sector AI development, this imbalance can manifest in several ways. Technologists, policymakers, and government officials often hold positions of authority, possess specialized knowledge, and control resources. This can inadvertently lead to a dynamic where their perspectives, priorities, and language dominate the co-design process, marginalizing the voices of community members, frontline workers, or individuals with lived experience of the issues the AI aims to address. For instance, during a co-design workshop focused on an AI system for optimizing social service delivery, a government official might champion a data-driven approach emphasizing efficiency metrics, while a community advocate might highlight the crucial need for human empathy and personalized support, a perspective that can be easily overlooked if not actively amplified. The language used can also perpetuate these imbalances; technical jargon, bureaucratic terminology, and abstract policy language can create barriers for those unfamiliar with these domains, rendering their contributions less impactful.

Another significant hurdle is the **communication barrier**, which extends beyond mere language differences. It encompasses disparities in understanding, varying communication styles, and the difficulty of translating complex technical concepts into accessible terms for non-experts, and vice-versa. Imagine a co-design session for an AI system intended to assist in urban planning.

A data scientist might present complex visualizations of predictive modeling outputs, assuming a shared understanding of statistical significance and algorithmic bias. Meanwhile, residents might express their concerns in more qualitative, narrative terms, focusing on their daily experiences of their neighborhood, which can be challenging to integrate directly into algorithmic parameters without careful facilitation. This gap in understanding can lead to misinterpretations, missed opportunities for valuable input, and a sense of alienation for participants who feel their contributions are not being understood or valued. Effective facilitation is key to bridging these gaps, requiring mediators who can reframe technical concepts and help articulate community concerns in ways that are actionable for the development team.

**Time constraints and resource limitations** present practical, yet substantial, challenges to achieving inclusive participation. Public sector projects often operate under tight deadlines and with limited budgets. Comprehensive participatory design, which requires dedicated time for meaningful engagement, consultation, and iterative feedback, can be perceived as a luxury that slows down development. Organizing workshops, compensating participants for their time and expertise, and ensuring accessibility for all individuals (e.g., providing childcare, accessible venues, translation services) all require resources that may not be readily available. For example, a community member who relies on their income for daily living expenses may not be able to afford to take unpaid time off work to attend multiple co-design sessions. Similarly, a small non-profit organization representing a marginalized community might lack the staff capacity to dedicate to extensive participation in a large government AI project. These constraints can lead to superficial engagement, tokenistic consultation, or the exclusion of those who cannot afford to participate, undermining the very principles of inclusivity.

Beyond these immediate practicalities, there are deeper challenges related to
**trust and historical context**. Many communities, particularly those that have experienced systemic discrimination or been negatively impacted by past government initiatives, may harbor a deep-seated distrust of public institutions and the technologies they deploy. They may view participatory processes with skepticism, perceiving them as mere public relations exercises or as opportunities for their data to be exploited without genuine benefit. If an AI system is being developed for areas with a history of biased policing or discriminatory housing practices, residents might understandably be wary of participating in its design, fearing that it will simply automate and amplify existing injustices. Building trust is a slow, deliberate process that requires transparency, genuine commitment, and a demonstrable track record of acting on community feedback.

Finally, ensuring
**genuine accessibility and adaptability** of participatory methods is a significant ongoing challenge. "Inclusivity" itself is a multifaceted concept. It means not only bringing diverse people to the table but also ensuring they can meaningfully contribute. This involves considering a wide spectrum of needs: individuals with disabilities (visual, auditory, cognitive, mobility), people with varying levels of digital literacy, those who speak languages other than the dominant one, and individuals in different geographic locations or with different access to technology. A workshop designed for digitally fluent participants using laptops will exclude those who lack reliable internet access or digital devices. A presentation heavy on visual data might alienate visually impaired participants. A purely online engagement strategy will miss those who are not digitally connected. Adapting methods to suit these diverse needs and contexts is essential but requires significant planning, flexibility, and creativity.

Despite these formidable challenges, there are effective strategies that can be employed to foster truly inclusive participation in public sector AI development. The overarching goal is to shift from a model of consultation to one of genuine collaboration and co-creation, ensuring that all voices are not only heard but also actively shape the outcome.

To address **power imbalances**, a conscious and consistent effort must be made to democratize the co-design process. This begins with the framing of the project itself, emphasizing that the AI is being built *for* and *with* the community, not *on* them. Facilitators play a pivotal role here, actively managing group dynamics to ensure equitable speaking time and encouraging participation from quieter individuals. Techniques like "round-robin" sharing, where each participant has an uninterrupted turn to speak, can be invaluable. Furthermore, deliberately including participants who are often marginalized – such as individuals with disabilities, low-income residents, or members of ethnic minorities – is paramount. This might involve targeted outreach through community organizations that already have established trust with these groups. In co-design workshops, explicitly acknowledging and discussing power dynamics can also be helpful, creating a shared understanding of the goal to level the playing field. For example, in a session about an AI for job matching, a facilitator might explicitly state that the aim is to ensure the AI doesn't perpetuate biases against certain demographics, and invite participants to share their experiences and concerns related to hiring discrimination.

**Fair compensation** for participants is not merely a matter of courtesy; it is a fundamental strategy for ensuring equitable participation, especially when addressing time constraints and resource limitations. Community members and frontline workers often possess invaluable lived experience and expertise that is critical to the success of the AI system. Their time and knowledge should be recognized and compensated accordingly, treating them as valuable collaborators rather than unpaid consultants. This compensation can take various forms: hourly stipends reflecting their professional or opportunity costs, gift cards, or even covering expenses such as childcare, transportation, and meals during workshops. For example, if a co-design workshop for an AI-powered healthcare navigation tool requires participants to attend for an entire day, offering a stipend equivalent to a day's wages, along with covering their travel and providing lunch, can make participation feasible for individuals who might otherwise be unable to attend. This approach acknowledges that their input has economic value and helps remove financial barriers that disproportionately affect lower-income individuals.

Overcoming
**communication barriers** requires intentional design of engagement processes and skilled facilitation. This involves a commitment to using clear, accessible language, avoiding jargon, and actively seeking clarification when ambiguity arises. Visual aids, analogies, and storytelling can be powerful tools for explaining complex concepts. For instance, when explaining the concept of algorithmic bias in an AI system for loan applications, a facilitator might use the analogy of a biased recipe: if the original ingredients (data) are skewed, the final dish (the AI's decision) will also be skewed. Translating community members' qualitative feedback into actionable design requirements is equally important. This can be achieved through collaborative synthesis sessions where facilitators help the development team understand the underlying needs and values expressed in community narratives, rather than simply treating them as abstract complaints. Employing multiple communication channels – written, verbal, visual – and offering opportunities for participants to express themselves in ways that feel

most comfortable for them (e.g., through drawings, role-playing, or written notes) can also enhance understanding and participation.

To mitigate the challenges posed by **time constraints and resource limitations**, organizations must integrate participatory design as a core component of the project from its inception, rather than an add-on. This means allocating dedicated budget and staff time for engagement activities from the outset. Furthermore, flexibility in engagement methods is crucial. Not all participation needs to occur in lengthy, in-person workshops. Shorter, more focused online sessions, asynchronous feedback mechanisms (e.g., dedicated online forums, comment periods on design drafts), and distributed participation models (e.g., engaging community representatives who then report back to their constituents) can all contribute to inclusive engagement. For example, instead of a single, all-day workshop, a project might utilize a series of brief, themed online meetings over several weeks, allowing participants to engage when it is most convenient. Mobile-friendly platforms and easily accessible feedback forms can also facilitate participation for those on the go. Strategic partnerships with community organizations can also leverage existing resources and trust networks, reducing the burden on the core project team and ensuring more effective outreach.

Building and maintaining **trust** is paramount, especially when working with communities that have historical reasons for skepticism. Transparency is the cornerstone of trust-building. This means being open about the project's goals, the intended use of the AI, how data will be collected and protected, and importantly, how community feedback will be incorporated into the design and decision-making process. It is vital to be upfront about what is and is not possible, managing expectations realistically. Demonstrating a commitment to action is also critical. If community feedback leads to specific design changes, these should be clearly communicated back to the participants and the wider community. For example, after a co-design session on an AI system for park safety, if residents express concern about the AI potentially leading to over-policing in certain areas, the project team should publicly commit to implementing strict oversight mechanisms, anonymizing data where possible, and establishing a community advisory board to review the AI's performance. Regular updates, accessible progress reports, and establishing clear channels for ongoing dialogue and grievance redressal further strengthen trust.

Ensuring
**accessibility and adaptability** requires a proactive and person-centered approach. This means going beyond basic compliance and actively designing engagement processes to be inclusive of diverse needs. For individuals with visual impairments, this might involve providing screen reader-compatible materials, large-print documents, or verbal descriptions of visual content. For those with auditory impairments, offering sign language interpretation or written transcripts of audio content is essential. Cognitive accessibility can be supported by breaking down information into smaller chunks, using clear and simple language, and providing opportunities for repetition and clarification. Physical accessibility of venues is a baseline requirement, but also considering remote participation options with robust digital accessibility features is key. Digital literacy can be addressed by providing training and support on how to use online participation tools, or by offering alternative, non-digital participation methods. Ultimately, adaptability means being prepared to tailor the participatory approach to the specific needs and contexts of the community being engaged, rather than expecting the community to adapt to a pre-defined, one-size-fits-all process. This might involve holding sessions in community centers rather than sterile government offices, conducting outreach through trusted local leaders, or designing materials in multiple languages.

The strategies for overcoming challenges in participatory design are interconnected. Fair compensation enables participation despite time constraints; clear communication bridges power imbalances; and building trust facilitates willingness to engage despite historical concerns. By thoughtfully implementing these strategies, public sector organizations can move beyond superficial engagement to cultivate genuine, inclusive participation, leading to the development of AI systems that are not only technically sound but also ethically responsible and truly beneficial to all members of society. This ongoing commitment to refining participatory practices is not an endpoint but a continuous process of learning, adaptation, and a testament to the belief that the future of public sector AI must be shaped by the diverse hands and minds of the people it serves.

### Measuring the Impact of Participatory Approaches

The preceding discussions have underscored the transformative potential and the intricate challenges inherent in integrating participatory design principles into public sector AI development. Having navigated the complexities of power dynamics, communication barriers, resource limitations, historical trust deficits, and the imperative for accessibility, the logical and critical next step is to establish a robust framework for evaluating the efficacy and impact of these participatory endeavors. Simply implementing co-design processes without a mechanism to measure their outcomes risks perpetuating a cycle of well-intentioned but potentially misdirected effort. A rigorous evaluation approach is not merely an academic exercise; it is a pragmatic necessity for demonstrating accountability, justifying investment, fostering continuous improvement, and, most importantly, ensuring that participatory AI development genuinely leads to more equitable, effective, and trusted public services.

Establishing metrics for measuring the impact of participatory approaches requires a multifaceted perspective, moving beyond simplistic notions of "did people show up?" to a more nuanced understanding of the depth, quality, and tangible influence of their engagement. This evaluation should encompass several key dimensions: the quality of the engagement itself, the extent to which community input is integrated into the AI system's design and functionality, and the ultimate impact on the AI's performance and its reception by the intended users and the public.

One crucial area for measurement is the **quality of engagement**. This goes beyond mere participation numbers. It delves into how well the participatory process fostered genuine understanding, equitable contribution, and a sense of agency among all stakeholders. Metrics here could include assessing the diversity of participants relative to the target population, ensuring representation across various demographics, socioeconomic backgrounds, and levels of digital literacy. We need to ask: were the voices of those historically marginalized amplified and heard? Were the methods employed truly inclusive, accommodating different communication styles and needs? Qualitative data, gathered through post-engagement surveys, focus groups with participants, or observations by trained facilitators, can provide rich insights. For instance, participants might be asked to rate their satisfaction with the co-design process, their perceived ability to influence decisions, and their comfort level in expressing their views. Indicators of high-quality engagement might include a high percentage of participants reporting feeling respected, understood, and empowered to contribute. Conversely, low scores in these areas would signal a need to refine the participatory methods, perhaps by enhancing facilitation skills, diversifying outreach strategies, or adjusting the timing and format of engagement activities.

Another vital dimension for evaluation is the **incorporation of user feedback into the AI system's design**. This is where the tangible influence of participatory approaches can be most directly observed. Metrics should track how the insights, suggestions, and concerns raised by participants are translated into concrete design choices, feature developments, or modifications to the AI's algorithms and user interfaces. This can be operationalized by creating a feedback log that documents all input received during participatory sessions, categorizing it, and assigning responsibility for its consideration. The log should then track the decision made for each piece of feedback – whether it was incorporated, why it was not incorporated (with clear justification), or if it led to further investigation. For an AI system designed to assist with public transportation routing, for instance, community feedback might highlight the need for real-time updates on accessibility features for individuals with mobility challenges. Measuring the impact would involve tracking whether this feedback led to specific technical changes in the AI's data inputs or display of information, and whether these changes were documented and communicated back to the community. A successful outcome would see a significant percentage of actionable feedback being demonstrably integrated into the final product or clearly addressed with well-reasoned explanations for deviations. The depth of integration can also be assessed; for example, was feedback simply accommodated as a minor tweak, or did it lead to a fundamental rethinking of a core AI functionality?

Furthermore, it is essential to assess the **influence of community input on the AI system's performance and its impact on users**. This moves the evaluation from the process to the product and its real-world consequences. Metrics here can be both quantitative and qualitative, focusing on whether the AI system, shaped by participatory design, actually performs better, is more user-friendly, and addresses the intended societal needs more effectively. For an AI designed to streamline the application process for social housing, for example, impact could be measured by reduced application processing times, decreased error rates in applications, improved user satisfaction with the platform, and, critically, a greater sense of fairness and transparency reported by applicants. Participatory design, if successful, should lead to AI systems that are not only technically sound but also align with community values and priorities, thereby enhancing trust and adoption. Evaluating this aspect often requires post-deployment studies, user surveys, pilot testing with diverse groups, and analysis of system usage data. We might compare the performance of a participatory-designed AI system against a baseline or a conventionally designed system, looking for statistically significant improvements in key performance indicators. Qualitative data from users, such as testimonials or in-depth interviews, can illuminate whether the AI feels like a helpful tool or an alienating obstacle, and whether it has genuinely improved their access to public services.

The process of establishing these metrics should itself be participatory. The very stakeholders whose engagement is being evaluated can offer invaluable perspectives on what constitutes meaningful impact. This ensures that the evaluation framework is aligned with community priorities and captures the aspects of AI development that matter most to them. For example, when developing metrics for an AI system to support disaster response, community members might emphasize the importance of the AI's ability to provide culturally sensitive information or facilitate equitable resource distribution, rather than solely focusing on speed of deployment. Their input would then shape the evaluation criteria, ensuring that the success of the participatory process is measured not just by technical efficiency but also by social equity and community resilience.

Beyond specific metrics, a meta-level evaluation of the **long-term value and sustainability of participatory approaches** is also crucial. This involves assessing whether the participatory processes built capacity within the community and the public sector organization, fostered ongoing relationships, and established mechanisms for continuous feedback and iterative improvement. Did the participatory design efforts empower community members with new skills or knowledge about AI? Did they foster stronger partnerships between government agencies and community groups? Are there established channels for community members to provide ongoing input on the AI system's evolution, or was the engagement a one-off event? Measuring this long-term impact might involve tracking the recurrence of participatory engagement in subsequent projects, the number of community partners actively involved in AI governance, or the extent to which lessons learned from participatory processes are codified in organizational policies and training. A successful participatory AI initiative should ideally create a virtuous cycle, where early successes build trust and capacity for deeper, more impactful engagement in the future.

Ultimately, the measurement of participatory AI's impact is not a final judgment but an ongoing diagnostic tool. It provides essential feedback to refine strategies, allocate resources effectively, and demonstrate the profound value that intentional, inclusive co-creation brings to public sector AI. By diligently measuring and reflecting on these outcomes, we can move beyond the idealism of participatory design to its practical realization, ensuring that AI in the public sector truly serves the people it is intended to benefit, in ways that are both effective and ethically grounded. This iterative process of design, engagement, and evaluation is the bedrock of responsible and impactful public sector AI development.

# Chapter 4: Establishing Robust Equity Metrics

The introduction of Artificial Intelligence (AI) into public services heralds a new era of potential for efficiency, personalization, and expanded reach. However, this promise is inextricably linked to a profound ethical imperative: the demand for algorithmic fairness. This is not a peripheral consideration, a mere technical add-on, or a box to be ticked; it is a foundational requirement for any AI system intended to serve the public good. The very essence of public service is to act in the best interest of all citizens, to uphold principles of justice and equity, and to ensure that resources and opportunities are distributed fairly. When AI systems are deployed within this domain, they inherit this profound responsibility. Failure to ensure algorithmic fairness means that these powerful tools can inadvertently, or even deliberately, undermine these core public service values, creating new forms of discrimination or amplifying existing ones. The stakes are simply too high to allow for any ambiguity or compromise on this front.

Algorithmic fairness, at its core, is the principle that AI systems should not produce outcomes that are systematically biased against certain groups of people, particularly those defined by protected characteristics such as race, gender, age, disability, or socioeconomic status. It is about ensuring that the decisions made or influenced by algorithms are just, equitable, and do not disproportionately disadvantage or disadvantage specific populations. This is not a purely abstract philosophical concept; it has tangible, real-world consequences. Consider an AI system used in hiring for public sector jobs. If this algorithm, due to biases in its training data or its design, consistently ranks candidates from underrepresented backgrounds lower, it directly perpetuates employment inequality. Similarly, an AI used to determine eligibility for social benefits, if it unfairly penalizes individuals with certain communication patterns or who live in specific geographic areas, can push vulnerable populations further into hardship. The fairness of the algorithm, therefore, directly translates into the fairness of the public service itself.

The complexity of achieving algorithmic fairness arises from the multifaceted nature of "fairness" itself. There is no single, universally accepted mathematical definition that perfectly captures what it means for an algorithm to be fair in all contexts. Different fairness metrics often make competing claims, and optimizing for one can sometimes lead to a decrease in another. This means that practitioners must engage in careful, context-specific deliberation to determine which definitions of fairness are most relevant and critical for a particular public service application.

One prominent definition is
**demographic parity**, also known as statistical parity. This metric aims to ensure that the proportion of individuals receiving a positive outcome (e.g., being approved for a loan, being flagged as high-risk for a disease, being selected for an interview) is the same across different groups. For instance, if an AI is used to recommend candidates for a public scholarship, demographic parity would dictate that the proportion of scholarships awarded to men and women, or to different racial groups, should be identical to the proportion of those groups in the applicant pool. While intuitively appealing for its simplicity and focus on equal representation, demographic parity can be problematic. It does not account for differences in true underlying qualifications or risks. For example, if an AI is used to predict recidivism rates for individuals accused of crimes, enforcing demographic parity might mean that the system flags individuals from a certain group as high-risk at the same rate as individuals from another group, even if, based on other legitimate factors, one group genuinely has a higher propensity for reoffending. This could lead to unjustifiable disparities in sentencing or parole decisions.

Another important set of fairness metrics falls under the umbrella of **equality of opportunity**, which focuses on ensuring that individuals with similar underlying qualifications or true risk levels have an equal chance of receiving a positive outcome. This is often operationalized through metrics like **equalized odds** and **predictive parity**. **Equalized odds** requires that the true positive rate (also known as sensitivity or recall) and the false positive rate are the same across different groups. In simpler terms, it means that individuals who truly deserve a positive outcome are equally likely to receive it, regardless of their group, and individuals who do not deserve a positive outcome are equally unlikely to receive it, regardless of their group. For example, in a system that flags individuals for additional screening at a border, equalized odds would mean that individuals who are genuinely carrying contraband are equally likely to be flagged, and individuals who are not carrying contraband are equally unlikely to be falsely flagged, regardless of their nationality or ethnicity.

**Predictive parity**, on the other hand, focuses on the accuracy of the predictions themselves. It requires that the precision of the algorithm (the proportion of individuals predicted to have a positive outcome who actually do) is the same across different groups. This is particularly relevant when the absolute value of the prediction matters. Consider an AI used to diagnose a rare disease. Predictive parity would ensure that when the AI predicts a patient has the disease, the likelihood that they actually have it is the same, whether the patient is from demographic group A or demographic group B. This is distinct from equalized odds. For instance, an algorithm might have equalized odds but still fail predictive parity if it is overly conservative for one group, meaning it rarely flags them as positive, but when it does, it is highly accurate (high precision), while for another group it is more liberal with positive flags, leading to a lower precision overall.

The tension between these different metrics is profound. Imagine an AI system designed to allocate scarce public housing. If the goal is demographic parity, the system might ensure that the proportion of housing units allocated to different racial groups mirrors their proportion in the general population, even if one group has demonstrably higher need based on objective criteria. If the goal is equalized odds in predicting need, the system might be more accurate in identifying truly needy individuals across all groups, but this could lead to a demographic distribution of housing that doesn't match the population proportion. Which is "fairer"? This is a question that cannot be answered by mathematics alone; it requires societal deliberation, ethical judgment, and a clear understanding of the policy objectives and values at stake.

Furthermore, the data used to train these AI systems are often a primary source of bias. Historical data reflects the societal inequalities and discriminatory practices of the past. If an AI is trained on arrest records that disproportionately feature individuals from certain minority communities due to biased policing, the AI may learn to associate those communities with higher crime rates, perpetuating a cycle of discriminatory surveillance and intervention. Similarly, if loan application data shows that historically marginalized groups have been denied loans at higher rates, an AI trained on this data might learn to discriminate against applicants from those groups, even if their current financial profiles are strong. This is often referred to as "bias amplification," where the AI not only reflects existing biases but also magnifies them.

The concept of

**counterfactual fairness** offers another perspective, asking: would the outcome for an individual have been the same if their protected attribute (e.g., race, gender) had been different, all other causally relevant factors remaining the same? This requires a deep understanding of the causal relationships within the data and the decision-making process. It's an aspirational goal, often difficult to achieve in practice due to the challenge of isolating the causal impact of a single attribute while controlling for all other relevant variables that might be correlated with it.

The challenge is further compounded by the fact that different stakeholders will have different interpretations of fairness, influenced by their own experiences and values. Community members who have historically faced discrimination may prioritize outcomes that directly address systemic inequalities, even if it means deviating from purely statistical parity. Public administrators may be concerned with efficiency and the accurate allocation of resources. Technologists might focus on the mathematical properties of fairness metrics. Navigating these divergent perspectives requires open dialogue, transparency, and a commitment to democratic deliberation, ensuring that the definition of fairness adopted for a particular AI system reflects the consensus and values of the community it serves, rather than being solely dictated by technical considerations or the perspectives of a select few.

The practical implementation of algorithmic fairness in public sector AI also necessitates robust mechanisms for
**auditing and monitoring**. An AI system is not static; its performance and fairness can change over time as new data becomes available or as the underlying societal conditions evolve. Therefore, continuous evaluation is critical. This involves regularly assessing the AI's outputs against chosen fairness metrics, identifying any drift or emerging biases, and having clear protocols for recalibrating or redesigning the system

when necessary. This might involve establishing independent oversight bodies, developing standardized testing procedures for algorithmic bias, and ensuring that the sources and nature of the data used for training and ongoing operation are transparent.

Moreover, the choice of fairness metric itself is not a neutral act. It represents a value judgment about what kind of fairness is most important in a given context. For example, in a criminal justice setting, prioritizing
**equalized odds** might aim to ensure that individuals who are truly innocent are equally unlikely to be falsely accused, and individuals who are truly guilty are equally likely to be identified, regardless of race. However, this might still allow for disparities in the overall number of arrests or convictions between racial groups if underlying crime rates or policing practices differ. Conversely, if the goal is to actively *reduce* disparities in outcomes, one might lean towards **demographic parity**, acknowledging that it might not perfectly reflect individual risk but aims to correct for historical systemic disadvantages. This is a complex trade-off that demands careful consideration and public debate.

The "fairness-accuracy trade-off" is another important consideration. In some cases, achieving perfect fairness according to certain metrics might lead to a slight reduction in the overall accuracy or predictive power of the AI system. For example, an AI designed to predict loan default might become more accurate if it heavily relies on factors that are highly correlated with race or socioeconomic status, even if those correlations are themselves a product of historical discrimination. However, enforcing fairness constraints might mean that the AI uses these factors less aggressively, potentially leading to a minor decrease in its ability to predict default accurately across the board, but significantly improving equity. Public sector AI must grapple with this trade-off, prioritizing equity and justice, even if it means accepting a slightly less "performant" system in purely technical terms, if that performance is achieved through discriminatory means. The ultimate goal of public service AI is not just predictive power, but serving the public equitably.

Ultimately, algorithmic fairness is not a technical problem with a purely technical solution. It is an ethical and societal challenge that requires ongoing vigilance, multi-stakeholder engagement, and a deep commitment to justice. It necessitates moving beyond simplistic, one-size-fits-all definitions of fairness and engaging in the nuanced, context-specific work of understanding what fairness means for each public service application. This involves not only mathematicians and computer scientists but also ethicists, social scientists, legal experts, policymakers, and, crucially, the communities that are most likely to be affected by these algorithms. Without this comprehensive and collaborative approach, the promise of AI in public service risks becoming a source of further division and injustice, rather than a tool for building a more equitable society. The imperative of algorithmic fairness demands that we confront these complexities head-on, ensuring that the algorithms we deploy are not just smart, but also just.

The preceding discussion has illuminated the multifaceted nature of algorithmic fairness and the inherent difficulties in applying universal definitions to complex real-world scenarios. It has become clear that what constitutes "fair" can differ dramatically depending on the specific context, the intended purpose of the AI system, and the societal values that are being upheld. Therefore, the crucial next step in establishing robust equity metrics is to move beyond abstract notions of fairness and engage in the practical, and often challenging, work of defining context-specific fairness goals. This endeavor is not a purely technical exercise; it is fundamentally an ethical and participatory process that demands deep collaboration with all relevant stakeholders.

To embark on this process, we must first acknowledge that each public sector AI application operates within a unique ecosystem. This ecosystem includes the sector itself—be it healthcare, criminal justice, employment services, housing, education, or social welfare—as well as the specific population it serves and the particular risks of harm that might arise from biased algorithmic decisions. For instance, an AI system designed to assist in allocating scarce public housing has a fundamentally different risk profile and set of ethical considerations than an AI tool intended to support medical diagnosis in a hospital setting.

In the context of public housing allocation, the primary concern might revolve around equitable access to a fundamental human need and the historical patterns of segregation and redlining that have disadvantaged certain communities. Here, fairness goals could be centered on ensuring that demographic representation in housing assignments reflects community proportions (demographic parity), or more proactively, ensuring that individuals with demonstrably higher need, regardless of their background, are prioritized. The potential harms of bias could include exacerbating existing housing shortages for marginalized groups, perpetuating residential segregation, or denying essential stability to vulnerable families. A fairness goal might therefore be to ensure that the algorithm's allocation recommendations do not disproportionately disadvantage applicants from historically underserved neighborhoods, even if this requires a careful calibration against metrics like predictive accuracy of income or employment stability, which themselves could be proxies for historical bias. The dialogue here would likely involve community housing advocates, residents of public housing, urban planners, and policymakers, all weighing in on what constitutes a just distribution of a vital resource. The focus might be less on perfectly predicting individual likelihood of rent payment and more on ensuring that the system actively works to counteract systemic disadvantages in housing access.

Contrast this with an AI system deployed in a healthcare setting, such as an algorithm designed to predict a patient's risk of developing a specific chronic condition or to assist in triaging emergency room patients. In this scenario, the paramount concern is patient well-being and the equitable provision of care. The risks of bias here are grave and immediate: misdiagnosis, delayed treatment, or unequal access to life-saving interventions. A fairness goal in this context might prioritize **equalized odds** or **predictive parity** for diagnostic accuracy across different demographic groups. For example, ensuring that the AI's sensitivity (true positive rate) and specificity (true negative rate) are consistent across racial, ethnic, and gender lines is crucial. This means that individuals who genuinely have the condition should have an equal chance of being correctly identified, and individuals who do not have the condition should have an equal chance of being correctly identified as healthy, regardless of their background. Similarly, if the AI is used to predict the likelihood of a positive test result, **predictive parity** would ensure that when the AI predicts a positive result for a patient, the probability that they actually have the condition is the same for all groups. The potential harms of failing these metrics could lead to Black patients being systematically undertreated for conditions like heart disease, or women's symptoms being dismissed as less serious than men's, with potentially fatal consequences. The stakeholders in this context would include clinicians, hospital administrators, patient advocacy groups, public health officials, and medical ethicists. They would grapple with questions like: Is it acceptable for an AI to be slightly less accurate overall if it dramatically reduces disparities in diagnostic accuracy for a specific underserved population? How do we define "true risk" or "true need" when historical data may already be biased by differential access to healthcare and preventative services?

The process of defining these context-specific goals requires a structured and inclusive approach. It begins with a thorough **risk and impact assessment** that goes beyond technical performance. This assessment should explicitly identify:

1. **The intended function of the AI system:** What specific decision or recommendation is the algorithm designed to support? What is the range of possible outcomes?
2. **The populations affected:** Who are the individuals or groups that will interact with or be impacted by the AI system? This includes direct users (e.g., job applicants, benefit claimants) and indirect beneficiaries or those affected by resource allocation decisions. Particular attention must be paid to groups with protected characteristics and those historically marginalized or underserved.
3. **Potential sources of bias:** Where might biases creep into the system? This could be in the data (historical, measurement, or representation bias), the algorithm design (feature selection, model architecture), or the deployment and human-AI interaction.
4. **Specific harms of bias:** What are the concrete negative consequences if the AI system exhibits bias against particular groups? These harms can range from economic disadvantage (denial of loans, employment) and social exclusion (denial of benefits, unequal access to services) to physical harm (misdiagnosis, wrongful arrest) and erosion of trust in public institutions. The severity and immediacy of these harms must be carefully considered. For example, a bias in a social media content moderation algorithm might lead to unfair censorship, while a bias in an AI-powered parole recommendation system could result in unjust imprisonment, a far more severe consequence.

Once these potential harms and biases are understood, the next critical step is

**stakeholder engagement**. This is not a perfunctory consultation but a genuine dialogue aimed at co-creating fairness objectives. The stakeholders should include not only the technologists and policymakers who are developing and deploying the system but, crucially, representatives from the communities most likely to be affected by its decisions. This might involve:

**Community members and advocacy groups:** Those with lived experience of systemic inequalities and discrimination can offer invaluable insights into what constitutes fairness and what harms are most salient. Their input is essential for ensuring that fairness goals are aligned with lived realities and aspirations for social justice, not just abstract principles.
**Domain experts:** Individuals with deep knowledge of the specific sector (e.g., doctors, educators, legal professionals) can help clarify the nuances of the decision-making process and the practical implications of different fairness metrics.
**Ethicists and social scientists:** These professionals can provide frameworks for ethical reasoning, help analyze power dynamics, and guide the interpretation of fairness concepts within their societal context.
**Legal and compliance officers:** Ensuring that fairness goals align with existing legal protections and regulatory requirements is paramount.

During these engagements, it is vital to translate technical fairness metrics into understandable terms that resonate with the lived experiences and concerns of diverse stakeholders. Instead of simply presenting options like "demographic parity" or "equalized odds," the discussion should focus on the desired
*outcomes*. For example:

"We want to ensure that the proportion of job opportunities

recommended by the AI to women is the same as their proportion in the qualified applicant pool." (Demographic parity)

"We need to guarantee that individuals who are truly qualified for the scholarship have an equal chance of being recommended, regardless of their racial background." (Equality of opportunity/Predictive parity depending on framing)
"It is critical that if the AI flags someone for further review, the likelihood that they actually require that review is consistent across all age groups." (Predictive parity)
"We must ensure that individuals who do not pose a genuine risk are equally unlikely to be wrongly flagged by the system, regardless of their ethnicity." (Equality of opportunity/Equalized odds focusing on false positive rates)

The process of selecting and prioritizing fairness goals is often characterized by
**trade-offs**. As previously discussed, different fairness metrics can conflict, and sometimes achieving perfect fairness according to one definition might necessitate a compromise on another, or even on overall predictive accuracy. This is where ethical deliberation becomes indispensable. The choice is not merely technical; it is a policy decision that reflects societal values. For instance, in a system that determines eligibility for a highly sought-after public service, should the priority be to ensure that the *proportion* of recipients from different groups matches their proportion in the general population (demographic parity), even if it means some individuals with very high need from a majority group are overlooked? Or should the priority be to maximize the accurate identification of *all* individuals with high need, even if this results in demographic outcomes that don't perfectly mirror the population (a form of equality of opportunity focused on need)?

The decision-making process for these trade-offs should be transparent and justifiable. It requires clear articulation of *why* a particular definition of fairness or a specific metric was chosen over others, acknowledging the potential downstream consequences. For example, a public employment agency might decide to prioritize **demographic parity** in its AI-driven candidate referral system, not because it perfectly captures individual merit, but as a deliberate intervention to counteract historical underrepresentation and systemic barriers that have prevented certain groups from accessing opportunities. The justification would be that in this specific context, achieving a more equitable distribution of opportunities is a primary societal good, even if it means the algorithm is less precise in identifying every single "best fit" candidate based on historical patterns alone. Conversely, an AI system for allocating critical medical supplies during a public health crisis might prioritize **predictive parity** to ensure that when the system flags an individual as high-risk for a life-threatening condition, there is a very high probability that the prediction is accurate, thereby ensuring that scarce resources are allocated to those most likely to benefit.

Furthermore, the chosen fairness goals must be **measurable and actionable**. This means that the abstract principles discussed with stakeholders need to be translated into concrete metrics that can be quantified and monitored over time. For each chosen fairness goal, the development team must define:

**The specific metric(s)** that will be used to assess fairness (e.g., difference in false positive rates between groups, ratio of true positive rates, statistical parity difference).

**The acceptable threshold or tolerance level** for bias. What degree of disparity is considered acceptable, and what triggers an intervention?

This threshold should be informed by the risk assessment and stakeholder input. For example, a difference of 0.5% in false positive rates might be acceptable for a low-stakes application, but a difference of 5% might be unacceptable for a system impacting critical healthcare access.

**The data required for measurement:** What data points are needed to calculate the chosen metrics, and how will this data be collected and protected, especially when dealing with sensitive personal information?

**The process for monitoring and evaluation:** How often will the AI system's performance against these fairness metrics be evaluated? Who will be responsible for this evaluation, and what are the established protocols for addressing detected biases (e.g., recalibration, re-training, human review, or even decommissioning the system)?

The definition of context-specific fairness goals is an iterative process. It is unlikely that the "perfect" set of goals will be established at the outset. As the AI system is deployed, new data will emerge, societal conditions may change, and unforeseen consequences may arise. Therefore, the framework for defining fairness must include mechanisms for ongoing review and adaptation, ensuring that the AI system remains aligned with evolving community needs and ethical standards. This dynamic approach is crucial for maintaining the integrity and trustworthiness of AI in public service, ensuring that these powerful tools serve to enhance equity and justice, rather than inadvertently undermining them. The commitment to defining context-specific fairness goals is not a one-time checkbox; it is an ongoing pledge to equitable stewardship of AI within the public sphere.

The previous section underscored the critical imperative of moving beyond abstract notions of fairness to define concrete, context-specific goals for AI systems in the public sector. This transition from philosophical ideals to practical application necessitates the careful selection of appropriate equity metrics – the quantifiable indicators that will allow us to measure progress towards our defined fairness objectives. This is not merely a technical exercise; it is the bedrock upon which trust and accountability in public sector AI are built. Without robust, well-chosen metrics, the most meticulously defined fairness goals remain aspirational rather than actionable. The process of selecting these metrics requires a nuanced understanding of both the available measurement tools and the specific context in which the AI system is deployed.

To begin, it is essential to categorize the types of metrics available, broadly distinguishing between quantitative and qualitative approaches. Quantitative metrics offer numerical measurements that allow for precise comparisons and statistical analysis, providing a clear picture of performance against predefined thresholds. Qualitative metrics, on the other hand, capture the nuances of user experience, perception, and impact, offering insights that numbers alone cannot fully convey. A comprehensive approach to equity assessment will invariably leverage both.

Within the realm of quantitative metrics, a fundamental starting point often involves examining overall accuracy alongside group-specific performance. While a system might boast high overall accuracy, this can mask significant disparities in its performance across different demographic groups. For instance, an AI tool designed to predict loan default might achieve an impressive 95% overall accuracy. However, if this accuracy is only 90% for a particular racial minority group, while being 98% for the majority group, the aggregate figure obscures a critical equity concern. This highlights the inadequacy of relying solely on global performance indicators.

A more granular approach necessitates the examination of various accuracy-related metrics for each defined subgroup. These include:

**True Positive Rate (TPR), also known as Sensitivity or Recall:** This metric measures the proportion of actual positive cases that are correctly identified by the AI. In a predictive policing context, a high TPR would mean that when a crime is predicted to occur in a certain area, it is indeed likely to occur. For equity, we would assess if the TPR is consistent across different neighborhoods or demographic compositions. A low TPR for a historically over-policed community, for example, could mean the system is failing to accurately predict genuine risks, potentially leading to under-resourcing or misplaced attention, or conversely, if the TPR is inflated, it could lead to over-surveillance. In a healthcare setting, a high TPR for detecting a disease means that individuals who actually have the disease are correctly identified. Ensuring equal TPR across racial groups is vital to prevent certain populations from being systematically missed when they are ill.

**True Negative Rate (TNR), also known as Specificity:** This metric measures the proportion of actual negative cases that are correctly identified by the AI. In a fraud detection system, a high TNR means that legitimate transactions are correctly identified as not fraudulent. For equity, we would examine if the TNR is consistent across different user groups. A low TNR for a particular demographic could lead to them experiencing more false alarms or incorrect negative classifications. In a criminal justice context, this translates to correctly identifying individuals who are not a flight risk or not likely to re-offend. Ensuring equal TNR across racial groups is critical to prevent individuals from being unfairly penalized due to a higher rate of false negatives, which could lead to inappropriate detentions or denials of parole.

**False Positive Rate (FPR):** This is the proportion of actual negative cases that are incorrectly identified as positive. It is calculated as 1 - TNR. A low FPR is desirable in most applications. In an AI system for flagging potentially fraudulent insurance claims, a high FPR would mean many legitimate claims are incorrectly flagged, leading to delays and frustration for policyholders. For equity, we must assess if this FPR is uniform across demographic groups. A higher FPR for a minority group might indicate a system that is more prone to unjustly scrutinizing their applications or activities. In a welfare benefits system, a high FPR for certain ethnic groups could mean more individuals are subjected to unnecessary investigations, causing undue stress and potential loss of essential support.

**False Negative Rate (FNR):** This is the proportion of actual positive cases that are incorrectly identified as negative. It is calculated as 1 - TPR. A low FNR is also crucial. In an AI used for medical screening, a high FNR means that individuals who actually have the condition are missed by the system, leading to delayed diagnosis and treatment. Equity demands that the FNR is comparable across all relevant groups. A higher FNR for a specific demographic group could mean they are systematically not receiving timely interventions when they need them, with potentially severe health consequences. In an employment application screening tool, a high FNR for qualified candidates from underrepresented backgrounds would mean the system is failing to identify promising individuals, thereby perpetuating workforce inequality.

These four metrics—TPR, TNR, FPR, and FNR—are fundamental for understanding how an AI system performs for different groups. They allow us to move beyond aggregated statistics and scrutinize potential biases in classification tasks. For instance, if an AI is used to determine eligibility for a public service, and it exhibits a significantly higher FPR for one racial group compared to another, it means that individuals from that group are more likely to be incorrectly denied service. Conversely, a higher FNR for a particular group might indicate they are more likely to be incorrectly deemed ineligible when they are actually qualified.

Beyond these basic accuracy metrics, several fairness-specific quantitative metrics have emerged, each offering a different perspective on equitable outcomes:

**Demographic Parity (also known as Statistical Parity):** This metric aims for equal selection rates across different groups. It asserts that the proportion of individuals receiving a positive outcome (e.g., being approved for a loan, being hired, being recommended for a program) should be the same for all protected groups, regardless of their underlying qualifications as defined by the model. For example, if an AI system is used to refer candidates for job interviews, demographic parity would dictate that the percentage of candidates referred from Group A is the same as the percentage referred from Group B, assuming both groups are represented in the applicant pool. The strength of demographic parity lies in its simplicity and its direct focus on proportional representation. It is often appealing because it directly addresses historical underrepresentation and aims to equalize outcomes. However, its major limitation is that it ignores the underlying qualifications or true need as predicted by the model. It can lead to selecting less qualified individuals from one group over more qualified individuals from another, which might be undesirable or even counterproductive in certain contexts. This metric is most appropriate when the goal is to ensure broad access or representation, and where disparities in underlying "qualifications" are themselves suspected to be a result of systemic bias that the AI should actively counteract, rather than simply reflect. For instance, in affirmative action policies, the aim is often to achieve demographic parity in educational institutions or workplaces, recognizing that traditional metrics of merit might be biased.

**Equality of Opportunity:** This metric focuses on ensuring that individuals who are truly qualified or deserving receive the positive outcome at equal rates across groups. It specifically looks at the True Positive Rate (TPR). Equality of opportunity requires that the TPR is the same for all groups. In simpler terms, if an individual is truly qualified for a loan, they should have the same probability of being approved by the AI, regardless of their demographic group. Similarly, if a patient truly has a disease, the AI should have the same probability of correctly identifying them, regardless of their race or gender. This metric is particularly relevant when the primary concern is ensuring that genuinely eligible individuals are not systematically overlooked. It addresses the risk of the AI systematically failing to recognize positive cases within certain groups. A limitation is that it does not constrain the False Positive Rate (FPR), meaning that while qualified individuals might be correctly identified, unqualified individuals from certain groups might also be more likely to be incorrectly approved.

**Equality of Odds:** This is a more stringent condition than equality of opportunity, as it requires both the True Positive Rate (TPR) and the True Negative Rate (TNR) to be equal across all groups. This means that the AI must perform equally well in correctly identifying positive cases (e.g., those who will default on a loan, those who are truly sick) and correctly identifying negative cases (e.g., those who will not default, those who are healthy) for all demographic groups. In essence, it aims for parity in both correctly identifying those who should receive a positive outcome and correctly identifying those who should not. Equality of odds is a powerful metric for ensuring overall fairness in classification, as it addresses biases in both false positives and false negatives. However, achieving equality of odds can be very challenging, as it often conflicts with other fairness definitions and might require sacrificing some level of overall accuracy. It is most suitable for applications where both types of errors have significant negative consequences and where there is a strong imperative to ensure consistent predictive performance across populations.

**Predictive Parity:** This metric focuses on the Precision or Positive Predictive Value (PPV). It requires that for individuals predicted to have a positive outcome, the probability that they actually have that outcome is the same across all groups. In other words, if the AI predicts that someone will default on a loan, the likelihood that they will actually *default should be the same, whether that prediction is made for a man or a woman, or for someone from Group A or Group B. This metric is crucial when the cost of a false positive is high, and we want to ensure that the predictions made by the AI are equally reliable across groups. For example, if an AI flags individuals for further investigation, predictive parity ensures that the likelihood of those flagged individuals* actually *warranting further investigation is consistent across demographics. This is distinct from equality of opportunity, which focuses on the rate of correct identification among those who* truly have the positive attribute.

**Conditional Use Accuracy Equality (also known as Equalized Predictive Accuracy):** This metric requires that the overall accuracy (proportion of correct predictions, both true positives and true negatives) is the same across different groups. It essentially ensures that the AI is equally good at making correct classifications for all populations. While seemingly straightforward, it can sometimes be a weaker guarantee than equality of odds or predictive parity, as improvements in overall accuracy might be achieved through different combinations of TPR, TNR, FPR, and FNR for different groups.

When selecting among these quantitative metrics, several critical considerations come into play. Firstly, the specific
**risks and harms** identified in the prior impact assessment must guide the choice. If the most severe harm is the systematic denial of essential services to a particular group (a false negative issue), then metrics focusing on the True Positive Rate (TPR), like Equality of Opportunity, would be paramount. If the primary concern is unwarranted suspicion or scrutiny leading to negative collateral consequences (a false positive issue), then metrics focusing on the False Positive Rate (FPR), like Predictive Parity or aspects of Equality of Odds, become more critical.

Secondly, the
**intended purpose and operational context** of the AI system are paramount. In scenarios where equitable distribution of scarce resources is the goal, and historical data suggests significant systemic disadvantages, Demographic Parity might be considered as a deliberate intervention to counterbalance past injustices, even if it means a deviation from perfect accuracy based on historical patterns. Conversely, in high-stakes medical diagnostics, where incorrect positive or negative predictions can have life-altering consequences, Equality of Odds or robust measures of Predictive Parity might be prioritized to ensure consistent reliability of diagnoses across patient demographics.

Thirdly, the
**interplay and potential conflicts between metrics** must be understood. As discussed in the previous section, it is often

mathematically impossible to satisfy all fairness metrics simultaneously, especially when base rates of outcomes differ across groups. For instance, achieving Demographic Parity might necessitate accepting disparities in predictive accuracy, while strictly enforcing Equality of Odds might lead to outcomes that do not reflect the demographic composition of the population. This reality necessitates ethical deliberation and policy decisions about which types of fairness are most important in a given context, and what trade-offs are acceptable. This decision should not be made in a vacuum by technologists alone, but through collaborative dialogue with affected communities and domain experts.

Fourthly, the
**data availability and quality** are crucial. Certain metrics require precise ground truth labels and detailed demographic information, which may not always be available or reliable. For example, measuring Predictive Parity requires accurate information on the actual outcome for all individuals, which can be challenging to obtain, especially for long-term predictions. Similarly, disaggregating performance by detailed demographic categories might be hindered by data limitations or privacy concerns. The chosen metrics must be calculable with the available data, or there must be a clear plan to acquire or improve the necessary data.

Fifthly, the
**actionability and interpretability** of the metric are important. A metric is more useful if it is clearly understood by all stakeholders and if deviations from the desired threshold can be readily translated into specific corrective actions. For instance, if a system shows a disparity in the False Positive Rate, it should be clear whether the intervention involves retraining the model, adjusting decision thresholds, or implementing human oversight.

Beyond quantitative metrics,

**qualitative assessment** plays an indispensable role in evaluating the equity of AI systems in the public sector. These methods capture aspects of fairness that are difficult, if not impossible, to quantify, such as user trust, perceived fairness, and the lived experience of those affected by the AI's decisions.

**User Satisfaction Surveys and Feedback Mechanisms:** Regularly soliciting feedback from individuals who interact with the AI system is crucial. This can be done through surveys, interviews, focus groups, or dedicated feedback channels. Questions should probe not only satisfaction with the outcome but also the perceived fairness of the process, the clarity of explanations (if provided), and the overall user experience. For example, a user applying for social benefits might be approved according to quantitative metrics, but if the interaction with the AI system felt impersonal, demeaning, or opaque, it can erode trust and create a sense of injustice. Qualitative feedback can reveal such issues.

**Interviews and Focus Groups with Affected Communities:** Direct engagement with individuals and communities most likely to be impacted by the AI system is essential for understanding the real-world implications of its deployment. These discussions can uncover harms or benefits that were not anticipated during the quantitative assessment. For instance, an AI used for resource allocation in public health might appear fair based on demographic parity, but interviews might reveal that the criteria it uses inadvertently disadvantage individuals with complex, intersecting needs that don't fit neatly into predefined categories. Community members can articulate what "fairness" means to them in their specific context, which might go beyond statistical parity.

**Audits of Decision-Making Processes and Outcomes:** Beyond examining the AI's output, qualitative audits can scrutinize the entire decision-making pipeline. This includes reviewing the data used for training and deployment, the algorithmic choices made, the human oversight mechanisms in place, and how the AI's recommendations are integrated into broader human decision-making processes. Such audits can identify potential points of bias or unfairness that might not be evident in the final statistical outcomes. For example, an audit might reveal that human reviewers are more likely to override the AI's recommendations for certain demographic groups, suggesting that the human operators themselves hold biases, or that the AI's outputs are not sufficiently trustworthy for all populations.

**Expert Reviews and Ethical Audits:** Independent reviews by ethicists, social scientists, and domain experts can provide an objective assessment of the AI system's fairness and its alignment with societal values. These experts can identify potential ethical blind spots, unintended consequences, and areas where the system might perpetuate systemic inequalities, even if it meets certain quantitative fairness metrics. They can offer a critical perspective that challenges assumptions and pushes for more robust equity considerations.

The integration of qualitative insights is vital because quantitative metrics, while powerful, represent a simplified model of reality. They capture specific dimensions of fairness but can miss the broader human experience. For instance, an AI system might achieve perfect Equality of Odds, meaning its error rates are the same across groups. However, if the system's output is communicated in a way that is biased, or if it leads to increased scrutiny or profiling in ways that cause distress, then despite meeting a technical fairness standard, it may still be perceived as unfair and lead to tangible harm. Qualitative data helps to contextualize quantitative findings and ensure that the AI system serves the public in a way that is not only statistically equitable but also genuinely just and trustworthy.

Ultimately, the selection of appropriate equity metrics is an ongoing, iterative process. It requires careful deliberation, robust stakeholder engagement, and a commitment to transparency. It is not about finding a single perfect metric or a universally applicable formula, but about making informed, context-aware choices that best align with the public sector's mandate to serve all citizens equitably and justly. The chosen metrics must be clearly documented, regularly monitored, and subject to periodic re-evaluation as the AI system evolves and societal understanding of fairness deepens. This diligent, multi-faceted approach to metric selection is fundamental to building AI systems that are not only effective but also ethical and aligned with the public good.

The effectiveness and, more crucially, the equity of any Artificial Intelligence system are intrinsically tied to the data upon which it is built and operated. In the public sector, where the stakes are particularly high and the potential for disparate impact on vulnerable populations is significant, a deep and critical examination of data is not merely a technical prerequisite but an ethical imperative. Biases embedded within datasets, often reflecting historical societal inequities, can be inadvertently amplified by AI algorithms, leading to outcomes that perpetuate and even exacerbate existing disparities. Therefore, establishing robust equity metrics necessitates a foundational understanding of data considerations, encompassing everything from how data is collected to how it is prepared for algorithmic consumption.

At the heart of data-driven inequity lies the concept of **data representativeness**. An AI model trained on data that does not accurately reflect the diversity of the population it will serve is inherently prone to bias. For instance, consider an AI system designed to predict the likelihood of a resident needing social services. If the historical data used for training disproportionately represents certain demographic groups or geographic areas while underrepresenting others, the model may develop blind spots or inaccurate predictive capabilities for the underrepresented populations. This could manifest as a failure to identify individuals in need, leading to a denial of crucial support, or conversely, an over-identification of individuals from overrepresented groups, potentially leading to unwarranted scrutiny or resource misallocation. Ensuring representativeness is a multifaceted challenge. It requires not only collecting data from all relevant demographic groups but also ensuring that the *quality* and *completeness* of data are consistent across these groups. For example, if a significant portion of data for a particular community is incomplete due to issues with data collection infrastructure or accessibility, the AI trained on this data will inherently be less reliable for that community. Proactive efforts to achieve representative data might involve targeted data collection initiatives, partnerships with community organizations to improve data access, and rigorous auditing of datasets to identify and quantify any representational gaps. This might mean actively seeking out data from historically marginalized communities, even if it requires more effort than drawing from readily available, but less representative, sources. The goal is to ensure that the training data is a mirror, albeit a carefully polished one, of the diverse populace the AI is intended to serve.

A related and often more insidious issue is the presence of **proxy variables**. These are seemingly neutral data points that, in reality, are highly correlated with sensitive attributes like race, gender, socioeconomic status, or geographic location, which themselves may be protected characteristics. An AI model might not explicitly use a protected attribute, but if it learns to rely on a strong proxy for that attribute, it can indirectly perpetuate discrimination. For example, in a housing allocation system, using data points such as "zip code" or "census tract" might seem innocuous. However, if historical patterns show that certain zip codes are overwhelmingly populated by specific racial or socioeconomic groups due to discriminatory housing policies or redlining, then using zip code as a predictive feature can inadvertently replicate those discriminatory patterns. The AI might learn to associate certain zip codes with a lower likelihood of receiving housing assistance, not because of any inherent characteristics of the individuals within those zip codes, but because the zip code itself is a proxy for protected characteristics. Identifying and mitigating the impact of proxy variables requires a deep understanding of the data's provenance and the social context in which it was generated. It involves thorough exploratory data analysis, correlation studies, and often, domain expertise to uncover these indirect discriminatory pathways. Strategies for mitigation can include removing the proxy variable, transforming it to obscure its correlation with sensitive attributes, or explicitly incorporating sensitive attributes into the model with fairness constraints during training. The latter, however, brings its own set of ethical considerations regarding the use of sensitive data.

The decision of whether and how to use **sensitive attributes** (e.g., race, ethnicity, gender, age, disability status) in AI models is one of the most ethically charged aspects of data consideration. On one hand, including sensitive attributes can be crucial for explicitly measuring and mitigating bias. By disaggregating performance metrics by these attributes, as discussed in the previous section, we can directly identify disparities and then use fairness-aware algorithms or post-processing techniques to equalize outcomes. For example, if an AI system for assessing risk in the criminal justice system shows a higher false positive rate for a particular racial group, explicit inclusion of race in the fairness-aware training process might help the model adjust its predictions to achieve parity in error rates. However, the use of sensitive attributes is fraught with challenges. It raises significant privacy concerns, as it involves collecting and processing highly personal information. There's also the risk that even with fairness interventions, the model might still learn spurious correlations or that the very act of using these attributes can lead to unintended consequences or stigmatization. Furthermore, in many jurisdictions, there are legal restrictions on the use of certain sensitive attributes. A pragmatic approach often involves a careful balancing act: use sensitive attributes strategically and transparently for the purpose of bias detection and mitigation, ensuring robust data security and anonymization where possible, and only when less direct methods (like identifying and mitigating proxies) prove insufficient. The decision to use sensitive attributes should be guided by a thorough ethical assessment, stakeholder consultation, and a clear justification of necessity.

Beyond the content of the data, the
**quality and integrity** of the data are paramount. Inaccurate, incomplete, or inconsistent data can lead to flawed models and inequitable outcomes. For instance, if data entry errors are more common for certain user groups due to issues with the user interface or data collection tools, the AI will learn from this noisy data, potentially leading to poorer performance for those groups. Data cleaning and validation processes are therefore critical. This involves identifying and correcting errors, handling missing values (e.g., through imputation methods that are themselves checked for bias), and ensuring consistency in data formats and definitions. For public sector AI, where data might be sourced from disparate legacy systems, the effort to achieve data quality and consistency can be substantial. It often requires significant investment in data governance, standardization, and automated data quality checks. Without this foundational work, any attempt to build an equitable AI system is built on shaky ground.

The
**temporal nature of data** also plays a critical role in equity. Societal norms, populations, and the very definition of fairness evolve over time.

AI models trained on historical data may become outdated and reflect past inequities that are no longer acceptable. For example, an AI system for job recruitment trained on data from decades ago might reflect hiring practices that were heavily biased against women in certain professions. If this data is not periodically updated and the model retrained, it will continue to perpetuate those outdated and discriminatory patterns. This necessitates a strategy for continuous monitoring and updating of the training data and the AI model itself. It also means being mindful of **data drift**, where the statistical properties of the data change over time, potentially rendering the model's predictions less accurate or fair. Regular re-evaluation of data representativeness and performance across demographic groups is essential to detect and address such drift. The public sector must establish mechanisms for ongoing data curation and model retraining that explicitly consider equity as data and the world change.

Furthermore, the
**origin and collection methods** of the data must be scrutinized. Was the data collected through voluntary participation, mandatory reporting, or surveillance? Were individuals informed about how their data would be used? For instance, if an AI system relies on data collected through mandatory compliance checks, and these checks are disproportionately applied to certain communities, the resulting data will inherently be biased against those communities. Transparency about data collection practices and the ethical justification for those practices are crucial for building public trust and ensuring that the data does not inherently disadvantage certain groups. This might involve undertaking detailed "data provenance" exercises, tracing the lineage of data from its source to its use in the AI system, and assessing the ethical implications at each step.

**Data preprocessing** techniques themselves can introduce or mitigate bias. Techniques like feature selection, feature engineering, and data augmentation must be applied with an equity lens. For instance, if during feature engineering, a process inadvertently creates new variables that are proxies for sensitive attributes, this can lead to downstream bias. Conversely, techniques like oversampling minority groups or undersampling majority groups in the training data (while controversial and requiring careful application to avoid creating artificial patterns) can sometimes help improve model performance for underrepresented populations. The choice of data imputation methods for missing values can also have an impact; for example, imputing missing income data for a particular group using a method that assumes a certain average might misrepresent individuals within that group. It is essential that these preprocessing steps are not treated as purely technical choices but as ethically-informed decisions, with potential impacts on fairness being carefully evaluated.

Finally, the
**ethical implications of data minimization and privacy-preserving techniques** must be considered in the context of equity. While collecting less data or using anonymization techniques can protect individual privacy, it can also obscure disparities. If sensitive attributes are removed or anonymized too aggressively, it may become impossible to measure or correct for bias in the AI system. This presents a delicate trade-off: how to protect individual privacy while still having the necessary information to ensure algorithmic fairness. Solutions might involve collecting aggregated demographic data rather than individual-level sensitive attributes, or using advanced privacy-preserving machine learning techniques that allow for fairness analysis without directly exposing sensitive information. The goal is to find methods that enable robust equity assessment without unnecessary compromise of individual privacy. This might involve sophisticated differential privacy mechanisms or federated learning approaches that allow models to be

trained on decentralized data without it ever leaving the local environment.

In summary, the journey towards equitable AI in the public sector begins with an unwavering commitment to understanding and critically evaluating the data. This involves ensuring data representativeness, diligently identifying and mitigating proxy variables, making informed and ethically grounded decisions about the use of sensitive attributes, prioritizing data quality and integrity, accounting for the temporal nature of data, scrutinizing data collection methods, applying preprocessing techniques with an equity lens, and navigating the complex interplay between privacy and fairness. Without this rigorous data foundation, even the most sophisticated fairness metrics will struggle to achieve their intended purpose, leaving the public sector vulnerable to deploying AI systems that inadvertently perpetuate the very inequalities they are meant to address.

The journey towards establishing robust equity metrics is not a destination, but a continuous voyage of assessment and adaptation. Once an initial set of metrics has been defined and implemented, the critical phase of iterative measurement and refinement begins. This is where the theoretical framework for fairness is tested against the practical realities of an AI system's deployment in the public sector. It is a dynamic process, demanding constant vigilance and a commitment to learning and evolving. The initial deployment of an AI system, no matter how meticulously planned and evaluated for fairness, marks only the beginning of its equitable operation. Real-world performance can diverge significantly from laboratory conditions, influenced by unforeseen user interactions, shifts in the external environment, and the inherent complexities of human behavior. Therefore, a robust system of continuous monitoring is not merely advisable; it is an ethical imperative. This monitoring must be structured to consistently track the performance of the AI against the predefined equity metrics, disaggregating these measures by relevant demographic groups and other intersectional identities identified during the design and planning phases.

The core of this iterative process lies in establishing a feedback loop. This loop connects the observed performance of the AI system back to the decision-making processes that govern its operation, maintenance, and potential modification. When the monitoring reveals deviations from the desired equity targets – for instance, a statistically significant disparity in service allocation, an elevated false positive rate for a particular community, or a failure to achieve equitable access to a resource – this information must trigger a structured response. This response should not be ad hoc but guided by predefined protocols designed to diagnose the root cause of the inequity and implement corrective actions. These actions could range from minor adjustments to algorithmic parameters, to significant retraining of the model, to revising the underlying data collection or preprocessing strategies, or even, in some extreme cases, to pausing or revoking the system's deployment. The key is that performance data, particularly when it highlights inequity, is not ignored but actively used to inform and drive improvements.

To facilitate this continuous improvement, several methodologies and practices are essential. One foundational element is the establishment of clear

**performance dashboards and reporting mechanisms**. These dashboards should provide real-time or near-real-time visibility into the AI system's performance across all defined equity metrics. They should present data in an easily digestible format, allowing stakeholders – from technical teams to policy makers to community representatives – to quickly understand the system's current state of fairness. The disaggregation of these metrics is paramount here. For example, if an AI is used for prioritizing public health interventions, the dashboard should not just show overall efficiency but also illustrate the rate of intervention across different racial, ethnic, socioeconomic, and geographic groups. Any widening of disparities must be immediately apparent. These reports should be generated regularly, and critically, they must be accessible to relevant parties, fostering transparency and accountability.

Beyond simple reporting,
**regular auditing and impact assessments** are crucial. These are deeper dives into the AI system's performance, conducted at predetermined intervals (e.g., quarterly, annually) or triggered by significant changes in the system or its operating environment. Audits should go beyond superficial checks; they should involve a thorough examination of the data feeding the system, the algorithms themselves, and the outcomes produced. This can involve employing independent third-party auditors with expertise in AI ethics and fairness to provide an objective assessment. Impact assessments, on the other hand, focus on the broader societal consequences of the AI system's deployment. They seek to understand not just whether the system is meeting its equity metrics, but whether it is contributing to broader societal goals of equity and justice, or inadvertently creating new forms of harm or disadvantage. For example, an impact assessment for an AI used in educational admissions might consider whether it is helping to diversify student bodies or inadvertently reinforcing existing educational access barriers.

A sophisticated approach to iterative refinement involves
**hypothesis testing and experimentation**. When an inequity is detected, it's rarely obvious what specific change will rectify it. This is where the scientific method becomes invaluable. Teams can formulate hypotheses about the causes of the observed inequity (e.g., "The current feature weighting for income is disproportionately penalizing low-income households due to the exclusion of crucial support program

data") and then design controlled experiments to test these hypotheses. This might involve retraining the model with modified features, adjusting algorithmic parameters, or augmenting the training data with more representative samples. Crucially, these experiments must be designed to isolate the impact of the change on the equity metrics. This often requires a period of parallel running where the existing system and the modified system operate simultaneously, allowing for direct comparison of their outputs and impacts on different demographic groups. Such experimentation allows for data-driven decision-making, moving away from guesswork towards evidence-based adjustments.

The
**dynamic nature of fairness definitions and societal values** also necessitates periodic re-evaluation of the equity metrics themselves. What was considered fair or equitable at the time of the initial system design might not reflect current societal understanding or evolving legal standards. For instance, the understanding of intersectionality – how multiple forms of disadvantage combine and interact – has deepened significantly in recent years. An AI system initially evaluated for fairness along single dimensions (e.g., race OR gender) might need to be re-evaluated for intersectional fairness (e.g., Black women). This requires ongoing engagement with ethicists, social scientists, legal experts, and crucially, the communities directly impacted by the AI system. These stakeholders can provide invaluable insights into whether the current metrics adequately capture the nuances of lived experience and whether the system's outcomes align with evolving notions of social justice. This might lead to the refinement, addition, or even removal of certain metrics, ensuring that the AI system remains aligned with contemporary ethical expectations.

Furthermore,
**proactive identification of drift** is a critical component of iterative refinement. AI models are susceptible to various forms of drift. **Data**

**drift** occurs when the statistical properties of the input data change over time. For example, if an AI system used for predicting unemployment relies on data that no longer accurately reflects the current job market due to economic shifts or policy changes, its predictions will become less accurate and potentially unfair. **Concept drift** happens when the relationship between the input features and the target variable changes. For instance, if the factors that predict loan default evolve due to new lending regulations or consumer behaviors, a model trained on old data will fail to capture these new relationships. To combat drift, continuous monitoring of input data distributions and model predictions is essential. When significant drift is detected, it should trigger a reassessment of the model and potentially a retraining process. This process itself must be equitable; retraining on new data should be performed with fairness considerations at the forefront, ensuring that the new model does not inherit or amplify new forms of bias.

The
**human element in the feedback loop** cannot be overstated. While automated monitoring and reporting are essential, human judgment and oversight are indispensable. Technical teams must be empowered to act on the insights generated by these systems. This includes having the resources and authority to implement necessary adjustments, retrain models, or even pause a system if an ethical breach is detected. Moreover, establishing clear **channels for public feedback and grievance redressal** is vital. Individuals who interact with or are affected by public sector AI systems should have a straightforward way to report perceived unfairness, errors, or harms. These reports should be taken seriously, investigated thoroughly, and used to inform the iterative refinement process. This not only helps identify issues that automated systems might miss but also builds trust and legitimacy for the AI's deployment. The public sector has a particular responsibility to ensure that these feedback mechanisms are accessible to all, including those with limited digital literacy or access.

The
**process of retraining and redeployment** itself requires careful consideration from an equity perspective. When a model is retrained, it's not simply a matter of running the same old code with new data. The retraining process must be guided by the lessons learned from the previous deployment and the ongoing monitoring. This might involve updating fairness constraints, adjusting regularization techniques, or employing different debiasing strategies. Furthermore, the redeployment of a retrained model should be carefully managed. It might be beneficial to deploy the new model in a shadow mode initially, running it in parallel with the existing system without making active decisions, to verify its performance and fairness before full cutover. This mitigates the risk of introducing new problems during the update process.

In essence, iterative measurement and refinement transform AI systems from static tools into dynamic, responsive partners in public service. It acknowledges that fairness is not a binary state but a continuous pursuit. It demands a culture of accountability, transparency, and learning within public sector organizations. The initial definition of equity metrics provides the compass, but it is the ongoing cycle of measurement, analysis, intervention, and re-measurement that keeps the AI system on course towards equitable outcomes. This iterative approach ensures that as society evolves, as data patterns shift, and as our understanding of fairness deepens, the AI systems deployed by the public sector remain not only effective but also just, serving all members of the community equitably and without discrimination. This commitment to ongoing adaptation is what distinguishes truly responsible AI implementation from a mere technical exercise, cementing its role as a tool for genuine societal benefit.

# Chapter 5: Conducting Thorough Impact Assessments

The deployment of artificial intelligence within the public sector, while promising unprecedented efficiencies and service enhancements, invariably introduces a landscape of potential unintended consequences. These are not merely abstract theoretical concerns but tangible risks that can disproportionately affect vulnerable populations, erode public trust, and ultimately undermine the very goals of equitable governance that AI is intended to serve. Recognizing this reality compels a fundamental shift in our approach, moving beyond the routine validation of an AI system's functional accuracy to a more profound and prescient task: the proactive identification of potential harms. This subsection delves into the imperative of such a forward-looking perspective, underscoring why anticipating and mitigating risks before they materialize is not an optional add-on but a cornerstone of responsible AI implementation.

Historically, technological adoption has often followed a pattern of reactive problem-solving. Issues are identified after they occur, leading to a period of diagnosis, blame, and eventual remediation. While this reactive stance might be acceptable for less consequential technologies, AI systems in the public sector operate on a different plane. Their decisions can directly impact individuals' access to essential services, their legal standing, their economic well-being, and even their personal safety. The potential for harm, therefore, carries a weight that necessitates a departure from the "fix it when it breaks" mentality. Proactive harm identification is about foresight, about embedding a culture of ethical vigilance that seeks to anticipate where and how an AI system might falter in its service to the public, and to build safeguards against these eventualities from the outset.

Consider an AI system designed to optimize the allocation of affordable housing units. Its primary function is to match eligible applicants with available properties based on a complex set of criteria. While functional testing might confirm that the system accurately processes applications and assigns units according to predefined rules, a proactive harm identification lens would ask: What if the data used to determine "eligibility" inadvertently disadvantages individuals with precarious employment histories, who are disproportionately represented in certain minority groups? What if the system, in its pursuit of efficiency, prioritizes applicants residing in areas with lower median incomes, inadvertently creating new forms of residential segregation, even if this was not an explicit objective? What if the algorithm fails to account for the nuances of family structures that deviate from the "nuclear family" model, leading to certain applicants being unfairly penalized? These are not questions that can be easily answered by merely testing if the system "works" as intended. They require a deliberate, imaginative, and critical exploration of potential negative externalities.

The challenge of proactive harm identification lies in its inherent difficulty. Harms are often subtle, emergent, and dependent on context. They can arise from the intersection of multiple factors, including biases in historical data, flawed assumptions embedded in the model's design, limitations in the proxy variables used, and the complex interplay between the AI's outputs and the real-world social, economic, and political systems it interacts with. Furthermore, the very definition of "harm" can be contested and may evolve over time, reflecting shifting societal values and understandings of fairness and equity. For instance, what was once considered an acceptable level of disparity might, with greater societal awareness of systemic inequities, be recognized as a significant harm.

To effectively identify potential harms proactively, a multi-pronged approach is essential. It begins with a deep understanding of the context in which the AI system will operate. This means going beyond the technical specifications of the system itself and immersing oneself in the social, economic, and political realities of the community it will serve. For the affordable housing example, this would involve engaging with community organizations, social workers, housing advocates, and residents themselves to understand their lived experiences, their existing challenges, and their concerns regarding housing access. Such engagement can surface potential harms that might not be apparent from an analysis of data alone. For example, community members might reveal that certain neighborhoods, while seemingly attractive from a data perspective (e.g., lower crime rates, good schools), are actually difficult to access for those without personal transportation, a factor an algorithm might overlook.

This contextual understanding must then be translated into a rigorous process of
**risk assessment**. This involves systematically identifying potential sources of harm, analyzing their likelihood and severity, and considering the populations most likely to be affected. This is not a one-time activity but an ongoing process that should be revisited throughout the AI lifecycle, from initial design and development through deployment and ongoing operation. Techniques such as "what-if" scenario planning, red-teaming (where teams actively try to "break" the system by exploiting potential vulnerabilities), and structured brainstorming sessions involving diverse stakeholders are invaluable tools in this regard. The goal is to cultivate a mindset that actively seeks out potential failures and negative consequences, rather than assuming they will not occur.

A critical element of proactive harm identification is the
**examination of data provenance and quality**. AI systems learn from data, and if that data reflects historical biases or systemic inequities, the AI will inevitably perpetuate and potentially amplify them. This requires

meticulous scrutiny of all data sources. Questions must be asked: Where did this data come from? How was it collected? What populations does it represent, and which are excluded or underrepresented? Are there known biases in its collection or recording? For example, if an AI system for predicting recidivism is trained on historical arrest data, it must be recognized that arrest rates can be influenced by policing practices that disproportionately target certain communities, even if those communities do not have higher rates of criminal activity. Proactively identifying these data-related biases is crucial for preventing the AI from unfairly penalizing individuals based on historical discrimination.

Beyond data, the
**algorithmic design itself** warrants deep scrutiny. Even with "clean" data, the choices made in designing an algorithm can introduce or exacerbate harms. This includes the selection of features, the definition of objective functions, and the choice of learning paradigms. For instance, an algorithm that optimizes for "efficiency" without explicit consideration for equity might inadvertently concentrate resources in already well-served areas, leaving marginalized communities further behind. Similarly, the use of proxy variables can be a significant source of unintended harm. If, for example, an AI system uses "zip code" as a proxy for socioeconomic status, it might inadvertently penalize individuals living in areas with high costs of living but who are otherwise financially stable, or conversely, it might group individuals with vastly different circumstances under a single socioeconomic label. Proactive identification involves questioning these proxy relationships and understanding their potential downstream consequences.

The concept of
**fairness metrics**, discussed in previous chapters, is intimately linked to proactive harm identification. While fairness metrics are often implemented to *measure* and *monitor* equity, their development process is also a crucial phase for *anticipating* potential harms. By defining what

constitutes fairness in a particular context, and by disaggregating these definitions by relevant demographic groups, we are inherently engaging in a process of identifying what kinds of disparities we want to avoid. For instance, if we decide that an AI system for loan applications must exhibit parity in approval rates across different racial groups (a specific fairness metric), we are implicitly acknowledging the potential harm of racial discrimination in lending and are proactively seeking to prevent it. However, it is vital to recognize that no single fairness metric can capture all forms of potential harm. A truly proactive approach requires considering a portfolio of metrics and, crucially, acknowledging that quantitative metrics alone may not suffice. Qualitative assessments and contextual understanding are also indispensable.

Furthermore,
**stakeholder engagement** is not just about understanding the community; it's also about tapping into the diverse expertise within the development and deployment teams. This includes ethicists, social scientists, legal experts, domain specialists, and even individuals with lived experience of the systems being developed. These individuals can bring different perspectives and can help identify potential harms that might be invisible to those with purely technical backgrounds. A diverse team is more likely to consider a wider range of potential negative outcomes. For example, a legal expert might identify potential violations of anti-discrimination laws that a computer scientist might overlook, while a social worker might highlight how a particular algorithmic decision could exacerbate existing social inequalities. Establishing mechanisms for these diverse voices to be heard and acted upon throughout the AI lifecycle is paramount.

The
**iterative nature of AI development** itself offers opportunities for proactive harm identification. During the prototyping and testing phases, it is crucial to go beyond simple accuracy checks and actively probe for

potential discriminatory outcomes or unintended consequences. This might involve designing test cases that specifically target known areas of vulnerability or bias. For example, if an AI system is being developed to assist in child welfare risk assessments, the testing phase should include scenarios involving families with specific cultural practices or socioeconomic challenges that might be misinterpreted by a biased algorithm. The goal is to uncover these potential issues in a controlled environment, allowing for their correction before the system is exposed to real-world populations where the consequences of failure could be severe.

A significant challenge in proactive harm identification is the **lack of definitive predictive power**. It is impossible to foresee every single negative outcome that an AI system might generate. The real world is complex and dynamic, and unforeseen interactions and emergent behaviors are inevitable. Therefore, proactive identification should not be seen as a guarantee of complete harm elimination, but rather as a strategy to significantly reduce the likelihood and severity of foreseeable harms. This means acknowledging the limitations of foresight and complementing proactive measures with robust mechanisms for **ongoing monitoring, rapid response, and continuous learning**, as discussed in the previous section. The journey of responsible AI deployment is one where anticipation is paired with vigilance, and where the commitment to minimizing harm is a perpetual endeavor.

In essence, the necessity of proactive harm identification stems from a profound ethical responsibility. It is about acknowledging the power of AI and the potential for that power to be misused, either intentionally or, more often, through oversight, flawed assumptions, or systemic blind spots. It is about shifting from a reactive posture to one of diligent anticipation, recognizing that the public sector's mandate is to serve and protect all citizens, and that AI systems must be designed and deployed with this fundamental principle at their core. By embedding foresight, critical inquiry, and diverse perspectives into every stage of the AI lifecycle, we can move closer to realizing the promise of AI for public good, while simultaneously safeguarding against its potential to inflict harm. This proactive stance is not merely good practice; it is an indispensable prerequisite for building and maintaining public trust in the age of artificial intelligence.

The imperative to rigorously assess the impact of artificial intelligence systems, particularly within the public sector, has given rise to a variety of structured methodologies designed to anticipate and mitigate potential harms. These frameworks serve as essential navigational tools, guiding practitioners and policymakers through the complex terrain of AI's societal implications. By providing systematic approaches, they aim to ensure that the deployment of AI aligns with public values, upholds ethical principles, and serves the broader public good without inadvertently causing damage or exacerbating existing inequities.

One of the most prominent and increasingly adopted frameworks is the **Algorithmic Impact Assessment (AIA)**. Originating from governmental and regulatory initiatives, particularly in Canada and more recently in other jurisdictions like the European Union, the AIA is designed to provide a standardized process for evaluating the potential risks and impacts of automated decision-making systems. At its core, an AIA seeks to answer critical questions about how an AI system might affect individuals and society, focusing on areas such as fairness, accountability, transparency, and safety.

The typical stages of an AIA often begin with a **screening process** to determine if a full assessment is necessary. This involves identifying the AI system, its intended purpose, and the types of data it utilizes. If the system processes personal information, makes decisions that significantly impact individuals' rights, freedoms, or well-being, or if it is deployed in a high-risk context, a more in-depth assessment is usually triggered. This initial step is crucial for resource allocation, ensuring that the most impactful systems receive the most thorough scrutiny.

Following the screening, the core of the AIA involves a detailed **impact evaluation**. This stage requires a comprehensive analysis of the AI system across several dimensions. A critical component is the examination of **bias and discrimination**. This involves scrutinizing the data used to train the AI for historical biases and ensuring that the algorithm itself does not perpetuate or amplify discrimination against protected groups. Methodologies here might include statistical tests for disparate impact, qualitative reviews of data sources, and an analysis of the model's decision-making logic for potential discriminatory patterns. The goal is to understand how the system might treat different demographic groups and to identify any unwarranted disparities.

Another vital dimension of the AIA is the assessment of **transparency and explainability**. Public sector AI systems, especially those influencing critical decisions, must be understandable to those affected by them, as well as to oversight bodies. This stage involves evaluating the extent to which the AI's decisions can be explained, documented, and communicated. For complex "black box" models, this might involve employing explainability techniques (like LIME or SHAP) or developing narrative explanations that, while not fully technical, convey the rationale behind a decision in an accessible manner. The ability to understand *why* a decision was made is fundamental for accountability and for enabling individuals to challenge outcomes they believe are unfair.

**Accountability and governance** are also central to the AIA framework. This involves clarifying who is responsible for the AI system's operation, its outcomes, and any negative impacts it may cause. It requires establishing clear lines of responsibility for monitoring, auditing, and rectifying issues. Furthermore, it necessitates the development of robust governance structures, including policies, procedures, and oversight mechanisms, to ensure that the AI system is used ethically and legally, and that recourse is available to those who are harmed.

The
**safety and security** of the AI system are also rigorously examined. This includes assessing the potential for technical failures, cyber-attacks, or unintended emergent behaviors that could lead to harm. Robust testing, validation, and security protocols are essential to ensure that the system operates reliably and securely, protecting both individuals and the integrity of public services.

Finally, an AIA typically concludes with the development of a **mitigation plan**. This plan outlines concrete steps to address any identified risks and impacts. It might involve modifying the AI system, improving data quality, enhancing transparency, establishing new oversight procedures, or providing additional training for human operators. The mitigation plan is not a one-time document; it forms the basis for ongoing monitoring and iterative improvement of the AI system throughout its lifecycle. The process of conducting an AIA often involves extensive **stakeholder consultation**, engaging with affected communities, domain experts, civil society organizations, and internal personnel to gain diverse perspectives and ensure that the assessment is comprehensive and reflects real-world concerns.

Closely related to the AIA, and sometimes used interchangeably or in conjunction with it, is the
**Ethical Impact Assessment (EIA)**. While an AIA might be more focused on regulatory compliance and specific algorithmic risks, an EIA often takes a broader, more philosophical approach to ethical considerations.

EIAs are designed to systematically identify, analyze, and evaluate the ethical implications of introducing a new technology or system, such as an AI application, into a specific context.

The components of an EIA typically include a **definition of ethical principles** relevant to the context. These principles can be drawn from established ethical theories, professional codes of conduct, or societal values, and might include principles like beneficence (doing good), non-maleficence (avoiding harm), justice (fairness), autonomy (respecting individual choice), and accountability. The assessment then examines how the AI system uphms or potentially violates these principles.

A key aspect of an EIA is the **stakeholder analysis**, which goes beyond simply identifying who is affected to understanding their values, interests, and potential ethical concerns. This often involves participatory methods, such as focus groups, interviews, and deliberative workshops, to ensure that the voices of those most likely to be impacted are heard and integrated into the assessment process. This approach recognizes that ethical considerations are not solely technical but are deeply embedded in social and human contexts.

EIAs also involve a thorough examination of the **intended and unintended consequences** of the AI system through an ethical lens. This means considering not only the direct outcomes of the AI's decisions but also the broader societal shifts it might instigate. For instance, an AI system that automates a public service might increase efficiency but could also lead to job displacement or a de-skilling of the human workforce, raising ethical questions about societal well-being and the future of work. Similarly, an AI used for predictive policing might be ethically scrutinized for its potential to reinforce existing social biases and lead to over-surveillance of marginalized communities, even if its immediate goal is crime reduction.

Another crucial element of EIAs is the exploration of **values conflicts**. AI systems often operate at the intersection of competing values. For example, a system designed to optimize resource allocation might prioritize efficiency, which could conflict with equity or individual liberty. An EIA aims to identify these conflicts and explore potential trade-offs, seeking solutions that minimize ethical compromises or offer justifiable resolutions. This might involve trade-off analysis, ethical reasoning, and the development of guidelines for navigating such conflicts.

The EIA process also emphasizes **reflexivity** and **critical self-examination** on the part of developers, deployers, and policymakers. It encourages a continuous questioning of assumptions, motivations, and the underlying ethical frameworks that inform the design and use of AI. This reflective practice is essential for identifying blind spots and ensuring that ethical considerations are not an afterthought but are integrated into the very inception of an AI project.

Similar to the AIA, an EIA culminates in recommendations for **ethical mitigation and governance**. These recommendations might include the adoption of specific ethical guidelines, the establishment of ethics review boards, the development of training programs for personnel, or the redesign of the AI system to better align with ethical principles. The emphasis is on creating robust ethical safeguards and fostering a culture of ethical responsibility within organizations deploying AI.

Beyond these established frameworks, a growing number of specialized

methodologies and approaches are emerging to address specific challenges posed by AI. For instance,
**AI Safety Impact Assessments** are gaining prominence, focusing specifically on the risks associated with advanced AI systems, including potential existential risks, unintended emergent behaviors in complex systems, and the alignment problem (ensuring AI goals remain aligned with human values). These assessments often draw from fields like formal verification, robust AI research, and scenario planning to evaluate and mitigate potential catastrophic failures.

Furthermore,
**human rights impact assessments** are being adapted for the AI context. These assessments focus on how AI systems might affect internationally recognized human rights, such as the right to privacy, freedom of expression, non-discrimination, and due process. By framing AI impacts through the lens of human rights, these assessments provide a strong legal and ethical foundation for evaluation, linking AI deployment to existing international legal obligations.

The development and adoption of these frameworks are not without their challenges. One significant challenge is the
**complexity and rapid evolution of AI technology**. Methodologies need to be flexible enough to adapt to new AI techniques, models, and applications. Another challenge is the **availability of expertise**. Conducting thorough impact assessments requires a multidisciplinary team with skills in AI, ethics, law, social science, and domain-specific knowledge. Furthermore, **data availability and quality** remain critical issues, as many assessments rely on data that may be biased, incomplete, or inaccessible.

The
**operationalization of these frameworks** also presents practical hurdles. Integrating impact assessments into existing bureaucratic

processes, securing adequate funding and staffing, and ensuring that assessments lead to meaningful changes rather than mere bureaucratic exercises are ongoing concerns. There is a risk that assessments can become tick-box exercises, performed to satisfy regulatory requirements without genuinely informing decision-making or leading to significant improvements in AI design and deployment.

However, the proliferation and refinement of frameworks like AIAs and EIAs represent a crucial step forward. They provide the public sector with structured, systematic ways to engage with the complex ethical and societal implications of AI. By offering concrete stages, key considerations, and recommended actions, these frameworks empower organizations to move beyond ad-hoc approaches and to proactively address potential harms. They foster a culture of responsibility, transparency, and accountability, which is indispensable for harnessing the transformative potential of AI for public good while safeguarding against its risks. The ongoing evolution and adoption of these assessment methodologies underscore a growing societal commitment to ensuring that AI serves humanity ethically and equitably.

The previous sections have established the foundational principles and overarching frameworks for conducting robust impact assessments of AI systems. We have explored methodologies like Algorithmic Impact Assessments (AIAs) and Ethical Impact Assessments (EIAs), highlighting their systematic approaches to identifying and mitigating risks. We've also touched upon specialized assessments like AI Safety Impact Assessments and Human Rights Impact Assessments, underscoring the need for context-specific scrutiny. However, a truly comprehensive impact assessment cannot remain at a general level; it must dive deep into the unique operational landscapes and societal considerations of each domain where AI is deployed. The risks and consequences associated with an AI system used in healthcare, for instance, will inherently differ from those of an AI deployed in the justice system or in urban planning. Therefore, a critical next step in our journey is to dissect these assessments across various sectors, recognizing that a one-size-fits-all approach to risk identification is insufficient.

When we consider the healthcare sector, the potential benefits of AI are immense, ranging from enhanced diagnostic accuracy and personalized treatment plans to streamlined administrative processes and accelerated drug discovery. However, these advancements are inextricably linked to a unique set of risks that demand careful consideration. One of the most significant concerns revolves around **diagnostic and treatment errors**. An AI system designed to interpret medical images, such as X-rays or MRIs, might, due to biased training data or algorithmic flaws, consistently misdiagnose certain conditions in specific demographic groups. For example, if the training dataset for a melanoma detection AI is disproportionately composed of lighter skin tones, it may perform poorly when identifying cancerous lesions on darker skin, leading to delayed or missed diagnoses and potentially severe health outcomes. Similarly, AI-powered treatment recommendation systems, if not rigorously validated for efficacy and safety across diverse patient populations, could suggest suboptimal or even harmful therapies. The stakes in healthcare are uniquely high; errors can directly translate into physical harm, suffering, and loss of life. Consequently, impact assessments in this domain must place a paramount emphasis on clinical validation, real-world performance monitoring, and the establishment of clear protocols for human oversight and intervention. The "explainability" of AI decisions becomes not just an ethical ideal but a medical necessity, allowing clinicians to understand the basis of an AI's recommendation and to override it when their professional judgment dictates.

Beyond direct clinical impacts, data privacy and security are amplified concerns in healthcare. AI systems often require access to vast amounts of sensitive personal health information (PHI). While this data is crucial for training and operation, its handling must adhere to stringent regulations like HIPAA in the United States or GDPR in Europe. An impact assessment must therefore scrutinize how PHI is collected, stored, processed, and shared. Are robust anonymization and de-identification techniques employed? What are the protocols for data breach response? Could the aggregation of data by an AI system inadvertently lead to re-identification of individuals, thereby compromising their privacy? The potential for discrimination based on health data, such as an AI system that predicts a higher likelihood of certain diseases for individuals from specific socioeconomic backgrounds or geographic locations, also presents a serious ethical and practical challenge. This could lead to discriminatory practices in insurance coverage, access to care, or even employment, creating a new layer of health inequity. Therefore, any impact assessment for AI in healthcare must include a thorough examination of data governance, ethical data use policies, and the potential for privacy violations or discriminatory profiling, ensuring that the pursuit of technological advancement does not erode fundamental rights to privacy and equitable treatment.

Turning our attention to the domain of housing and urban planning, AI is being increasingly leveraged for tasks such as property valuation, tenant screening, loan application processing, and even resource allocation for public services. Here, the risks are deeply intertwined with historical patterns of discrimination and systemic inequities. A primary concern is the perpetuation or exacerbation of **bias in allocation and decision-making**. AI algorithms used for tenant screening, for example, might inadvertently learn to associate certain zip codes or demographic characteristics with higher eviction rates or lower creditworthiness, even if these correlations are rooted in systemic disadvantages rather than individual culpability. This can lead to AI systems effectively replicating historical practices of redlining, excluding marginalized communities from rental opportunities or access to housing finance. An impact assessment must therefore meticulously audit the data used to train these algorithms, looking for proxies for protected characteristics and evaluating the fairness of the algorithm's outcomes across different racial, ethnic, and socioeconomic groups.

Furthermore, the deployment of AI in areas like predictive policing for resource allocation within a city or in the planning of public services can lead to
**disproportionate surveillance and resource diversion**. If an AI system is trained on historical crime data that reflects biased policing practices, it might direct more law enforcement resources to already over-policed minority neighborhoods, creating a feedback loop of increased arrests and further data reinforcing the bias. Conversely, AI systems used to optimize the placement of new public facilities, such as schools or community centers, could inadvertently disinvest from historically underserved areas if the underlying data does not adequately account for community needs and historical inequities. Impact assessments in this sector must therefore critically examine the data sources, the algorithmic objectives, and the potential for reinforcing spatial or social segregation. Beyond bias, there are also concerns about **transparency and accountability** in AI-driven housing decisions. When an AI system denies a loan application or rejects a housing application, individuals have a right to understand the reasons behind that decision. If the AI's decision-making process is opaque, it becomes exceedingly difficult for individuals to challenge potentially unfair outcomes or for regulators to identify and rectify systemic problems. Therefore, an assessment must consider the mechanisms for providing meaningful explanations to affected individuals and establish clear lines of accountability for algorithmic decisions.

In the realm of transportation, AI holds the promise of revolutionizing mobility through autonomous vehicles, optimized traffic management,

and enhanced public transit systems. However, the introduction of AI into this domain introduces a distinct set of risks, primarily centered around
**safety and accessibility**. The development of self-driving cars, for instance, requires AI systems capable of making split-second decisions in complex, unpredictable environments. While extensive testing is conducted, the possibility of unforeseen scenarios or algorithmic failures leading to accidents remains a significant concern. An impact assessment must delve into the robustness of the AI's perception systems, its decision-making logic under various conditions (e.g., adverse weather, complex intersections), and its fail-safe mechanisms. The "edge cases"—rare but critical scenarios that the AI may not have encountered during training—are particularly important to analyze. For example, how does an autonomous vehicle respond to an unexpected pedestrian darting into traffic, or to a sudden obstruction on the road? Ensuring that these systems are not only functional but demonstrably safer than human drivers across a wide spectrum of conditions is paramount.

Beyond safety,
**accessibility** is a crucial consideration. AI-powered transportation solutions must be designed to serve all members of society, including individuals with disabilities, the elderly, and those in low-income or rural communities. Autonomous vehicles, for example, must be designed with features that accommodate diverse needs, such as wheelchair accessibility and intuitive interfaces for individuals with cognitive or sensory impairments. Similarly, AI used to optimize public transportation routes should not inadvertently lead to service reductions or exclusion of certain neighborhoods, perpetuating existing transportation deserts. An impact assessment in this sector must therefore evaluate the AI system's design from an equity and inclusion perspective, ensuring that it enhances, rather than diminishes, mobility for all. Furthermore, the cybersecurity of transportation AI systems is critical. A compromised traffic management system could lead to

widespread chaos and accidents, while a hacked autonomous vehicle could be weaponized. Therefore, the assessment must rigorously examine the security protocols, the potential attack vectors, and the resilience of the AI systems against malicious actors.

The application of AI in the **criminal justice system** presents a particularly sensitive area, where the consequences of flawed or biased systems can have profound and life-altering impacts on individuals and communities. AI is increasingly being used for risk assessment tools to inform decisions about pretrial detention, sentencing, and parole. These tools are designed to predict an individual's likelihood of reoffending or failing to appear in court. However, a significant body of research has demonstrated that many of these tools exhibit **racial and socioeconomic bias**. They often rely on historical arrest and conviction data, which are themselves products of systemic biases in policing and judicial practices. As a result, individuals from minority groups may be unfairly assigned higher risk scores, leading to harsher pretrial detention decisions, longer sentences, or denial of parole, even when their individual circumstances do not warrant such outcomes. An impact assessment in this domain must therefore focus intensely on the fairness and validity of these risk assessment tools. This requires not only scrutinizing the algorithms themselves for disparate impact but also critically examining the data sources and the very definitions of "risk" that the AI is trained to identify. Are these definitions neutral, or do they implicitly encode societal biases?

Beyond risk assessment, AI is also being explored for predictive policing, facial recognition for identification, and even for analyzing evidence. Predictive policing algorithms, when fed biased data, can lead to the over-surveillance and over-policing of certain neighborhoods, disproportionately impacting minority communities and eroding trust between law enforcement and the public. Facial recognition technology, while useful for identification, has been shown to have significantly higher error rates for women and people of color, leading to a substantial risk of false arrests and wrongful accusations. An impact assessment must therefore rigorously evaluate the accuracy and reliability of these technologies across different demographic groups and consider the potential for misuse. The **transparency and explainability** of AI decisions in the criminal justice system are also paramount. When an AI tool influences a decision about someone's liberty, that person, their legal counsel, and the court have a right to understand how the decision was reached. If the AI is a "black box," it undermines due process and the ability to challenge potentially erroneous or discriminatory outcomes. Therefore, impact assessments must prioritize the development and deployment of explainable AI systems in this sector, alongside robust mechanisms for human review and override. The fundamental principles of justice—fairness, impartiality, and due process—must be the guiding stars for any AI deployed within this sensitive domain.

In the **education sector**, AI offers promising avenues for personalized learning, automated grading, and administrative efficiency. However, risks emerge concerning **equity in access and outcomes, data privacy, and the potential for deskilling educators**. AI-powered adaptive learning platforms, while designed to tailor instruction to individual student needs, can inadvertently create or widen achievement gaps if the underlying algorithms are biased or if access to the technology is inequitable across socioeconomic strata. Students in under-resourced schools may not have the same access to sophisticated AI tutors or personalized learning tools as their peers in affluent districts, exacerbating existing educational disparities. An impact assessment must therefore consider the equitable distribution of AI educational tools and scrutinize the algorithms for biases that might disadvantage certain student populations. For instance, an AI tutor that is optimized for a particular learning style might fail to adequately support students with different cognitive approaches.

Furthermore, AI systems in education often collect vast amounts of student data, including performance metrics, learning patterns, and even behavioral observations. The **privacy and security of this sensitive data** are of paramount importance. Impact assessments must address how this data is collected, stored, used, and protected, ensuring compliance with privacy regulations and preventing its misuse for commercial purposes or discriminatory profiling. There is also the risk that an over-reliance on AI for tasks such as grading could lead to the **deskilling of educators**, diminishing their pedagogical judgment and their ability to provide nuanced feedback and mentorship. The human element of teaching—empathy, inspiration, and the fostering of critical thinking and socio-emotional skills—is difficult, if not impossible, for AI to replicate. Therefore, impact assessments in education should not only focus on the technical performance of AI tools but also on their impact on the role of educators and the holistic development of students. The goal should be to augment, not replace, the crucial human interaction in the learning process.

Finally, considering the deployment of AI in **financial services**, beyond housing and general credit, risks center on fairness, market stability, and consumer protection. AI is widely used in algorithmic trading, fraud detection, personalized financial advice, and credit scoring. In algorithmic trading, the speed and complexity of AI systems can potentially contribute to **market volatility and systemic risk**. Flash crashes, where market prices plummet rapidly due to automated trading, highlight the potential for AI to amplify market instability. An impact assessment in this context needs to consider the interconnectedness of AI trading systems and their potential for cascading failures. For fraud detection and credit scoring, the same concerns about **bias and discrimination** that we've seen in other sectors are highly relevant. An AI system trained on historical loan default data might unfairly penalize applicants from marginalized communities or those with non-traditional financial histories, limiting their access to essential financial products and services. The fairness of these algorithms must be rigorously assessed to ensure that they do not perpetuate or create new forms of financial exclusion. Moreover, personalized financial advice delivered by AI must be both accurate and in the best interest of the consumer, raising questions about **fiduciary duty and accountability** when AI is making recommendations that could significantly impact an individual's financial well-being. Ensuring transparency in how these AI systems operate and providing clear avenues for recourse for consumers who believe they have been unfairly treated are critical components of any responsible impact assessment.

Across all these domains, a common thread emerges: the imperative to move beyond generic risk checklists and engage in a deep, sector-specific understanding of how AI interacts with existing societal structures, human vulnerabilities, and established regulatory landscapes. Each sector has its own unique set of ethical considerations, legal obligations, and potential for harm. Therefore, impact assessments must be tailored, drawing on domain expertise, engaging with affected communities, and employing methodologies that are sensitive to the specific risks at play. Only through this nuanced, context-aware approach can we hope to harness the transformative potential of AI for public good while diligently safeguarding against its myriad risks.

The preceding sections have underscored the necessity of granular, sector-specific impact assessments for AI systems. We've delved into the unique challenges and risks presented by AI deployment across domains like healthcare, urban planning, transportation, criminal justice, education, and finance, highlighting how a one-size-fits-all approach is insufficient. However, even the most sophisticated, domain-aware assessment framework will fall short if it does not actively and meaningfully incorporate the voices and experiences of the individuals and communities who will be most directly impacted by the AI system under scrutiny. This subsection shifts our focus to the critical practice of integrating stakeholder perspectives, recognizing that quantitative data and algorithmic analyses, while indispensable, often fail to capture the full spectrum of potential harms, particularly those that are qualitative, nuanced, or deeply rooted in lived experiences.

To truly understand the impact of an AI system, we must move beyond abstract risk matrices and theoretical frameworks to engage with the concrete realities of those who will interact with, be subjected to, or be otherwise affected by its deployment. These stakeholders are the ultimate arbiters of an AI system's true societal impact. They possess invaluable insights into how an AI might manifest existing inequalities, create new forms of discrimination, erode trust, or disrupt established social fabrics in ways that may not be immediately apparent through purely technical or quantitative evaluations. For instance, an AI system designed for loan applications might be deemed statistically fair according to certain metrics, but individuals from historically marginalized communities might articulate how the system's opacity or reliance on proxy variables still leads to disproportionate rejections, reflecting a deeper, experiential bias that quantitative measures alone cannot fully articulate. Similarly, an AI used for public service allocation could appear efficient on paper, but residents of affected neighborhoods might describe how it bypasses crucial community knowledge, leading to the neglect of genuine, pressing needs that were not quantifiable inputs into the algorithm.

Therefore, a cornerstone of any robust impact assessment process must be the systematic and inclusive engagement of diverse stakeholders. This is not merely a matter of public relations or compliance; it is an ethical imperative and a practical necessity for identifying and mitigating risks effectively. The "who" of stakeholder engagement is as crucial as the "how." It must extend beyond easily accessible groups or vocal advocates to include those who are most vulnerable, least likely to have their voices heard through conventional channels, and most likely to bear the brunt of any negative consequences. This includes, but is not limited to, individuals who belong to protected characteristics groups (such as racial and ethnic minorities, women, individuals with disabilities, LGBTQ+ individuals), low-income populations, the elderly, geographically isolated communities, and any other groups that may be disproportionately affected by algorithmic decision-making. Furthermore, engagement should encompass a range of perspectives, including end-users of the system, those who are subject to its decisions, domain experts who understand the nuances of the application area, civil society organizations, community leaders, and even those who might be indirectly affected.

The methods employed for stakeholder engagement must be as varied and adaptable as the communities they aim to reach. Traditional public forums or online surveys, while useful, may not be sufficient. More proactive and participatory approaches are often required. One powerful method is the use of
**community workshops and focus groups**. These settings allow for in-depth discussions, the exploration of complex issues, and the surfacing of nuanced concerns. Facilitators must be skilled in creating safe spaces for open dialogue, ensuring that all participants feel comfortable sharing their perspectives without fear of reprisal or dismissal. It is essential that these workshops are designed with cultural sensitivity, taking into account language barriers, accessibility needs, and local customs. For example, when assessing an AI system for use in a community with a significant non-English speaking population, providing interpretation services and materials in multiple languages is non-negotiable.

Moreover, the timing and location of these sessions should be convenient for participants, perhaps holding them during evenings or weekends in accessible community centers, rather than during standard business hours in formal government buildings.

Another vital approach is **participatory design and co-creation**. In this model, stakeholders are not just consulted but actively involved in the design, development, and evaluation phases of the AI system. This can involve forming advisory boards composed of representatives from affected communities, or even embedding community members within the development team. For instance, if an AI is being developed to assist with urban park maintenance and resource allocation, involving local residents, community garden organizers, and neighborhood association representatives in defining the system's objectives, identifying key performance indicators, and even testing its outputs can ensure that the system addresses actual community needs and priorities, rather than just abstract efficiency metrics. This collaborative approach fosters a sense of ownership and ensures that the AI is more likely to be accepted and effective. It also provides a continuous feedback loop, allowing for iterative refinement of the system based on ongoing stakeholder input.

**Storytelling and qualitative data collection** are also crucial complements to quantitative assessments. While metrics can tell us *what* is happening, stories can often illuminate *why* and *how*. Encouraging individuals to share their personal experiences with existing systems or their concerns about proposed AI applications can reveal subtle forms of bias, exclusion, or harm that might otherwise go unnoticed. For example, a personal narrative from someone who was denied a crucial service due to an AI-driven decision, detailing the emotional toll, the perceived unfairness, and the lack of recourse, can be far more persuasive and informative than a statistical report on denial rates. Impact assessments should therefore incorporate methods for systematically gathering and analyzing such qualitative data, such as in-depth interviews, ethnographic studies, and the collection of testimonials. These qualitative insights can help to contextualize quantitative findings, identify emergent risks, and inform the development of more sensitive and equitable AI systems.

The principle of
**"nothing about us without us"** must be central to the entire process. This means that decisions regarding the deployment and governance of AI systems should not be made in isolation by technologists or policymakers without genuine input from the people who will be most affected. This principle requires a commitment to transparency throughout the assessment process itself. Stakeholders should be informed about the purpose of the assessment, the AI system being evaluated, the methodologies being used, and how their input will be incorporated into the final assessment and subsequent decision-making. They should also be given access to summaries of the assessment findings and have opportunities to comment on them.

Furthermore, effective stakeholder engagement necessitates a recognition of power imbalances. Historically marginalized communities often face systemic barriers to participation, including lack of time, resources, technical literacy, and access to decision-making forums. Impact assessments must proactively address these barriers. This could involve providing compensation for participants' time and expertise, offering training to enhance understanding of AI concepts and assessment methodologies, and utilizing accessible communication channels and formats. It also means actively seeking out and amplifying the voices of those who are often excluded, rather than relying on them to find their own way to the table. This might involve partnering with trusted community organizations, conducting outreach in familiar settings, and employing culturally competent engagement strategies.

The findings from stakeholder engagement must then be demonstrably integrated into the impact assessment report and, more importantly, into the recommendations and mitigation strategies developed. It is insufficient to simply collect input; the assessment must articulate how this input has informed the analysis of risks, the identification of potential harms, and the proposed solutions. For example, if community members express deep concerns about the potential for an AI-powered surveillance system to exacerbate profiling and harassment in their neighborhood, the impact assessment must not only acknowledge these concerns but also analyze the specific mechanisms through which this harm could occur and propose concrete mitigation strategies, such as strict limitations on data retention, independent oversight bodies, or even a moratorium on deployment until robust safeguards are proven effective. The assessment should clearly document how stakeholder input has shaped the risk assessment and what specific mitigation measures have been developed in response to their feedback.

Moreover, the integration of stakeholder perspectives should extend to the ongoing monitoring and evaluation of AI systems post-deployment. AI systems are not static; they evolve, and their impacts can change over time. Establishing mechanisms for continuous feedback and ongoing dialogue with affected communities is essential for ensuring that the AI system remains aligned with societal values and continues to meet its intended objectives without causing undue harm. This might involve periodic reviews involving community representatives, mechanisms for reporting and addressing emergent issues, and a commitment to adapting the system or its governance based on ongoing feedback and observed impacts. By embedding stakeholders throughout the lifecycle of an AI system, from its inception through its deployment and evolution, we can move towards a more responsible, equitable, and truly beneficial use of artificial intelligence. This sustained commitment to inclusive participation transforms impact assessments from mere technical exercises into dynamic, community-centered processes that foster trust, promote accountability, and ultimately lead to AI systems that serve the public good in a meaningful and equitable manner.

Having diligently navigated the complexities of AI impact assessment, from understanding the specific vulnerabilities of various sectors to ensuring the inclusion of diverse stakeholder voices, we now arrive at the crucial juncture: translating these insights into tangible action. The preceding discussions have underscored that an impact assessment is not an end in itself, but rather a foundational step—a diagnostic tool that illuminates potential risks and harms. Without a robust framework for mitigation and remediation, even the most thorough assessment risks becoming a purely academic exercise, failing to deliver on its promise of responsible AI deployment and citizen protection. This final stage of the impact assessment process is where the commitment to ethical AI is truly tested, moving from analysis to proactive intervention and prepared response.

The core of this actionable phase lies in the development of concrete **mitigation strategies**. These are deliberate measures designed to prevent, reduce, or eliminate the negative consequences identified during the assessment. Mitigation is not a one-size-fits-all endeavor; its efficacy hinges on its specificity, its feasibility, and its direct alignment with the risks identified. For an AI system determining eligibility for social housing, for instance, if the impact assessment reveals a disproportionate denial rate for single-parent households due to the algorithm's weighting of income stability metrics, a mitigation strategy could involve introducing an explicit adjustment factor for family structure or a mandatory human review for all applications flagged with specific demographic indicators. This strategy directly addresses the identified harm by recalibrating the system's decision-making process.

Developing effective mitigation strategies requires a multi-pronged approach, considering technical, procedural, and organizational adjustments. On the technical front, this might involve modifying algorithms, refining datasets, or implementing bias detection and correction mechanisms *within* the AI system itself. For an AI used in predictive policing, if the assessment highlights the risk of reinforcing existing racial biases due to historical over-policing in certain neighborhoods being used as training data, mitigation could involve actively re-balancing the training data, employing fairness-aware machine learning techniques, or implementing a strict oversight mechanism that scrutinizes and overrides any recommendations that show signs of algorithmic bias. This might also entail developing simpler, more transparent models where complex, opaque ones pose an unacceptable risk, prioritizing explainability and auditability over marginal gains in predictive accuracy.

Procedural mitigation strategies focus on the processes surrounding the AI system's deployment and operation. This could include establishing clear guidelines for human oversight, defining appeal processes for individuals affected by AI decisions, and implementing regular audits. If an AI system for university admissions is found to inadvertently disadvantage applicants from under-resourced schools due to its reliance on standardized test scores that correlate with access to expensive test preparation, a procedural mitigation could be to mandate that admissions committees always consider a holistic review of each application, with specific attention paid to the context of the applicant's educational background. Furthermore, it might involve developing a clear, accessible pathway for applicants to provide additional context or challenge aspects of their application that they believe have been unfairly assessed by the AI. The design of these appeal mechanisms is paramount; they must be understandable, accessible, and provide a genuine opportunity for recourse, not merely a superficial nod to due process.

Organizational mitigation strategies involve changes in governance, training, and accountability structures. This includes training personnel who interact with the AI system on its limitations, potential biases, and the ethical implications of its use. For an AI deployed in healthcare to assist with diagnosis, if the assessment reveals concerns about potential racial disparities in diagnostic accuracy for certain conditions, organizational mitigation could involve mandatory training for clinicians on recognizing and mitigating implicit biases, establishing a clear protocol for when and how to override the AI's recommendations based on clinical judgment, and creating a cross-disciplinary ethics review board to monitor the system's performance and address emerging concerns. This board might include clinicians, data scientists, ethicists, and patient advocates, ensuring a range of perspectives inform ongoing governance.

The process of identifying and selecting appropriate mitigation strategies should itself be iterative and informed by the stakeholder engagement that forms a cornerstone of the impact assessment. Community members, domain experts, and civil society organizations often possess invaluable practical knowledge about the real-world implications of identified risks and can propose solutions that might not be apparent from a purely technical standpoint. For example, if an AI system designed to manage public transportation routes is found to inadequately serve elderly residents with mobility issues, community feedback might reveal that simply adjusting bus schedules is insufficient. The actual mitigation might need to involve integrating the AI with on-demand accessible transport services or developing a parallel AI-assisted dispatch system specifically for paratransit. This co-creation of solutions ensures that mitigation efforts are not only technically sound but also socially relevant and practically implementable.

A critical element in effective mitigation planning is the establishment of clear
**accountability mechanisms**. Who is responsible for implementing and monitoring these strategies? What are the performance indicators for their success? Without defined roles and measurable outcomes, mitigation plans can easily languish. For an AI system used in hiring, if a mitigation strategy involves implementing bias detection filters, the accountability should clearly lie with the technology team to ensure these filters are operational and effective, while the HR department might be accountable for monitoring the diversity metrics of the candidate pool and hiring outcomes, and escalating any concerning trends. Regular reporting on the effectiveness of mitigation strategies should be a standard practice, with findings made accessible to relevant oversight bodies and, where appropriate, to the public.

Beyond proactive prevention, a comprehensive approach to AI impact assessment necessitates
**remediation planning**. This is the crucial component that addresses what happens when, despite best efforts, adverse impacts do occur. Remediation acknowledges that AI systems, like any complex technology, can have unforeseen consequences, and that a robust governance framework must include provisions for addressing these failures effectively and equitably. It is about establishing clear protocols for response, correction, and, where necessary, compensation, ensuring that individuals and communities are not left to bear the brunt of algorithmic harm without recourse.

Remediation planning begins with the establishment of a

**clear incident response framework**. This framework should outline the steps to be taken immediately upon the identification of an adverse impact. This includes defining who is responsible for detecting and reporting incidents, how they will be investigated, and who has the authority to initiate corrective actions. For an AI system used to detect fraudulent insurance claims, if an investigation reveals that the system has erroneously flagged a significant number of claims from a specific demographic group, leading to undue scrutiny and delays for legitimate claimants, the incident response framework would dictate immediate steps such as temporarily disabling the suspicious detection module, initiating an urgent audit of the flagged claims, and informing the affected claimants about the situation and the steps being taken.

Central to remediation is the concept of **correction**. This involves not only addressing the immediate harm but also rectifying the underlying cause within the AI system or its operational processes. In the insurance fraud example, correction would involve identifying the specific data patterns or algorithmic logic that led to the erroneous flagging, such as a spurious correlation with specific zip codes or historical claims patterns unique to that demographic. The system would then need to be retrained or reconfigured to eliminate this discriminatory or erroneous logic. This might also extend to correcting any misinformation or reputational damage that the erroneous flagging might have caused to individuals or groups.

Furthermore, remediation planning must include provisions for **redress and compensation**. When individuals suffer tangible harm—financial loss, denial of essential services, reputational damage, or significant emotional distress—due to the actions of an AI system, there must be a mechanism for making them whole. This could range from simple apologies and reversal of decisions to financial compensation for documented losses. For an AI system used in credit scoring that leads to an incorrect denial of a mortgage, remediation might involve correcting the credit score, re-evaluating the mortgage application without the erroneous data, and providing compensation for any demonstrable financial losses incurred due to the delay or denial. The complexity and scale of compensation would, of course, depend on the severity and nature of the harm. Establishing a clear process for individuals to claim redress is vital; it should be accessible, timely, and free from undue burden.

The development of these remediation protocols should also be informed by stakeholder input. Communities that have historically been marginalized or have experienced harms from technological systems can offer crucial insights into what constitutes effective and just remediation. Their lived experiences can help define what constitutes 'harm' in practical terms, what forms of redress are meaningful, and what processes are most likely to be trusted. For instance, if an AI system used in the justice system leads to wrongful arrests or unfair sentencing, community representatives can advise on how to ensure that any subsequent compensation or review processes are conducted with sensitivity and do not further re-traumatize those affected.

Crucially, remediation planning should not be viewed as a purely reactive measure. It is also a critical component of **continuous improvement and accountability**. By meticulously documenting incidents, investigations, and remediation actions, organizations can gain invaluable insights into the persistent vulnerabilities of their AI systems and governance structures. This data can then feed back into the impact assessment and mitigation planning cycles, leading to more robust systems and more effective preemptive measures in the future. A detailed log of AI-related incidents and their resolutions can serve as a powerful audit trail, demonstrating a commitment to learning from mistakes and fostering public trust.

Consider an AI system deployed for optimising traffic flow in a city. If an incident occurs where the AI prioritises emergency vehicle routes so aggressively that it causes significant gridlock in residential areas, leading to missed school buses and delayed commutes for thousands, the remediation plan would involve not just re-calibrating the traffic optimisation algorithm to balance emergency response with general traffic flow, but also establishing a clear protocol for compensating affected residents for demonstrably late arrivals at work or school, and for proactively communicating with communities about how such an event will be prevented in the future. This communication might include public meetings with traffic engineers and AI developers, and the establishment of a community feedback channel specifically for the traffic AI. The process of documenting this incident, analyzing its root causes, implementing technical fixes, providing compensation, and communicating changes to the public reinforces the accountability loop.

The challenge with remediation often lies in the difficulty of definitively attributing harm solely to the AI. In complex socio-technical systems, AI rarely operates in isolation. Its outputs are often mediated by human decision-makers, existing organizational policies, and broader societal factors. Therefore, remediation planning must include strategies for **establishing causality and responsibility** in these multi-layered environments. This might involve developing sophisticated forensic tools for auditing AI decision trails, establishing clear lines of responsibility among AI developers, system operators, and end-users, and creating independent bodies capable of investigating complex AI-related incidents. The aim is to move beyond simply identifying that something went wrong, to understanding precisely *how* and *why* it went wrong, and who or what should be accountable for rectifying it and preventing recurrence.

Moreover, remediation planning must consider the **long-term implications** of AI failures. An AI system designed for job recruitment that exhibits bias may not just deny opportunities to individuals in the short term; it can also perpetuate cycles of disadvantage, impacting career trajectories, economic stability, and social mobility for entire cohorts of people. Remediation in such cases may need to extend beyond immediate corrections to encompass broader interventions, such as affirmative action policies, targeted skills development programs, or even systemic reviews of the industries or sectors where the biased AI was deployed. This holistic view of remediation ensures that we are not just patching individual failures but are actively working to undo systemic inequities that AI might exacerbate.

Ultimately, the development of robust mitigation strategies and comprehensive remediation planning transforms the impact assessment from a retrospective analysis into a forward-looking, dynamic process. It imbues the deployment of AI with a sense of proactive stewardship, acknowledging that the pursuit of innovation must be balanced with an unwavering commitment to protecting individuals and communities from potential harms. By integrating these actionable phases, organizations can move beyond simply identifying risks to actively managing them, building trust, ensuring accountability, and steering the development and deployment of artificial intelligence towards outcomes that are not only efficient and effective, but also equitable and just. This dedication to translating assessment into action is paramount for fostering a future where AI serves as a genuine force for public good.

# Chapter 6: The Art and Science of Co-Creation

The preceding discussion has laid the groundwork for understanding that the impact assessment of AI systems is not merely an analytical exercise but a springboard for proactive measures. We've explored the crucial steps of developing concrete mitigation strategies – technical, procedural, and organizational – and the equally vital planning for remediation, encompassing incident response, correction, redress, and continuous improvement. This journey from identification to action is where the promise of responsible AI is truly tested. However, a critical component is missing if we rely solely on internal assessments and planned interventions without deeply integrating the very people who will be affected by these technologies. This brings us to a fundamental shift in how AI systems should be conceived and built: moving beyond traditional feedback loops to embrace genuine co-creation.

Co-creation, in the realm of AI development, signifies a profound departure from conventional methodologies. It is not simply about soliciting opinions or conducting surveys after a system has been largely designed. Instead, it represents a paradigm shift where the communities affected by an AI system are not just recipients of information or end-users providing feedback, but active, equal partners throughout the entire lifecycle of the AI. This means bringing community members, domain experts, advocacy groups, and those most vulnerable to potential harms to the table from the very inception of an idea, through the conceptualization, design, development, testing, deployment, and ongoing refinement of AI systems. It is an iterative, dynamic process built on mutual respect, shared learning, and a commitment to collective ownership of the AI's outcomes.

Imagine an AI system designed to allocate limited public resources, such as grants for community development projects. In a traditional approach, developers might build a system based on predefined criteria, deploy it, and then collect feedback from grant applicants and administrators about its usability or perceived fairness. This feedback might lead to adjustments, but the core architecture and decision-making logic would likely remain largely intact, potentially perpetuating underlying biases or failing to capture nuanced community needs. In a co-creative model, however, the process would begin with intensive dialogue with community leaders, local organizations, and residents who understand the unique socio-economic fabric of the area. These conversations would not be about explaining a pre-existing system, but about jointly exploring the community's priorities, identifying critical needs that AI *could* address, and collaboratively defining what constitutes a "successful" allocation of resources from the community's perspective. This might involve identifying specific qualitative factors that are vital for local impact but difficult to quantify, or understanding historical injustices that need to be actively counteracted by the AI's design.

This deep engagement at the outset ensures that the AI is not an imposition but a tool forged in partnership. Developers gain invaluable insights into the lived realities, cultural contexts, and specific vulnerabilities of the intended beneficiaries. This understanding allows them to design algorithms that are not only technically sound but also ethically grounded and socially relevant. For instance, if the community highlights the critical importance of supporting micro-enterprises in underserved neighborhoods, the co-creation process might lead to the development of AI features that prioritize applications from these areas, perhaps by incorporating measures of local economic impact or by building in a weighting system that acknowledges the challenges faced by small businesses in accessing traditional funding channels. This is fundamentally different from an algorithm that might, by default, favour larger, more established entities with a track record of quantifiable success.

The "user feedback" model, while valuable, often operates with an inherent power imbalance. Developers create, and users react. Information flows primarily in one direction, and the community's role is often passive, reactive, or limited to identifying problems rather than shaping solutions. Co-creation, conversely, actively dismantles this imbalance. It positions the community as co-architects, bringing their expertise – often tacit knowledge and lived experience – to the forefront. This collaborative approach fosters a sense of shared ownership. When communities are involved in the creation process, they are more likely to trust, adopt, and advocate for the AI system, seeing it as a reflection of their collective aspirations rather than an external imposition. This shared ownership is a powerful antidote to the mistrust and resistance that can plague AI deployments, especially in public sector contexts where equity and public good are paramount.

The practical implementation of co-creation requires a commitment to new ways of working. It necessitates breaking down traditional silos between technical teams, policy makers, and community stakeholders. This involves establishing dedicated forums for dialogue, workshops, and collaborative design sessions. For example, a city government might convene regular "AI co-design labs" bringing together data scientists, urban planners, community organizers, and residents to brainstorm and prototype AI solutions for issues like public safety, waste management, or transportation. These labs would be structured to facilitate open communication, encourage diverse perspectives, and ensure that the technical feasibility of AI solutions is always balanced against community needs and ethical considerations. Prototypes might be developed and tested iteratively, with feedback loops that are immediate and deeply integrated, allowing for rapid adaptation and refinement based on communal input.

Furthermore, co-creation demands transparency and accountability that goes beyond typical corporate or governmental disclosure. It requires developers to be open about the data being used, the algorithms being developed, and the limitations of the AI. This transparency is not just about publishing documentation; it is about engaging in ongoing, accessible conversations where community members can ask questions, voice concerns, and receive understandable answers. When an AI system is being developed to assist in the allocation of affordable housing, for instance, the co-creation process would involve explaining to community representatives how the algorithm works, what data points it uses, and how decisions are made. If concerns arise about potential biases, such as an over-reliance on credit scores that might disadvantage low-income applicants, the community members would be empowered to suggest alternative data sources or modified weighting mechanisms, and the development team would be obligated to investigate and incorporate these suggestions where feasible and ethically sound. This level of openness builds trust and allows for a more robust challenge to potentially harmful assumptions embedded within the technology.

The ethical implications of AI are often best understood through the lens of lived experience. A technical team might analyze data and identify a statistical bias, but a community member experiencing the direct consequences of that bias – perhaps being unfairly denied a service or facing increased scrutiny – can articulate the human cost in ways that data alone cannot convey. Co-creation provides a structured pathway for this experiential knowledge to inform AI development. It moves the conversation from abstract ethical principles to concrete, actionable insights derived from real-world impacts. For example, an AI system designed to predict school truancy might inadvertently flag students from disadvantaged backgrounds who face systemic barriers to attendance, leading to punitive measures rather than supportive interventions. Through co-creation, educators, parents, and students themselves can highlight these risks, guiding developers to build in features that prioritize understanding underlying causes and connecting families with resources, rather than simply identifying absence.

This collaborative approach extends to the testing and validation phases. Instead of solely relying on internal testing or controlled user studies, co-creation involves the community in rigorously testing the AI system in real-world scenarios. This might include pilot programs where community members actively use the system, provide feedback on its performance, and flag any unexpected or harmful outcomes. For an AI system designed to help small businesses navigate regulatory compliance, community business owners could be part of the testing group, identifying points of confusion, suggesting improvements to the user interface, and flagging instances where the AI's advice might be technically correct but practically unfeasible or misleading in their specific local context. This type of real-world validation is invaluable for uncovering edge cases and ensuring that the AI functions effectively and equitably for its intended users.

Moreover, co-creation is not a one-time event; it is a continuous commitment. As AI systems evolve and the contexts in which they operate change, ongoing collaboration is essential. This means establishing mechanisms for continuous dialogue, feedback, and iterative refinement. It acknowledges that AI development is not a linear process with a definitive endpoint but a dynamic journey. For an AI-powered public health platform designed to provide personalized health advice, co-creation would involve ongoing engagement with patients, healthcare providers, and public health officials to ensure the platform remains relevant, accurate, and sensitive to evolving health needs and societal concerns. This could include regular advisory board meetings, user forums, and participatory design workshops to adapt the AI's recommendations and functionalities.

The integration of co-creation into the AI lifecycle fundamentally alters the relationship between technology developers and the communities they serve. It shifts from a paternalistic or purely transactional model to one of genuine partnership. This partnership is built on the recognition that the most effective and ethical AI systems are those that are built *with* people, not just *for* them. By embedding community participation at every stage, from conceptualization to ongoing improvement, organizations can move beyond simply mitigating potential harms to proactively building AI systems that are truly beneficial, equitable, and aligned with the diverse needs and values of society. This deepens the commitment to responsible AI, transforming it from a set of compliance requirements into an intrinsic aspect of the development process itself. Bridging the divide between the sophisticated world of artificial intelligence development and the rich tapestry of community lived experience is not a matter of chance, but of deliberate, strategic effort. It requires a conscious and sustained commitment to building robust bridges of communication and collaboration. This section delves into the practicalities of establishing these vital connections, acknowledging the inherent challenges and proposing concrete strategies to foster mutual understanding and respect.

One of the most immediate and pervasive barriers is the difference in language. Developers often operate within a lexicon of technical jargon – algorithms, datasets, neural networks, APIs, hyperparameter tuning – that can be utterly opaque to individuals outside their field. Conversely, community members speak the language of lived reality, of social dynamics, of cultural nuances, and of practical needs that may not be easily translated into technical specifications. Imagine a data scientist discussing "feature engineering" to identify patterns in urban traffic flow, while a community organizer is concerned with how the resulting AI might affect pedestrian safety in their neighborhood, particularly for elderly residents or children. Without intentional effort, these distinct vocabularies can create an unbridgeable chasm, leading to misunderstandings, frustration, and a sense of exclusion.

To overcome this, the first step is a deliberate act of translation and simplification. Developers must cultivate the ability to explain complex technical concepts in clear, accessible language, avoiding jargon wherever possible. This isn't about dumbing down the technology, but about making its essence comprehensible to a wider audience. Analogies can be powerful tools here. For instance, explaining an algorithm as a recipe, where the data are the ingredients and the model is the cooking process that produces a specific dish, can demystify the underlying logic. Similarly, when discussing data, developers can move beyond mentioning "training data" to explaining it as "examples we're showing the AI so it can learn."

Conversely, community members need to be encouraged and supported to articulate their needs and concerns in ways that developers can understand and act upon. This might involve facilitated workshops where community representatives are guided through structured exercises to describe their challenges, priorities, and desired outcomes. For example, if a community group is concerned about an AI system for crime prediction, they might be prompted to discuss not just the fear of increased policing, but also specific instances where such systems have led to profiling or the disproportionate surveillance of certain neighborhoods. This shifts the conversation from an abstract concern about "bias" to concrete examples of how the AI's predictions might manifest in daily life, providing developers with actionable insights.

Beyond language, cultural norms and communication styles can present significant hurdles. Technical development often thrives on directness, rapid iteration, and a focus on problem-solving. Community engagement, however, may prioritize relationship-building, consensus-seeking, and a more circuitous approach that allows for reflection and the airing of a broader range of perspectives. A developer might feel impatient with lengthy discussions that don't immediately yield a technical solution, while a community member might feel steamrolled if their concerns are not given adequate space and time to be fully explored.

Fostering empathy and active listening is paramount in navigating these differing styles. For developers, this means consciously stepping outside their technical frame of reference and making a genuine effort to understand the social, economic, and cultural context in which the AI will be deployed. It involves recognizing that community members are not just "users" but stakeholders with valuable expertise derived from lived experience. Active listening here means not just hearing the words spoken, but understanding the underlying emotions, values, and priorities being expressed. This might involve paraphrasing what has been heard to confirm understanding, asking clarifying questions without judgment, and demonstrating through body language and verbal cues that their contributions are valued.

For community members, active listening involves approaching the technical team with an openness to understand the constraints and possibilities of AI development. It means recognizing that developers are working within a framework of technical feasibility, ethical guidelines, and resource limitations. While their concerns must be prioritized, there's also a need to understand the realities of building and deploying technology. This does not diminish the importance of their input, but rather frames it within a collaborative problem-solving dynamic.

Power dynamics are another significant challenge that requires careful attention. In traditional development settings, developers and the organizations they work for often hold a disproportionate amount of power due to their technical expertise, access to resources, and decision-making authority. This can lead to situations where community input is sought but ultimately sidelined if it conflicts with the developers' initial plans or perceived efficiencies.

To counter this, deliberate efforts must be made to level the playing field. This can begin with establishing a shared understanding of roles and responsibilities from the outset. A co-creation charter, developed collaboratively, can outline expectations for all parties, including how decisions will be made, how disagreements will be resolved, and how feedback will be incorporated. This document serves as a foundational agreement, ensuring that everyone is working from the same set of principles.

Furthermore, providing adequate resources and support for community participation is crucial. This might include compensating community members for their time and expertise, offering childcare or transportation to facilitate attendance at meetings, and providing training to help them effectively engage with the technical aspects of the project. When community members are not burdened by practical barriers and are recognized for their valuable contributions, their ability to participate meaningfully is significantly enhanced.

Creating dedicated, safe, and inclusive spaces for dialogue is also essential. This could take the form of regular community advisory board meetings, participatory design workshops, or online forums designed for accessible engagement. These spaces must be carefully moderated to ensure that all voices are heard, that discussions remain respectful, and that dominant perspectives do not overshadow marginalized ones. The goal is to create an environment where open and honest feedback can be shared without fear of repricthe or dismissal.

For example, consider an AI system being developed to assist in the allocation of affordable housing. A traditional approach might involve developers creating an algorithm based on factors like income, credit history, and existing housing status, and then presenting this system to housing authorities and tenant advocacy groups for review. In a co-creative model, however, the process would begin much earlier. Developers, urban planners, and representatives from housing advocacy groups, as well as residents of affordable housing, would convene to discuss the fundamental goals.

During these initial sessions, the community representatives might highlight historical patterns of discrimination in housing allocation, the inadequacy of traditional credit scores for assessing the financial stability of low-income families, and the importance of factors like proximity to social services, schools, and employment opportunities. They might also articulate the need for a system that actively works to *correct* historical injustices, rather than merely perpetuate them. This could lead to discussions about incorporating qualitative data, such as community impact statements or references from local support organizations, into the AI's decision-making process.

The developers, in turn, would need to explain the technical challenges and possibilities of incorporating such qualitative data. They might discuss methods for sentiment analysis on written statements, the difficulty of standardizing subjective assessments, and the potential for introducing new forms of bias if these qualitative inputs are not carefully managed. This back-and-forth is where the bridging happens. Developers gain a deeper understanding of the complex human factors that traditional metrics miss, and community members gain insight into the technical realities and limitations.

As the development progresses, this collaboration would continue. For instance, when discussing the specific features of the AI, community members might express concerns about how the system prioritizes applications. They might advocate for a weighted system that gives additional points to families who have been on waiting lists for extended periods, or to those moving from precarious living situations. Developers would then translate these desires into algorithmic parameters, explaining the trade-offs involved. They might demonstrate how different weighting schemes affect the outcomes, perhaps showing that an overemphasis on one factor inadvertently disadvantages another equally important group.

Testing phases are also critical opportunities for bridge-building. Instead of solely relying on internal testing, the AI system would be piloted with diverse groups of prospective users. This could involve community members actively using a simulated version of the AI to apply for housing, providing feedback on the clarity of the application process, the fairness of the questions asked, and the transparency of the outcomes. A community resident might point out that a question about employment history is phrased in a way that is confusing for individuals who work in the informal economy. Another might highlight that the AI's assessment of "housing stability" doesn't account for the reality of temporary shelter arrangements for those experiencing homelessness.

These real-world insights are invaluable. They allow developers to refine the AI's logic, improve its user interface, and ensure that it aligns with the lived experiences of those it is intended to serve. This iterative process of testing, feedback, and refinement, driven by both technical expertise and community wisdom, is the hallmark of effective co-creation. It moves beyond simply soliciting feedback to actively shaping the technology based on collective understanding and shared goals.

Moreover, the process of building these bridges requires an ongoing commitment to transparency. Developers should be prepared to explain not only how the AI works but also its limitations, the data it uses, and the assumptions it makes. This transparency should be ongoing and accessible, not confined to a single technical document. For example, if the AI is being used to identify individuals at risk of requiring social services, developers might hold public sessions to explain how risk factors are identified, what data sources are consulted, and how the AI's predictions are used to inform interventions. They would also be open to discussing the ethical implications, such as the potential for over-identification or stigmatization, and how these risks are being mitigated.

This commitment to transparency fosters trust. When communities understand the mechanics of the AI and are confident that their concerns are being heard and addressed, they are more likely to embrace the technology and its potential benefits. It transforms the AI from a black box into a tool that is understood and, ideally, trusted.

Ultimately, building bridges between developers and communities is an investment in the ethical and effective deployment of AI. It is a recognition that the most impactful and responsible technologies are not those created in isolation by technical experts, but those that are forged in partnership, drawing on the diverse knowledge and lived experiences of all those who will be affected. This sustained effort in communication, empathy, and shared decision-making is not a secondary consideration; it is fundamental to the very art and science of co-creation.

The transition from understanding the *why* of co-creation to implementing the *how* involves a deep dive into the practical aspects of shaping an AI system. This is where the abstract principles of shared vision and mutual respect begin to manifest as concrete features, intuitive interfaces, and responsive user experiences. Co-designing the AI's functionality and user experience is not merely about adding a layer of user-friendliness; it is about imbuing the technology with the intelligence and understanding derived from the very communities it is intended to serve. This iterative process ensures that the AI is not only technically sound but also culturally relevant, ethically aligned, and genuinely useful.

One of the most impactful areas for co-creation is the definition of the AI's core functionalities. Often, developers approach a problem with a predefined set of capabilities in mind, driven by technical feasibility or perceived efficiency. However, community members, experiencing the problem firsthand, can offer a richer, more nuanced understanding of what functionalities are truly necessary and how they should operate. For instance, consider the development of an AI-powered platform designed to connect individuals with social services. While developers might focus on creating an algorithm that efficiently matches service requests with available providers based on capacity and eligibility criteria, community members might highlight the critical need for features that address trust, transparency, and accessibility. They might advocate for a system that clearly explains *why* a particular service is recommended, rather than simply presenting a list. They might also emphasize the importance of features that allow users to provide feedback on their experience with a service provider, creating a feedback loop that enhances accountability. Furthermore, they could point out the necessity of multilingual support or features that accommodate low digital literacy, such as voice-based input or simplified visual interfaces. These are functionalities that might not arise from a purely technical brainstorming session but are vital for ensuring the AI's actual adoption and effectiveness in the real world.

The user experience (UX) of an AI system is equally, if not more, important than its underlying functionality. An AI that is technically brilliant but confusing or off-putting to interact with will inevitably fail to achieve its goals. Co-design allows community members to actively participate in shaping how the AI communicates, how it presents information, and how it guides users through tasks. This means moving beyond generic interface design and creating experiences that resonate with the specific cultural contexts and expectations of the intended users. For a public health AI aimed at providing personalized health advice, community input could be invaluable in determining the appropriate tone of communication. Should the AI be authoritative, friendly, or empathetic? What kind of language should it use to explain complex medical information without causing alarm or confusion? Should it use emojis or visual aids? Community members can also offer crucial insights into how the AI should present data. For example, when an AI is designed to help residents understand local environmental risks, community members might suggest that raw data on pollutant levels is less useful than contextualized information about *what* those levels mean for their daily lives – for example, how they might affect children's play or outdoor activities. They might also advocate for clear calls to action, such as recommendations for specific protective measures or information on how to report concerns.

Another crucial aspect of co-designing user experience relates to error handling and unexpected situations. AI systems, by their nature, can sometimes produce erroneous outputs or fail to understand user input. How the AI handles these situations can significantly impact user trust and satisfaction. Community members can provide vital perspectives on how errors should be communicated and what recourse users should have. For example, if an AI system for managing public transportation schedules provides incorrect information, a community member might suggest that the AI should not just state it made an error but offer alternative solutions, such as suggesting the next available bus or providing a direct link to customer support. They might also emphasize the importance of a clear apology and a straightforward process for reporting inaccuracies, rather than a generic error message that leaves the user feeling frustrated and unassisted. This focus on graceful failure and transparent recovery mechanisms is something that often emerges from lived experience rather than theoretical design.

The process of co-designing these elements often involves specific participatory design techniques. For instance, persona development, a common UX tool, can be greatly enhanced by including community members. Instead of developers creating hypothetical user personas based on assumptions, community representatives can help create detailed, realistic personas that reflect the diversity of experiences, needs, and digital literacy levels within their communities. These personas can then serve as powerful touchstones throughout the design process, ensuring that decisions are made with real users in mind. Storyboarding and scenario mapping are other valuable techniques where community members can illustrate how they would interact with the AI in various real-world situations, highlighting potential pain points and opportunities for improvement. Imagine a scenario where an AI is designed to assist elderly individuals with daily tasks. Community members might use storyboarding to show how an individual might struggle with a voice command due to hearing difficulties, or how they might feel confused by an automated notification that lacks personal context. These visual narratives provide concrete examples that are far more compelling than abstract discussions of potential usability issues.

Furthermore, the co-design of AI functionality and UX is an ongoing process, not a one-time event. As AI systems are deployed and used, new challenges and opportunities for improvement will invariably arise. Establishing mechanisms for continuous feedback and iterative refinement is therefore essential. This could involve setting up user panels, conducting regular usability testing with diverse groups, or creating dedicated channels for users to report issues and suggest enhancements. The key is to create a feedback loop that empowers the community to actively participate in the evolution of the AI system over time. This ensures that the technology remains relevant, responsive, and continues to meet the evolving needs of its users.

The implications of this collaborative approach are far-reaching. When communities are involved in shaping AI functionality and user experience, the resulting systems are more likely to be adopted, trusted, and ultimately successful. They are less prone to unintended consequences, ethical missteps, and outright failure. This is because the AI is not designed *for* a community in a vacuum, but *with* a community, drawing on their collective wisdom and lived experience. This is the essence of creating AI that truly serves humanity, moving beyond the purely technical to embrace the deeply human.

Consider a practical example in the realm of public safety. An AI system might be proposed to help emergency services allocate resources more effectively. Developers might initially conceive of features focused on predicting the likelihood of an incident occurring in a specific location based on historical data, weather patterns, and social media trends. However, during a co-design workshop, community representatives, including local activists and residents from historically underserved neighborhoods, might highlight the critical need for a feature that flags areas where residents have historically experienced over-policing or discriminatory application of emergency services. They might argue that an AI solely focused on prediction, without accounting for these societal dynamics, could inadvertently exacerbate existing inequalities. They might propose a "community trust index" as a crucial input, which would factor in historical data on community-police relations, the presence of community support networks, and the perceived legitimacy of emergency response in different areas. This is not a readily quantifiable metric that developers might find on their own; it requires a deep understanding of social justice issues.

Further co-design of the user interface for such a system could involve discussions about how information is presented to dispatchers. Instead of just a heat map of predicted incidents, community members might advocate for layered information. For instance, when an AI flags a high-risk area, the interface could simultaneously display information about local community outreach programs, established trusted community leaders, or specific communication protocols that have been effective in that neighborhood. This ensures that resource allocation is not just about raw prediction but also about informed, context-aware deployment that builds rather than erodes community trust. They might also insist on the inclusion of a clear disclaimer acknowledging the potential for bias in the predictive models and a mechanism for dispatchers to override AI recommendations based on their local knowledge. This level of detail and human-centric consideration is what transforms a potentially problematic AI into a more equitable and effective tool.

In another scenario, imagine an AI designed to assist small businesses in navigating complex regulatory landscapes. Developers might focus on building a comprehensive database of regulations and an algorithm to identify potential compliance issues. However, community engagement with small business owners could reveal a pressing need for an AI that not only identifies regulations but also provides practical, actionable advice on how to comply, especially for businesses with limited resources or technical expertise. Co-design could lead to features such as simplified checklists, interactive guides that break down complex requirements into manageable steps, and even AI-powered tools that help businesses generate necessary documentation. The user experience would need to be tailored to individuals who are often time-poor and may not be digitally native. This might involve designing a mobile-first interface, offering voice-based support, and ensuring that the language used is accessible and avoids legalistic jargon. Community members could help develop scenarios demonstrating how a business owner might use the AI at different stages of their operation – from initial setup to ongoing compliance – and identify where the AI's guidance is most needed and where it could be most effective.

The "black box" problem, where the inner workings of AI are opaque even to its creators, is a significant concern for ethical AI deployment. Co-designing the functionality and UX provides a crucial opportunity to demystify the AI. By involving users in defining what information the AI provides and how it is presented, developers can be pushed to make the AI's reasoning more transparent. For example, if an AI is recommending a particular course of action, community input might lead to a requirement that the AI explicitly state the factors it considered in making that recommendation and perhaps even provide links to supporting data or evidence. This transparency is not just about technical documentation; it's about building the AI's interface and conversational logic in a way that naturally promotes understanding and accountability.

Furthermore, the process of co-designing the AI's interaction style is paramount. This goes beyond the visual layout of an interface to the very nature of the AI's dialogue. Should it be conversational, formal, or instructive? How should it handle ambiguity or requests for clarification? For an AI assisting with job applications, for example, community members might emphasize the need for an AI that is encouraging and supportive, especially for individuals who may have faced barriers to employment. They might suggest that the AI should offer to elaborate on terms, provide examples of strong answers, or even connect users with resources for resume building or interview preparation. The AI's persona and conversational flow are powerful tools for building rapport and ensuring that users feel comfortable and empowered.

Ultimately, the co-design of AI functionality and user experience is a testament to the belief that technology should be a tool that amplifies human capabilities and addresses human needs, rather than an abstract construct imposed upon society. By actively involving communities in defining what an AI does and how it behaves, we move towards creating AI that is not only intelligent but also wise, not only efficient but also equitable, and not only functional but also profoundly human. This collaborative shaping process ensures that AI development is guided by a compass of real-world impact and shared values, paving the way for technologies that truly benefit everyone.

The preceding discussion has illuminated the vital role of co-creation in shaping the outward expression of an AI – its functionalities and user experience. However, the deeper, more intricate work of co-creation lies in translating community values into the very core of the AI's operational framework: its logic. This involves moving beyond mere user interface design and functional specification to influence how the AI *thinks*, *prioritizes*, and *makes decisions*. It is about ensuring that the algorithmic underpinnings of the system are not solely dictated by technical efficiency or data patterns but are instead informed by the ethical principles and lived realities of the communities it serves.

One of the most profound areas where community values can and must be embedded is in the definition and implementation of fairness. What constitutes "fairness" is rarely a universally agreed-upon metric, especially in AI. Developers might initially lean towards statistical notions of fairness, such as ensuring equal accuracy across different demographic groups or equal false positive rates. However, community engagement can reveal that fairness extends far beyond these quantitative measures. For instance, in the development of an AI system for loan application assessment, developers might focus on ensuring that applicants from different racial or gender groups have an equal chance of their applications being approved, based on historical data and financial risk factors. But community members, particularly those from marginalized groups, might argue that true fairness also requires accounting for systemic disadvantages and historical redlining that may have prevented certain communities from accumulating wealth or building credit histories. This could translate into a need for the AI's logic to incorporate factors that mitigate past discrimination, perhaps by giving more weight to alternative credit assessment methods or by adjusting risk profiles based on socioeconomic context. This is not about lowering standards but about creating a more equitable playing field where past disadvantages do not perpetually dictate future opportunities.

The process of defining fairness might involve detailed discussions within co-creation workshops. Participants could be asked to consider hypothetical scenarios and articulate what an equitable outcome would look like. For an AI designed to allocate public housing, for example, community members might express that while waiting list order is a common criterion, fairness also demands prioritizing families experiencing homelessness, individuals with disabilities who require specific accessible units, or those displaced by natural disasters. These are qualitative considerations that need to be translated into quantifiable parameters or rules within the AI's decision-making logic. This might involve developing a multi-criteria decision-making framework where elements like urgency, vulnerability, and specific housing needs are weighted alongside standard eligibility criteria. The community's input ensures that the AI’s prioritization system reflects genuine societal needs and values, rather than simply optimizing for bureaucratic efficiency.

Beyond fairness, co-creation can also guide the AI's prioritization of outcomes. Many AI systems are designed to optimize for a specific objective function. However, this objective function might not fully capture the multifaceted goals of a community. Consider an AI designed to manage traffic flow in a city. A purely technical objective might be to minimize overall travel time. Yet, community members might argue that minimizing travel time for private vehicles should not come at the expense of pedestrian safety, air quality in residential areas, or accessibility for public transportation. Co-design sessions could lead to a revised objective function that balances these competing priorities. This might involve incorporating constraints into the AI's logic, such as mandating reduced speed limits in school zones, prioritizing bus signal preemption over car signal preemption, or dynamically rerouting traffic away from sensitive ecological areas. Translating these community priorities into algorithmic constraints requires a deep understanding of how different objectives interact and how to balance them without compromising the overall effectiveness of the system.

The setting of boundaries for algorithmic behavior is another critical area where community values can be embedded. AI systems can sometimes operate in ways that, while technically permissible, are ethically questionable or socially undesirable. For example, an AI used for predictive policing might identify patterns that suggest increased criminal activity in certain neighborhoods. Without clear boundaries, this could lead to over-surveillance and harassment of residents in those areas. Community input can help establish ethical guardrails. This might involve defining specific types of data that the AI is not allowed to use, such as protected characteristics or sensitive personal information, unless absolutely necessary and with strict oversight. It could also involve setting thresholds for action; for instance, an AI might be programmed to flag a potential issue but require human verification and additional corroborating evidence before any intervention is initiated.

Furthermore, community values can inform the AI's approach to uncertainty and risk. AI systems often operate with a degree of uncertainty. How the AI communicates this uncertainty and how it manages risk can have significant implications. For an AI designed to assist with medical diagnoses, for instance, developers might focus on maximizing diagnostic accuracy. However, community members, particularly patients and their families, might emphasize the importance of cautiousness. They might advocate for an AI that errs on the side of recommending further testing or consultation with a human specialist when its confidence level is below a certain threshold, even if this means a slightly lower overall accuracy rate. This reflects a value placed on patient safety and trust, where avoiding a misdiagnosis is prioritized over achieving the highest possible statistical accuracy. This would require embedding probabilistic thresholds and decision rules into the AI's logic that reflect this community-defined risk tolerance.

The implementation of these embedded values often requires specialized techniques in AI development. Techniques like "value alignment" and "ethical AI frameworks" are not just abstract concepts but concrete methodologies that can be guided by community input. Value alignment seeks to ensure that an AI's goals and behaviors are consistent with human values. Co-creation provides the raw material for this alignment by surfacing what those values actually are for the specific community. Ethical AI frameworks, on the other hand, provide structured approaches to identifying, assessing, and mitigating ethical risks. Community participation is crucial for identifying the relevant ethical risks from the perspective of those most affected by the AI.

Consider the development of an AI to assist in the administration of social welfare benefits. A purely data-driven approach might identify individuals who meet certain eligibility criteria based on income and employment status. However, community members might highlight that "eligibility" can be more complex. They might point out situations where individuals are temporarily out of work due to caregiving responsibilities, recovering from illness, or participating in essential community activities that do not generate immediate income. The AI's logic would need to be adjusted to accommodate these nuanced understandings of an individual's circumstances. This could involve developing algorithms that can interpret qualitative data, accept attestations from community organizations, or incorporate a "human review" step for edge cases that fall outside standard quantifiable parameters. This ensures that the AI does not rigidly enforce rules in a way that creates undue hardship or overlooks genuine need.

Another important aspect is the AI's ability to learn and adapt while respecting established values. As an AI system interacts with the world, it may encounter new situations or data that could lead it to revise its internal models. Co-creation can help define the boundaries within which this learning should occur. For an AI used in educational settings to personalize learning paths, community members might agree on the AI's ability to adapt the pace and content of instruction based on a student's performance. However, they might also stipulate that the AI should not steer students away from certain subjects based on perceived aptitude or background, especially if this could reinforce existing stereotypes. This would translate into the AI's learning algorithms being constrained to prevent it from making such discriminatory recommendations, even if the data might suggest otherwise. The AI's learning process itself must be aligned with the community's vision of equitable opportunity.

The challenge of explainability in AI is also deeply intertwined with embedding community values. When an AI makes a decision, especially one that has a significant impact on an individual's life, the ability to explain
*why* that decision was made is paramount for trust and accountability. Co-creation can push for the development of AI architectures that are inherently more interpretable or for the creation of "explanation modules" that translate complex algorithmic processes into understandable terms for the community. For example, if an AI denies a permit application for a small business, community members might demand that the AI not only state the reason for denial but also provide

specific guidance on what steps the applicant can take to meet the requirements in the future. This moves beyond a simple "no" to a constructive and transparent process, reflecting a value of empowering individuals to understand and navigate systems.

Furthermore, the very notion of "risk" can be defined differently by a community than by a technologist. For an AI assessing risk in urban planning, developers might focus on economic risk or infrastructure risk. Community members, however, might highlight social risks, such as the potential for displacement of long-term residents, the disruption of cultural heritage sites, or the increased burden on essential services in underserved areas. These are legitimate risks that need to be factored into the AI's decision-making logic, perhaps by introducing new metrics or by adjusting the weighting of existing ones to give greater consideration to social and cultural impacts. This requires a shift in how the AI models and evaluates potential futures, ensuring that it accounts for the holistic well-being of the community.

In essence, embedding community values into AI logic is a continuous process of negotiation and translation. It requires a willingness from AI developers to move beyond purely technical considerations and to engage deeply with the ethical, social, and cultural dimensions of their work. Co-creation provides the essential bridge for this translation, transforming abstract principles of fairness, equity, and well-being into concrete rules, parameters, and objective functions that govern the AI's behavior. This is where the art of AI development truly meets its ethical and societal purpose, ensuring that the intelligence we create is not only powerful but also profoundly aligned with the human values it is meant to serve.

The journey of co-creation does not end with the successful deployment of an AI system. In fact, the initial development phase, while crucial, represents only the foundational stage of a dynamic and evolving relationship. Sustaining co-creative partnerships over time is paramount to ensuring that AI systems remain responsive, ethical, and beneficial throughout their operational lifespan. This ongoing collaboration is not a mere post-deployment formality; it is an intrinsic necessity for the continued relevance and responsible evolution of AI, particularly in public sector applications where societal impact is direct and profound. Without sustained engagement, even the most thoughtfully designed AI risks becoming a static artifact, out of sync with changing community needs, emerging societal values, or unforeseen consequences.

One of the most fundamental aspects of sustaining these partnerships is establishing robust and accessible feedback mechanisms. These channels must be designed not as optional add-ons but as integral components of the AI's governance structure. This means moving beyond the traditional customer service hotline or bug reporting system. Instead, it requires creating platforms where community members can continuously voice their experiences, concerns, and suggestions regarding the AI's performance and impact. For an AI system used in urban planning, for instance, feedback mechanisms could include participatory budgeting tools where residents can propose modifications to AI-driven resource allocation, or digital forums dedicated to discussing the AI's impact on local neighborhoods, alongside traditional comment periods for proposed policy changes informed by the AI. These mechanisms should be multilingual, easy to navigate across various digital literacy levels, and clearly communicated to all affected communities. The feedback collected should not be passively received; it needs to be actively analyzed, categorized, and acted upon, with clear communication back to the community about how their input has been considered and what actions, if any, have been taken. This iterative loop of feedback, analysis, and action builds trust and demonstrates a genuine commitment to a co-creative ethos.

Regular co-creation forums, akin to the initial workshops but adapted for an operational context, are also vital. These might take the form of periodic review meetings, community advisory boards, or even virtual town halls focused on specific aspects of the AI's performance. For an AI system deployed in a healthcare setting, such as one assisting with appointment scheduling or providing basic health information, these forums could convene patients, healthcare providers, and community health advocates to discuss the AI's accuracy, accessibility, and any instances of bias or miscommunication. The agenda for these meetings should be collaboratively developed, ensuring that the issues most pertinent to the community are addressed. The frequency of these gatherings will vary depending on the AI's complexity and the pace of change in its operational environment. The key is consistency and predictability, allowing community partners to plan and prepare for their involvement. These sessions serve not only to gather feedback but also to foster a shared understanding of the AI's limitations and capabilities, promoting a more informed and engaged community.

Transparency regarding the AI's performance and decision-making processes is another cornerstone of sustained co-creation. While the internal workings of complex AI algorithms can be difficult to fully explain, a commitment to transparency means making as much information available as is feasible and understandable to the community. This could involve publishing regular performance reports that detail key metrics, identify areas where the AI is underperforming, and outline any known issues or biases. For an AI system managing public transit schedules, for example, transparency would involve sharing data on on-time performance, identifying routes with persistent delays, and explaining the factors contributing to these issues, such as infrastructure problems or unexpected demand. Explanations of how community feedback has led to system adjustments are also crucial for reinforcing the co-creative relationship. This transparency extends to outlining the AI's current limitations and potential risks, managing expectations and preventing misunderstandings. When the community understands how an AI works and is informed about its performance, they are better equipped to provide meaningful input and to hold stakeholders accountable.

Moreover, it is essential to establish clear governance structures that embed community representation and participation in the ongoing monitoring and iteration of AI systems. This might involve formalizing the role of community advisory boards, granting them a consultative or even decision-making capacity on certain AI-related matters. In the context of an AI used for managing public resources, such as water allocation during droughts, a community governance council comprising representatives from agricultural, environmental, and residential sectors could be empowered to review the AI's operational parameters and recommend adjustments based on evolving needs and environmental conditions. The individuals serving on these boards or committees should be representative of the diverse communities impacted by the AI, and mechanisms should be in place to ensure their insights are effectively integrated into the AI's development and deployment lifecycle. This formalizes the co-creative partnership, moving it from an ad-hoc engagement to a structured and institutionalized process.

The iterative refinement of AI models based on ongoing feedback is a core aspect of sustained co-creation. AI systems are not static entities; they learn and evolve. This evolution, however, must be guided by the ethical principles and community values that were established during the initial co-design phase. When new data becomes available, or when the AI encounters novel situations, its responses and decision-making logic may need to be updated. This process of updating should involve the community. For an AI designed to assist in disaster response, for instance, after an event, community members could participate in post-event reviews to assess the AI's effectiveness, identify any unintended consequences, and provide input on how its algorithms should be adjusted for future scenarios. This ensures that the AI's learning process remains aligned with the community's priorities, preventing drift towards outcomes that may be technically optimal but socially undesirable. It requires a commitment to continuous improvement that is informed by lived experience.

Capacity building within the community is also a critical, albeit often overlooked, component of sustaining co-creative partnerships. For community members to effectively engage in ongoing dialogue about complex AI systems, they need access to relevant information, training, and resources. This might involve workshops to explain AI concepts in accessible terms, providing stipends to compensate community members for their time and expertise, or facilitating access to technical expertise to help them understand the implications of different AI configurations. For an AI system used in the justice system, such as one predicting recidivism risk, capacity building could involve training community advocates and legal aid organizations on how to interpret the AI's outputs and challenge its assessments. Empowering communities with knowledge and resources ensures that their participation is meaningful and that they can contribute to the AI's evolution from a position of informed agency, rather than simply being recipients of decisions.

The sustainability of co-creative partnerships also hinges on establishing clear lines of accountability. When an AI system produces an undesirable outcome, or when its performance falls short of community expectations, it must be clear who is responsible for addressing the issue and how.

This involves defining roles and responsibilities among AI developers, deployers, community representatives, and oversight bodies. For an AI used in public education to personalize learning, accountability would involve identifying which entity is responsible for ensuring the AI does not perpetuate bias, and what recourse is available to parents or students if such issues arise. This clarity prevents finger-pointing and ensures that there is a structured process for remediation and improvement. Accountability mechanisms should be transparent and accessible to the community, fostering trust and reinforcing the commitment to responsible AI deployment.

Furthermore, the evolution of AI technology necessitates a proactive approach to anticipating future needs and potential challenges. Co-creative partnerships should extend to forward-looking exercises, such as scenario planning and foresight workshops. These activities can help communities and AI developers collaboratively imagine future possibilities, identify emerging ethical dilemmas, and proactively develop strategies to address them. For an AI system designed to manage smart city infrastructure, for example, foresight workshops might explore the implications of increasingly sophisticated AI on privacy, employment, and social equity, and collaboratively develop guidelines for future AI development and deployment that mitigate potential negative impacts. This anticipatory approach ensures that the co-creative process remains agile and capable of addressing challenges before they become crises.

It is also important to recognize that co-creative partnerships are not always smooth sailing. There will inevitably be disagreements, moments of frustration, and differing perspectives on what constitutes the "best" course of action. Sustaining these partnerships requires a commitment to constructive conflict resolution. This involves creating safe spaces for dialogue, employing skilled facilitators, and adhering to established protocols for addressing disputes. For an AI system that influences the distribution of public services, disagreements might arise between different community groups advocating for competing priorities. A well-structured conflict resolution process, perhaps involving mediation by a neutral third party, would be essential to finding common ground and moving forward collaboratively. The ability to navigate conflict constructively is a hallmark of mature and resilient co-creative relationships.

Finally, the long-term success of co-creative partnerships is deeply intertwined with the broader policy and regulatory landscape. As AI technology matures and its societal impact becomes more pronounced, governments and regulatory bodies play a crucial role in fostering an environment that supports sustained co-creation. This could involve developing policies that mandate community engagement in AI development, establishing ethical guidelines for AI deployment, and providing funding for community-led AI initiatives. For instance, legislation that requires public sector AI projects to undergo community impact assessments and establish ongoing community oversight mechanisms can significantly bolster the sustainability of co-creative efforts. When policy frameworks recognize and actively encourage ongoing collaboration, it signals a societal commitment to ensuring that AI serves the public good, not just technical advancement. This overarching support system empowers and validates the efforts of communities and developers working together to shape the future of AI in a responsible and equitable manner.

# Chapter 7: Ethical Procurement and Contracting for AI

The act of procuring technology, especially sophisticated systems like those powered by artificial intelligence, represents far more than a simple financial transaction. For public sector organizations, it is a fundamental and potent mechanism for shaping the very nature of AI deployment, acting as an initial, critical juncture where ethical considerations can be deeply embedded or regrettably overlooked. Procurement, in this context, transcends the procurement officer's desk; it becomes a strategic tool wielded by the public sector to steer innovation towards societal benefit, thereby influencing not just the technology itself but also the vendors who develop and supply it. By thoughtfully structuring procurement processes, public bodies can proactively establish the ethical guardrails, transparency mandates, and fairness requirements that will govern AI systems from their inception. This proactive stance is not merely about mitigating future risks; it is about actively cultivating a vendor ecosystem that prioritizes responsible AI practices, thereby fostering a more trustworthy and equitable technological future for all.

When a public sector entity decides to acquire an AI system, it is implicitly setting a precedent for the type of technology it values and the standards it upholds. This decision point is fertile ground for embedding ethical considerations directly into the contract. Instead of viewing procurement as a reactive measure to acquire a finished product, it can and should be reimagined as a proactive process that co-designs the ethical framework for that product. This involves moving beyond generic service level agreements to demand specific ethical performance indicators. For instance, when procuring an AI system for citizen service delivery, a public agency could include clauses that require the vendor to demonstrate how the system has been tested for bias against protected demographic groups, and to outline the mitigation strategies employed. The procurement document becomes a blueprint for ethical AI, not just a shopping list. This approach shifts the burden of ethical compliance from an afterthought to a prerequisite, thereby guiding vendors toward developing AI that is inherently more responsible.

The language and stipulations within procurement documents hold immense power in shaping vendor behavior. By explicitly defining what constitutes ethical AI—requiring transparency in data sources, mandating explainability for decision-making processes, and demanding robust mechanisms for redress—public sector organizations can signal their commitment to these principles. This sends a clear message to potential vendors: that ethical considerations are not negotiable but are integral to the evaluation and selection process. For example, a request for proposal (RFP) for an AI-powered recruitment tool could stipulate that vendors must provide detailed documentation on their algorithms' fairness metrics, including measures for disparate impact across different racial, gender, and age groups. Furthermore, the RFP might require vendors to describe their internal processes for identifying and mitigating algorithmic bias throughout the AI's lifecycle, from data collection to model deployment. Such specific requirements force vendors to invest in ethical AI development from the ground up, rather than attempting to retrofit ethical considerations after the fact.

Moreover, procurement processes can be designed to incentivize innovation in ethical AI. Instead of solely focusing on technical specifications and cost, RFPs can incorporate evaluation criteria that reward vendors who demonstrate a strong commitment to ethical principles. This could include giving higher scores to proposals that offer more comprehensive plans for data privacy, enhanced user explainability features, or robust mechanisms for human oversight. For example, when procuring an AI system to optimize traffic flow, a city government might award bonus points to vendors who propose solutions that not only improve efficiency but also incorporate features to ensure equitable access to transportation for all neighborhoods, or that provide clear explanations for how traffic patterns are being managed. This encourages vendors to compete not just on performance but on the ethical integrity of their solutions, fostering a market where responsible AI becomes a competitive advantage.

The power of procurement also lies in its ability to enforce accountability. Contracts are legally binding documents, and well-drafted procurement agreements can include provisions for ongoing monitoring, auditing, and recourse in the event of ethical breaches. This means that if an AI system developed by a vendor fails to meet the agreed-upon ethical standards—perhaps by exhibiting discriminatory behavior or compromising user privacy—the public sector organization has a contractual basis for demanding corrective action or even terminating the contract. For instance, a contract for an AI system used in social service allocation could include clauses that allow for independent audits of the AI's decision-making processes and outcomes. If these audits reveal evidence of bias, the contract could stipulate penalties for the vendor and require them to implement remedial measures within a defined timeframe. This contractual leverage ensures that vendors remain accountable for the ethical performance of their AI systems long after the initial deployment.

Public sector procurement can also serve as a catalyst for advancing the broader understanding and adoption of ethical AI practices within the technology industry. By setting high standards and demanding specific ethical commitments, government agencies can influence market trends and encourage a broader shift towards responsible AI development. When public sector entities, which often have significant purchasing power, prioritize ethical AI, it signals to the market that there is demand for such systems. This can inspire other organizations, both public and private, to adopt similar procurement practices, thereby creating a ripple effect that elevates ethical standards across the board. For example, the procurement guidelines established by a national government for AI in healthcare could serve as a benchmark for other countries or for private healthcare providers, driving a collective improvement in the ethical landscape of AI in this critical sector.

Furthermore, procurement processes can be instrumental in fostering collaboration and knowledge sharing regarding ethical AI. By requiring vendors to share information about their AI development methodologies, data governance practices, and bias mitigation strategies, public sector organizations can gain valuable insights into the challenges and best practices in ethical AI development. This information can then be used to refine future procurement processes, inform policy development, and even contribute to public knowledge about responsible AI. For example, a procurement process for an AI system used in environmental monitoring could involve vendor workshops where best practices for data anonymization and privacy protection are discussed and shared among all participating companies. This collaborative approach ensures that the procurement process itself becomes a learning opportunity, benefiting all stakeholders.

The challenge, however, lies in crafting procurement documents that are both robust and practical. They must be specific enough to be meaningful but flexible enough to accommodate the evolving nature of AI technology. This requires a multidisciplinary approach, involving not only procurement specialists but also legal experts, AI ethicists, data scientists, and community representatives. Their collective expertise is crucial for identifying potential ethical pitfalls, defining clear and measurable ethical standards, and ensuring that the contractual language effectively translates ethical principles into actionable requirements. For instance, when developing an RFP for an AI system that will assist in judicial sentencing recommendations, the procurement team would need to consult with legal scholars, civil rights advocates, and AI fairness experts to ensure that the RFP adequately addresses concerns about fairness, due process, and the potential for reinforcing existing societal biases.

In essence, procurement is not merely a transactional exercise; it is a powerful policy instrument. When wielded with foresight and a deep understanding of ethical AI principles, it can transform the way AI is developed and deployed. It offers a concrete and actionable pathway for public sector organizations to steer technological advancement towards a future that is not only innovative but also equitable, transparent, and aligned with societal values. By embracing procurement as a strategic lever for ethical AI, public bodies can ensure that the AI systems they commission serve the public good, fostering trust and ensuring that technology works for everyone. This deliberate integration of ethical considerations at the earliest stages of acquisition is fundamental to building a responsible AI ecosystem.

Translating abstract ethical principles into concrete contractual obligations within procurement documents, particularly Requests for Proposals (RFPs), is a cornerstone of responsible AI acquisition for public sector entities. This process moves beyond aspirational statements to establish legally binding requirements that vendors must meet. The goal is to ensure that AI systems procured not only perform their intended functions but also align with societal values of fairness, equity, and transparency. Crafting these requirements demands a precise understanding of both the technology and the ethical implications, transforming vague notions of "fairness" or "privacy" into specific, measurable, and verifiable contractual terms.

A critical component of defining ethical requirements in RFPs revolves around establishing robust fairness criteria. This involves moving past a general desire for an "unbiased" system to specifying *how* fairness will be defined and measured. For instance, an RFP for an AI system intended to assist in parole decisions could mandate that the system be evaluated against specific fairness metrics, such as demographic parity, equalized odds, or predictive parity. Each of these metrics addresses different facets of fairness. Demographic parity, for example, would require the proportion of positive outcomes (e.g., parole granted) to be the same across different racial or gender groups. Equalized odds, on the other hand, would focus on ensuring that false positive rates and false negative rates are consistent across these groups. Predictive parity aims to ensure that the probability of a positive outcome given a predicted positive score is the same across groups. The RFP should explicitly state which metrics are prioritized, why, and how the vendor will be expected to demonstrate compliance. This might involve requiring vendors to submit a detailed fairness report based on independent audits or to provide access to the underlying algorithms and evaluation datasets for scrutiny.

Beyond fairness metrics, data governance standards are paramount. AI systems are only as good as the data they are trained on and the data they process. Therefore, RFPs must articulate clear expectations for how data will be collected, stored, used, and protected. This includes requirements related to data provenance, ensuring that the origins and history of the training data are well-documented. Vendors should be required to disclose the sources of their training data, the methods used for data collection and labeling, and any potential biases inherent in those processes. Furthermore, stringent data privacy and security protocols must be specified. For AI systems handling sensitive personal information, such as those used in healthcare or social services, RFPs should mandate compliance with relevant privacy regulations (e.g.,

GDPR, HIPAA) and outline specific security measures, including encryption, access controls, and data anonymization or pseudonymization techniques. The contract could stipulate that data used for training an AI model must be de-identified to the highest possible standard, and that any personally identifiable information processed by the AI during operation must be handled with strict confidentiality and deleted once its purpose is fulfilled, with clear audit trails to verify compliance.

Transparency obligations are another vital element. While the "black box" nature of some AI models can be a technical challenge, procurement documents can demand levels of transparency appropriate to the risk and impact of the system. For AI systems making decisions that significantly affect individuals, such as those used in loan applications, hiring, or criminal justice, RFPs should require a degree of explainability. This doesn't necessarily mean demanding a full breakdown of every algorithmic step, which might be technically infeasible or proprietary. Instead, it can mean requiring vendors to provide clear, understandable explanations of
*why* a particular decision was made. This could involve providing summaries of the key factors that influenced a decision, the confidence level of the AI's prediction, and information about the data points that were most influential. The RFP could also mandate that vendors provide documentation on the model's architecture, its intended use cases, its limitations, and its known failure modes. Furthermore, transparency extends to ongoing performance monitoring. Vendors might be required to provide regular reports on the AI system's performance, including metrics related to fairness, accuracy, and any instances of unexpected behavior or errors.

Mandatory impact assessments and community engagement processes are also crucial for ensuring that AI systems are developed and deployed in a manner that respects human rights and societal well-being. RFPs should stipulate that vendors must conduct thorough ethical and societal impact assessments *before* deployment. These assessments should proactively identify potential risks, such as discriminatory outcomes, privacy violations, or unintended consequences, and propose mitigation strategies. For AI systems that will interact directly with the public or influence significant decisions about individuals, the RFP should also require mechanisms for meaningful community engagement. This could involve mandating that vendors consult with affected communities during the development phase to gather feedback, understand their concerns, and incorporate their perspectives into the design. Such engagement can help identify potential harms that might not be apparent from a purely technical or legal perspective. The RFP could even specify that a portion of the vendor's proposal evaluation will be based on the quality and inclusiveness of their engagement plan and the extent to which community feedback has been integrated into their proposed solution.

To make these requirements actionable and verifiable, RFPs need to translate abstract ethical principles into concrete, contractual terms. For example, instead of stating "the system must be fair," an RFP might include a clause such as: "The vendor shall ensure that the AI system's performance in identifying eligible candidates for housing assistance demonstrates a disparity ratio of no more than 0.8 between any two demographic groups as defined by the U.S. Census Bureau, as measured over a six-month period post-deployment. The vendor shall provide quarterly reports detailing these performance metrics and the methodology used for their calculation. Any disparity ratio exceeding 0.8 will trigger a contractual obligation for the vendor to submit a remediation plan within 30 days and implement approved mitigation strategies within 90 days." This level of specificity leaves little room for ambiguity.

Another example could relate to transparency for an AI-powered facial recognition system intended for public safety. An RFP might stipulate: "The vendor shall provide a detailed technical specification of the facial recognition algorithm, including its architecture, training data characteristics (e.g., demographic composition, data sources, anonymization methods), and documented accuracy rates across different lighting conditions, angles, and demographic groups. Furthermore, the vendor must provide a clear explanation of the system's limitations, including known biases and potential error rates, which shall be prominently displayed to any operator using the system. Any deployment of this system must be accompanied by clear public signage informing individuals that AI-powered surveillance is in use." This requirement aims to ensure that both the procuring agency and the public understand the capabilities and limitations of the technology.

In the context of AI for hiring, an RFP might require vendors to detail their processes for mitigating algorithmic bias. The clause could read: "The vendor must describe its methodology for identifying and mitigating bias in its AI hiring tool throughout the system's lifecycle. This description must include steps taken during data collection and preprocessing, model development and validation, and ongoing monitoring. Specifically, the vendor shall provide evidence of testing for disparate impact on protected classes (as defined by Title VII of the Civil Rights Act) and outline the specific technical or procedural adjustments made to reduce such impact to statistically insignificant levels. The proposal must also include a plan for periodic re-evaluation and recalibration of the model to address emerging biases."

Moreover, RFPs can incorporate requirements for auditability and accountability. This means ensuring that the AI system's operations and decisions can be audited by independent third parties or the procuring agency. For an AI system used to manage citizen requests for services, an RFP might require that "all AI-driven decisions regarding the prioritization or allocation of services must be logged with sufficient detail to enable post-hoc review. These logs must include the input data, the AI's recommendation, the final decision made (whether by AI or human override), and the rationale for any human override. The vendor shall grant the procuring agency and its designated auditors full read-access to these logs, and shall cooperate fully with any audits conducted." This creates a clear path for understanding how the system is functioning and for identifying any deviations from expected or ethical behavior.

The inclusion of specific clauses related to data minimization and retention is also crucial. For instance, an RFP for an AI-powered customer service chatbot could state: "The vendor's AI system shall be designed to collect only the minimum data necessary to fulfill the user's request and improve service quality. All personally identifiable information collected during a user interaction shall be retained for no longer than 90 days, unless legally required otherwise. Following this period, all PII shall be securely purged. The vendor shall provide a data retention policy and schedule for all data processed by the AI system, subject to agency approval." This proactive approach limits the potential for data misuse and reduces the risk associated with data breaches.

Furthermore, RFPs can address the issue of vendor lock-in and promote interoperability, which has indirect ethical implications related to long-term control and flexibility. By requiring vendors to use open standards or provide mechanisms for data portability and system migration, public agencies can ensure they are not beholden to a single vendor indefinitely. This can be framed as an ethical requirement by stating:

"The vendor shall ensure that all data generated, processed, or stored by the AI system is exportable in a common, machine-readable format (e.g., JSON, CSV) at any time, and that the vendor will provide necessary tools or APIs to facilitate such export. The vendor shall also provide documentation for the system's architecture and data schemas to facilitate potential future integration with other systems or migration to alternative solutions."

The inclusion of provisions for continuous improvement and adaptation also plays a role in ethical procurement. AI systems are not static; they learn and evolve. RFPs can mandate that vendors have a robust process for updating and retraining models to address drift, emerging biases, or changing requirements. This might involve requiring vendors to propose a schedule for regular model re-evaluation and retraining, along with the criteria that would trigger such interventions. For example, an RFP for an AI system used in predictive maintenance for public infrastructure could include: "The vendor shall propose a lifecycle management plan for the AI model, including regular (e.g., annual) reviews of its predictive accuracy and potential biases. The plan must outline procedures for retraining the model using new data and for updating algorithms to maintain optimal performance and fairness. The vendor must provide an annual report detailing the model's performance, any updates made, and the rationale behind them."

Finally, the language used in RFPs must be accessible and understandable not only to technical experts and legal counsel but also to procurement officers and, ideally, to the public. While technical detail is necessary, it should be framed within a clear understanding of the desired ethical outcomes. The goal is to create procurement documents that serve as a blueprint for responsible AI development and deployment, ensuring that public sector investments in AI are aligned with the public good and contribute to a more just and equitable society. This requires ongoing dialogue between technologists, ethicists, legal professionals, and community stakeholders to ensure that the requirements are both technically sound and ethically robust.

Before any contract is signed, and as a continuous part of the vendor relationship, a rigorous process of vendor due diligence is not merely advisable; it is an ethical imperative. Public sector entities venturing into AI procurement must look beyond the technical specifications and the bottom-line cost. The trustworthiness of the AI solution is intrinsically linked to the trustworthiness of the vendor providing it. This necessitates a deep dive into their ethical foundations, their governance structures, and their demonstrated commitment to responsible AI practices.

Evaluating a vendor's ethical track record requires a multi-faceted approach. This begins with researching their public statements, published policies, and any documented instances of their engagement with ethical AI principles. Have they proactively published an AI ethics framework or principles? Do they participate in industry initiatives focused on responsible AI development? Publicly available information can offer initial insights, but it is rarely sufficient on its own. A more thorough assessment involves probing for concrete actions and evidence. This might include requesting information about their internal review processes for AI systems, their training programs for employees on AI ethics, and how they handle ethical challenges or dilemmas that arise during development or deployment. For example, a vendor that can readily provide documentation of an internal ethics review board, complete with meeting minutes and recorded decisions on AI projects, demonstrates a commitment to operationalizing ethical considerations. Conversely, a vendor that struggles to articulate their ethical framework or point to any specific internal mechanisms for addressing ethical concerns should raise a significant red flag.

Furthermore, understanding the vendor's approach to bias mitigation and fairness is paramount. This goes beyond simply asking if they "test for bias." The due diligence process should aim to uncover *how* they identify, measure, and mitigate bias throughout the AI lifecycle. This can involve requesting detailed reports from previous projects, where they outline the specific fairness metrics used, the demographic groups considered, the datasets employed for testing, and the mitigation strategies implemented. For instance, a vendor proposing an AI system for credit scoring might be asked to provide evidence of how they have addressed potential biases related to race, gender, or socioeconomic status in previous credit modeling projects. Did they use differential fairness, equalized odds, or another metric? What was the outcome of their bias mitigation efforts, and how was this measured? A vendor who can present robust methodologies, demonstrate tangible improvements in fairness metrics, and articulate the trade-offs they considered during the mitigation process is far more likely to deliver an ethically sound solution. The absence of such detailed evidence, or vague assertions about having "handled bias," should be a cause for concern and a prompt for further investigation.

A critical aspect of vendor due diligence also involves scrutinizing their internal AI governance policies and practices. This refers to the internal rules, procedures, and organizational structures that guide how the vendor develops, deploys, and manages AI systems. A mature AI governance framework typically includes defined roles and responsibilities for ethical oversight, processes for risk assessment and management, mechanisms for accountability, and protocols for continuous monitoring and improvement. When evaluating potential vendors, public sector entities should inquire about the existence and implementation of such frameworks. This could involve asking for their AI governance charter, details on their data stewardship policies, their approach to intellectual property related to AI, and how they ensure compliance with relevant regulations and ethical standards. For example, if a vendor is developing an AI system that will handle sensitive personal data, their data governance policies should clearly outline how data privacy is protected, how data quality is maintained, and how data access is controlled. Their response to such inquiries should be specific and demonstrate a systematic approach, rather than a collection of ad-hoc practices.

The concept of "AI ethics audits" emerges as a crucial tool within this due diligence framework. These audits are independent, objective assessments of an AI system's ethical compliance, fairness, transparency, and safety. They can be integrated into the procurement process as a mandatory step before contract award, or as a recurring requirement throughout the contract lifecycle to ensure ongoing adherence to ethical standards. Designing an effective AI ethics audit requires careful consideration of its scope, methodology, and independence.

The scope of an AI ethics audit should be tailored to the specific AI system and its intended use case. For a high-risk AI system, such as one used in criminal justice or healthcare, the audit would need to be comprehensive, covering aspects like algorithmic bias, data privacy, security, robustness, explainability, and societal impact. For lower-risk applications, the scope might be narrower, focusing on a few key ethical dimensions. A robust audit would typically involve a review of the AI system's design documentation, training data, algorithms, testing procedures, and deployment records. It might also include interviews with the development team, end-users, and potentially affected stakeholders.

The methodology for an AI ethics audit should be clearly defined and scientifically sound. This often involves employing a combination of quantitative and qualitative techniques. Quantitative methods could include the statistical analysis of model performance across different demographic groups to identify disparities, vulnerability testing to assess robustness, and data analysis to detect potential privacy leaks. Qualitative methods might involve expert review of code and documentation, scenario-based testing to evaluate decision-making processes, and interviews to understand user experience and potential unintended consequences. A critical element of the methodology is the establishment of clear, measurable benchmarks against which the AI system is evaluated. These benchmarks should be derived from the ethical principles and contractual obligations established in the procurement process.

Independence is perhaps the most critical factor for the credibility of an AI ethics audit. The auditors must be external to both the vendor and the procuring public entity to ensure objectivity. This could involve engaging specialized third-party auditing firms, academic institutions, or non-profit organizations with expertise in AI ethics and auditing. The selection process for auditors should be transparent, and their qualifications and experience thoroughly vetted. It is also important to ensure that the auditors have no conflicts of interest that could compromise their impartiality. The audit report should be comprehensive, detailing the findings, the methodologies used, any identified risks or non-compliance issues, and actionable recommendations for remediation.

Public sector entities can stipulate in their RFPs that vendors must undergo such independent AI ethics audits as a prerequisite for contract award. This shifts the burden of proof to the vendor to demonstrate ethical compliance. For example, an RFP for an AI-powered facial recognition system for law enforcement might require that the vendor submit an independent audit report attesting to the system's accuracy rates across diverse demographic groups, its susceptibility to adversarial attacks, and its adherence to privacy principles, prior to any contract being considered. This audit would need to be conducted by a pre-approved, independent body.

Beyond initial procurement, ongoing AI ethics audits are crucial for maintaining ethical standards over the lifecycle of an AI system. AI models can degrade over time, develop new biases, or be subjected to changing operational environments. Therefore, contracts should include provisions for periodic re-audits. These might be triggered by significant model updates, changes in data inputs, or the occurrence of unexpected system behavior. A contract for an AI-driven recruitment tool, for instance, might mandate an annual independent ethics audit to re-evaluate its fairness metrics and identify any emerging biases that may have arisen from new applicant data or changes in the job market. The findings of these ongoing audits should inform decisions about system updates, retraining, or even decommissioning.

The process of developing clear, actionable requirements for vendor due diligence and AI ethics audits requires collaboration among various stakeholders within the public sector. Procurement officers need to work closely with legal counsel, IT departments, data scientists, and ethicists to define the specific criteria and standards for evaluation. This cross-functional approach ensures that the due diligence process is both technically sound and ethically robust. It also helps in crafting clear contractual clauses that translate these requirements into legally binding obligations for vendors. For instance, a clause might state: "Prior to contract execution, the selected vendor shall provide a comprehensive AI Ethics Audit Report, conducted by an independent, NASBA-certified audit firm specializing in AI, detailing the system's compliance with the fairness, transparency, and accountability standards outlined in Appendix B of this RFP. This report must include evidence of bias testing using the demographic categories specified and demonstrate adherence to the mitigation strategies detailed in the vendor's proposal. Failure to provide an audit report that meets the minimum compliance criteria, as determined by the procuring agency, will result in disqualification."

In conclusion, thorough vendor due diligence and the systematic application of AI ethics audits are indispensable components of responsible AI procurement in the public sector. They serve as vital mechanisms for assessing the ethical integrity of AI solutions and the vendors who provide them, ensuring that public investments in AI align with societal values and uphold public trust. By scrutinizing a vendor's ethical track record, internal governance, and bias mitigation strategies, and by mandating independent audits, public entities can significantly de-risk AI deployments and pave the way for AI systems that are not only effective but also ethical, fair, and beneficial to society. This diligent approach moves beyond mere contractual compliance to foster a culture of ethical AI development and deployment within the public sphere.

When public sector entities embark on the procurement of artificial intelligence systems, the initial stages of vendor vetting and due diligence, as previously discussed, lay a crucial foundation. However, the ethical imperative does not cease once a contract is signed. To ensure that AI systems continue to operate in a manner that is aligned with public values and safeguards citizen rights throughout their lifecycle, contractual clauses specifically designed for ongoing transparency and accountability are indispensable. These clauses transform ethical aspirations into legally binding obligations, creating a framework for continuous oversight and recourse.

One of the most fundamental aspects of ensuring transparency is the right to access and understand the AI system's inner workings. This translates into contractual provisions that grant the procuring entity appropriate access to model documentation. This documentation should go beyond superficial descriptions, delving into the specifics of the AI model's architecture, the algorithms employed, and the underlying logic that drives its decision-making processes. For instance, a contract for an AI-powered system used in evaluating social welfare applications might require the vendor to provide detailed documentation on the features used in the model, the weights assigned to these features, and the thresholds or rules that lead to a particular outcome (e.g., approval or denial of benefits). This level of detail is essential for enabling auditors, internal review teams, and potentially oversight bodies to scrutinize the AI's fairness, identify potential biases, and understand the rationale behind its outputs. The contract should clearly define the scope of this documentation, specifying its required level of technical detail and ensuring that it is updated concurrently with any model modifications. Furthermore, clarity on the intellectual property rights associated with this documentation is vital, ensuring that the public sector entity can utilize it for its oversight purposes without infringing on proprietary claims, perhaps through licensing agreements that permit review and audit by authorized personnel.

Beyond model documentation, transparency in AI procurement necessitates comprehensive logging and auditing of data usage. AI systems are data-hungry, and understanding how data is accessed, processed, and utilized is critical for privacy protection and accountability. Contractual clauses should therefore mandate the logging of all data interactions with the AI system. This would include records of which data inputs were used for specific predictions or decisions, when these interactions occurred, and by which user or system process. For an AI system deployed in a healthcare setting, for example, to assist in diagnostic processes, logs would need to meticulously record patient data inputs, the corresponding AI output (e.g., a diagnosis recommendation), and the timestamp of this interaction. These logs serve as an immutable audit trail, allowing for the reconstruction of events, the identification of anomalies, and the investigation of any alleged misuse of data or erroneous outcomes. The contract should specify the retention period for these logs, the format in which they should be stored to ensure integrity and ease of access, and the security measures that must be in place to protect this sensitive information from unauthorized access or tampering. Moreover, clauses might stipulate that the vendor must provide regular summaries or access to these logs for review by the procuring agency, perhaps on a quarterly basis, or immediately upon suspicion of any data breach or system malfunction.

Performance monitoring reports are another cornerstone of ongoing accountability. AI systems are not static; their performance can change over time due to shifts in the data they encounter or the evolving nature of the problems they are designed to solve. Contracts must therefore require vendors to provide regular, detailed performance monitoring reports. These reports should go beyond generic metrics, focusing on key performance indicators (KPIs) that are directly relevant to the AI's intended purpose and its ethical implications. For an AI system designed to optimize traffic flow in a city, for instance, reports might include metrics on traffic congestion reduction, travel time improvements, and, crucially, any differential impact on travel times for different neighborhoods or demographic groups. If the AI is used for predictive policing, the reports must include metrics on accuracy, false positive/negative rates, and crucially, disparities in predictions across different racial or ethnic groups. The contract should specify the frequency of these reports (e.g., monthly, quarterly), the format of their presentation to ensure clarity and comparability, and the benchmark against which performance is being measured. These reports should be made available to internal review teams and potentially made public, depending on the sensitivity of the application, to foster public trust and enable external scrutiny. The contract should also empower the procuring entity to request ad-hoc reports if specific concerns arise regarding system performance or ethical adherence.

Furthermore, contractual clauses must establish clear mechanisms for addressing grievances and rectifying unforeseen harms caused by AI systems. It is a stark reality that even well-intentioned AI systems can produce unintended negative consequences. A citizen might feel unfairly treated by an AI-driven decision, or an organization might identify a systemic bias that was not apparent during initial testing. The contract should outline a structured process for individuals or groups to raise complaints or report issues related to the AI's operation. This grievance mechanism should be accessible, clearly communicated to the public, and ensure that all complaints are logged, investigated, and responded to in a timely manner. For example, if an AI system used for loan applications denies a loan to an applicant, the applicant should have a clear avenue to appeal this decision, potentially through a human review process facilitated by the procuring entity, supported by the AI's output and supporting documentation. The contract should specify the timelines for acknowledgement of grievances, investigation, and resolution, and what remedies are available, which could include manual review of decisions, re-evaluation of the AI's parameters, or compensation for proven damages. This process not only provides recourse for affected parties but also serves as a crucial feedback loop for improving the AI system and its governance.

The contractual framework should also address the concept of "red teaming" or adversarial testing throughout the AI's operational life. Just as cybersecurity threats evolve, so too can malicious actors seek to exploit vulnerabilities in AI systems. Proactive, ongoing adversarial testing by independent parties or by the vendor under strict supervision can help identify and mitigate such risks before they are exploited. Contracts can mandate that vendors conduct periodic red teaming exercises on their AI systems, simulating attacks designed to cause the AI to misbehave, generate biased outputs, or compromise sensitive data. The findings from these exercises, along with the vendor's remediation plans, should be reported to the procuring entity. For an AI system managing critical infrastructure, such as an energy grid, regular red teaming could identify vulnerabilities that might lead to widespread power outages or security breaches. The contract should define the scope and frequency of these exercises, the criteria for acceptable performance under adversarial conditions, and the reporting requirements for findings and mitigation efforts.

Moreover, provisions for model updates and retraining require careful ethical consideration. As data evolves and new societal norms emerge, AI models may need to be updated or retrained. Contracts should stipulate that any significant model updates or retraining must be preceded by a formal ethical review and impact assessment. This would involve re-evaluating the model for bias, fairness, and unintended consequences, using the same rigorous standards as during the initial procurement. The results of this assessment should be documented and, where appropriate, shared with the procuring entity before the updated model is deployed. For an AI system used in hiring, for instance, retraining on a new dataset of applicants could inadvertently reintroduce or amplify existing biases if not carefully managed. The contract should therefore require the vendor to provide assurance that all retraining and update processes are accompanied by a new round of bias and fairness testing, with detailed reports submitted to the procuring agency. The contract could also specify that certain types of updates, particularly those that might significantly alter the AI's decision-making logic or its potential impact, require explicit approval from the procuring entity before implementation.

Transparency also extends to the understanding of data provenance and quality. AI systems are only as good as the data they are trained on and process. Contracts should require vendors to provide clear documentation regarding the origin, quality, and potential limitations of the data used by the AI system. This includes information about how the data was collected, any preprocessing steps undertaken, and known biases or gaps in the dataset. For an AI system used to assess agricultural yields, for example, understanding the geographical distribution and temporal coverage of the historical weather and crop data used for training is crucial for interpreting its predictions. If the data primarily comes from one region, the AI's performance in other regions might be compromised. The contract should mandate that vendors disclose any known data quality issues or limitations and outline how these might affect the AI's performance and reliability. This information is vital for the procuring entity to contextualize the AI's outputs and make informed decisions about its deployment and limitations.

In addition to these specific clauses, overarching accountability frameworks are essential. This can include requirements for regular audits, independent verification of performance claims, and mechanisms for contract renegotiation or termination in cases of persistent ethical breaches or system failures. The contract should clearly define what constitutes a material breach of ethical obligations, outlining the steps for remediation and the consequences of non-compliance, which could range from financial penalties to the termination of the contract. For instance, if an AI system repeatedly demonstrates discriminatory outcomes despite repeated efforts at mitigation, the contract should provide a clear pathway for the public sector entity to terminate the agreement and seek alternative solutions.

Finally, the concept of "explainability" in AI, while technically challenging, should be a contractual aspiration. While full transparency into every algorithmic step may not always be feasible or necessary, contracts can require vendors to provide explanations for the AI's decisions that are understandable to humans. This might involve providing simplified explanations, identifying the key factors that influenced a decision, or offering alternative outputs based on slightly modified inputs. For an AI system that flags potential fraud in financial transactions, for example, a simple "fraud flagged" output is insufficient. The contract could require that the system also provides the top three indicators that contributed to the flag, making it easier for human analysts to investigate further. The level of explainability required will naturally vary depending on the risk and impact of the AI system, but the principle of striving for interpretable AI outputs should be embedded within contractual agreements. This commitment to explainability fosters trust and allows for more effective human oversight and intervention when necessary, ensuring that AI serves as a tool to augment human judgment, rather than replace it blindly.

The practice of building ethical AI ecosystems is not solely a matter of internal policy or post-deployment monitoring; it is profoundly shaped by the initial and ongoing decisions made during procurement. When public sector entities strategically leverage their purchasing power, they become powerful catalysts for responsible AI innovation and adoption across the entire industry. This section explores how procurement acts as a fundamental lever for fostering an environment where ethical AI development is not an afterthought, but a core differentiator, ultimately shaping market behavior and influencing public perception of artificial intelligence.

Public sector procurement, by its very nature, carries significant weight. The sheer scale of government purchasing, coupled with the imperative to serve the public good, means that the standards set by these entities ripple outwards. When a government agency specifies rigorous ethical requirements – be it for data privacy, fairness, transparency, or accountability – it effectively raises the bar for all vendors seeking to do business with it. This incentivizes not only the direct vendors to comply but also encourages their subcontractors and partners to adopt similar ethical considerations. Over time, this can lead to a market where ethical AI is no longer a niche offering but a standard expectation. Consider, for example, a city government seeking to procure an AI-powered system for optimizing public transportation routes. If the procurement documents explicitly demand that the AI must demonstrate equitable service delivery across all neighborhoods, irrespective of socioeconomic status, and that any potential disparities must be rigorously identified and mitigated, vendors will be compelled to invest in developing and deploying solutions that meet these criteria. This demand signals to the market that fairness in service provision is a critical component of successful AI deployment, pushing vendors to refine their algorithms, data selection, and bias detection methodologies. Without such explicit contractual mandates, vendors might prioritize cost-effectiveness or raw performance metrics, potentially overlooking the equitable distribution of benefits.

Furthermore, ethical procurement serves as a vital mechanism for disseminating best practices. When public agencies clearly articulate their ethical expectations and the rationale behind them, they are, in effect, educating the market. Vendors learn what constitutes acceptable

AI behavior within the public sphere. This knowledge transfer is invaluable, especially for smaller or less experienced AI developers who may not have established robust ethical frameworks within their organizations. By encountering these demands in tender documents and contract negotiations, they are exposed to crucial considerations that they might not have otherwise encountered. This can foster a culture of responsible innovation, where ethical thinking is integrated into the design and development process from the outset. Imagine a scenario where a national agency procures an AI system for managing citizen service centers. If the procurement requires vendors to outline their approach to ensuring data minimization, explaining AI outputs in plain language, and providing mechanisms for human oversight and intervention, this sets a precedent. Other agencies looking to procure similar systems will benefit from the clarity and detail provided in these earlier tenders, and vendors will begin to see these requirements as standard components of their offerings, rather than unique demands. This shared understanding reduces friction and promotes a more cohesive and ethical AI landscape.

The long-term perspective in procurement is crucial for cultivating an ethical AI ecosystem. It moves beyond single transactions to consider how these decisions shape the future development and deployment of AI technologies. Public sector entities can, and should, think about how their procurement strategies can support the broader societal goals related to AI. This might involve actively seeking out vendors who demonstrate a commitment to open-source AI development, contribute to public AI ethics research, or engage in community consultations regarding AI deployment. It's about fostering partnerships with organizations that are not just providing a service but are actively contributing to a more responsible and trustworthy AI future. For instance, a research institution procuring AI tools for scientific discovery could prioritize vendors who commit to making their AI models or datasets available for public scrutiny or further research, under appropriate ethical guidelines. This not only advances scientific progress but also builds trust in the AI tools being used. This proactive approach moves procurement from a transactional function to a strategic one, where each contract is an opportunity to invest in a better AI future.

Building such an ecosystem also requires a willingness to engage in ongoing dialogue and collaboration with vendors, researchers, and civil society. Procurement processes should not be rigid, one-way dictates. Instead, they can be platforms for co-creation and learning. This might involve pre-tender consultations where potential vendors can provide input on technical feasibility and ethical considerations, or post-contract reviews that allow for sharing lessons learned and refining best practices. This collaborative approach ensures that ethical standards are not only ambitious but also practical and adaptable to the rapidly evolving AI landscape. For example, when a public utility company procures an AI system to predict and manage infrastructure maintenance needs, an initial workshop involving potential vendors could explore the ethical implications of predictive algorithms, particularly in relation to potential biases that might disproportionately affect certain communities or infrastructure types. This dialogue can lead to more robust ethical considerations being embedded in the final request for proposals.

Moreover, public sector procurement can play a significant role in fostering transparency and accountability not just within specific projects but across the entire AI development lifecycle. By requiring vendors to demonstrate robust data governance, explainable AI practices, and mechanisms for independent audit and oversight, public agencies can champion principles that are essential for public trust. This commitment to transparency in procurement sends a powerful message to the public: that their governments are serious about ensuring AI systems are developed and used responsibly and ethically. This, in turn, can influence public perception, shifting it from one of apprehension or skepticism towards one of informed optimism. When citizens see that their public institutions are demanding ethical AI, they are more likely to accept and benefit from the technologies that are deployed. Consider the procurement of an AI system for use in the justice system. If the tender explicitly mandates that the AI's decision-making processes must be explainable to legal professionals and that its performance must be subject to regular, independent audits for bias, this demonstrates a commitment to fairness and due process. Such transparency reassures the public that the technology is being deployed with a focus on justice and equity, rather than as a black box decision-maker.

The development of ethical AI ecosystems through procurement also necessitates a careful consideration of market dynamics. Public sector entities often have the leverage to shape these dynamics. By consistently demanding higher ethical standards, they can create a competitive advantage for vendors who prioritize ethics. This can lead to a market where ethical AI development becomes a key selling point, driving innovation in areas such as bias mitigation, privacy-preserving techniques, and human-AI collaboration. Conversely, if procurement remains focused solely on cost and functionality, it can inadvertently perpetuate a race to the bottom, where ethical considerations are sidelined. Therefore, procurement strategies should be designed to reward ethical innovation. This might involve evaluation criteria that give significant weight to a vendor's ethical track record, their proposed AI ethics framework, and their commitment to ongoing ethical review. For instance, when procuring an AI system for managing public health data, a government agency might assign points not only for technical accuracy but also for the vendor's demonstrated expertise in secure data handling, anonymization techniques, and adherence to public health ethics guidelines. This incentivizes vendors to build capabilities and expertise in these ethically critical areas.

Furthermore, procurement can be a vehicle for encouraging diversity and inclusion within the AI field. By actively seeking out and supporting vendors from underrepresented groups, or those that demonstrate a commitment to diverse development teams, public agencies can help foster a more inclusive AI ecosystem. A diverse workforce is more likely to identify and address a broader range of potential biases and unintended consequences in AI systems. This can lead to the development of more robust and equitable AI solutions that serve the needs of all members of society. For example, a procurement process for an AI system designed to assist in educational assessments could actively encourage participation from vendors with diverse teams that reflect the student population. This inclusion can lead to the identification of culturally relevant factors and potential biases that might be missed by a less diverse development team, resulting in a fairer and more effective assessment tool.

The long-term impact of ethical procurement extends to building institutional capacity within the public sector itself. Engaging in the procurement of AI, especially with an ethical lens, requires government agencies to develop new skills and expertise. This includes understanding AI technologies, assessing ethical risks, and formulating complex contractual requirements. As agencies gain experience in procuring and overseeing ethical AI systems, they build internal knowledge and capabilities that enable them to make more informed decisions in the future. This continuous learning cycle strengthens the public sector's ability to govern AI effectively. It means that procurement departments, legal teams, and program managers become more adept at navigating the ethical complexities of AI, rather than relying solely on external expertise. This internal capacity building is fundamental to creating a sustainable ethical AI ecosystem.

Finally, the ultimate goal of ethical procurement in fostering AI ecosystems is to ensure that AI is developed and deployed in a manner that aligns with societal values and enhances public well-being. It is about using the power of public purchasing to steer the trajectory of AI development towards beneficial outcomes. By setting clear ethical expectations, incentivizing responsible innovation, promoting transparency, and fostering collaboration, public sector entities can play a pivotal role in shaping an AI future that is not only technologically advanced but also ethically sound and socially beneficial. This requires a sustained commitment to integrating ethical considerations into every stage of the procurement process, from initial planning and vendor selection to contract management and ongoing oversight. It is an investment in a future where AI serves humanity, rather than the other way around.

## Chapter 8: Establishing Clear Lines of Accountability

In the realm of artificial intelligence, particularly as these systems become more sophisticated and integrated into critical public services, a significant challenge emerges: the accountability gap. This gap refers to a situation where, when an AI system malfunctions, produces harmful outcomes, or fails to meet its intended objectives, it becomes exceedingly difficult to pinpoint a single entity or individual responsible. This ambiguity can arise from several interconnected factors, all of which contribute to a diffusion of responsibility that can leave individuals and communities without recourse when they are negatively impacted.

One of the primary drivers of this accountability gap is the inherent autonomy of advanced AI systems. As AI moves beyond simple automation and into domains requiring complex decision-making, it begins to operate with a degree of independence. Algorithms learn, adapt, and make choices based on vast datasets and intricate internal logic that may not be fully transparent even to their creators. When such a system, for example, an AI-powered tool used in criminal justice that influences sentencing recommendations, disproportionately flags individuals from certain demographics as high-risk, who is to blame? Is it the data scientists who trained the model, the engineers who deployed it, the policymakers who authorized its use, or the system itself, if its emergent behaviors were not explicitly programmed? The traditional models of accountability, which often rely on identifying a clear human error or intentional wrongdoing, struggle to map onto these scenarios. The AI acts, and its actions have consequences, but the chain of command and responsibility becomes fragmented and obscured by the complexity of the technology.

Furthermore, the intricate and often globalized supply chains involved in AI development and deployment significantly contribute to this accountability deficit. An AI system deployed by a government agency might have been developed by a private company, which in turn relied on open-source components, licensed data from a third party, and utilized cloud infrastructure managed by another entity. Each of these actors plays a role in the final product and its performance. If an AI-driven traffic management system causes widespread gridlock due to an unforeseen interaction between its predictive algorithms and real-time sensor data, tracing the root cause of the failure becomes a monumental task. Was the error in the initial algorithm design? Was it in the quality or labeling of the training data? Did a change in the cloud provider's infrastructure introduce an instability? Or was it the integration of disparate systems that created an emergent flaw? In such a complex ecosystem, each party can point to another, and the end user or affected citizen is left in a position of seeking redress from a nebulous collective rather than a clearly identifiable responsible party. This distributed nature of AI development means that responsibility is not easily contained or assigned.

The distributed nature of AI development and deployment extends beyond just the commercial supply chain. It also encompasses the various stakeholders involved in the lifecycle of an AI system. This can include the researchers who explore novel algorithms, the developers who code the software, the product managers who define its features, the deployment teams who integrate it into existing workflows, and the end-users who interact with it on a daily basis. Each of these groups has a hand in shaping the AI's behavior and outcomes. Consider an AI-powered diagnostic tool used in healthcare. If this tool misdiagnoses a patient, leading to delayed or incorrect treatment, accountability becomes a thorny issue. The researchers might argue their algorithm was sound in a controlled environment. The developers might claim they implemented the research faithfully. The product managers might state the tool was deployed within its specified parameters. The clinicians using the tool might argue they relied on its output as intended. In this scenario, the intended beneficiary of the AI – the patient – can be left without a clear avenue for recourse, as responsibility is so widely dispersed across numerous individuals and organizational units, each with their own justifications and defenses.

Moreover, the rapid pace of AI innovation and the evolving nature of the technology itself can outpace the development of legal and ethical frameworks for accountability. By the time a specific AI-related harm is identified and a legal case is considered, the technology may have already evolved, making the original design choices or deployment decisions seem less relevant or harder to prove as the direct cause of the harm. This temporal disconnect further complicates the assignment of blame. For instance, if an AI system designed for predictive policing is found to perpetuate historical biases in arrests, and a lawsuit is filed years after its initial deployment, the original datasets and training methodologies might have been updated, or the algorithm itself modified in ways that obscure the initial discriminatory patterns. The original developers or policymakers might argue that the current version of the system is not the same one that caused the harm, thus creating a moving target for accountability.

The "black box" nature of some sophisticated AI models exacerbates this problem. When an AI's decision-making process is opaque and its internal workings are not readily explainable, it becomes incredibly challenging to understand
*why* a particular outcome occurred. This lack of transparency makes it difficult to identify a specific failure point or a responsible human agent. If an AI trading algorithm makes a series of disastrous trades, leading to significant financial losses, but no one can fully explain the sequence of decisions that led to those trades, how can accountability be assigned? Was it a flaw in the algorithm's learning process, an unexpected market

event that the AI interpreted in a novel way, or a human oversight in setting risk parameters? Without the ability to interrogate the AI's reasoning, attributing responsibility becomes a matter of speculation rather than concrete evidence.

This accountability gap is not merely an abstract theoretical problem; it has tangible consequences for individuals and society. When AI systems fail to provide equitable services, discriminate against certain groups, or cause physical or financial harm, the lack of clear accountability mechanisms can lead to a sense of injustice and powerlessness among those affected. Without a clear path to seek redress, trust in both AI technologies and the institutions that deploy them erodes. This erosion of trust can hinder the adoption of beneficial AI applications and foster widespread skepticism, ultimately undermining the potential positive impact of AI on society. The public sector, in particular, bears a significant responsibility to ensure that AI systems deployed in its name are accountable. Citizens expect their governments to uphold justice and provide recourse when harm occurs, regardless of whether that harm is caused by human error or by the complex machinations of artificial intelligence. Therefore, understanding and actively working to close this accountability gap is a fundamental prerequisite for the responsible and ethical deployment of AI in public services.

In confronting the pervasive issue of the AI accountability gap, a foundational step involves the meticulous and proactive definition of roles and responsibilities for every entity and individual implicated in the artificial intelligence lifecycle. This is not a retrospective exercise undertaken after an incident, but rather a deliberate and ongoing process that must be embedded from the very inception of any AI initiative within the public sector. Without this clarity, the amorphous nature of AI development and deployment will inevitably lead to a diffusion of ownership, leaving a vacuum where accountability should reside. This section delves into the critical need to establish these clear lines of responsibility, exploring how to assign ownership for various facets of AI, from initial concept and design through to ongoing operation and eventual decommissioning.

The journey begins with a comprehensive mapping of all stakeholders. In the context of public sector AI implementation, this typically includes a diverse array of actors: the government agency initiating the AI project, specific departments or directorates within that agency, individual public officials who oversee or utilize the AI, the external technology providers or vendors who develop and supply the AI solutions, and crucially, independent oversight bodies or external auditors tasked with ensuring compliance and ethical operation. Each of these parties brings a unique perspective and set of capabilities, and their contributions, potential liabilities, and oversight functions must be explicitly delineated. For instance, a government agency might be responsible for defining the overarching policy goals and ethical boundaries for an AI system, while a specific department might be tasked with integrating it into existing workflows and ensuring data privacy standards are met. A technology provider, in turn, will hold responsibility for the technical integrity and performance of the AI model itself, subject to contractual obligations and regulatory compliance.

To achieve this clarity, it is imperative to develop robust frameworks for assigning accountability. These frameworks should not be one-size-fits-all but should be tailored to the specific nature and risk profile of each AI application. For a low-risk AI system, such as a simple chatbot for public information dissemination, the accountability structures might be relatively straightforward. However, for a high-stakes AI system, such as one used in predictive policing, healthcare diagnostics, or critical infrastructure management, the frameworks must be far more detailed and layered. A tiered approach to accountability can be highly effective, wherein different levels of responsibility are assigned for different types of AI failures. For example, the most severe failures might trigger accountability at the highest levels of agency leadership, while more minor operational issues might be addressed at the departmental or project management level.

Consider the initial stages of AI development. Here, clarity regarding who is accountable for the problem definition and the ethical considerations is paramount. Is it the policy-makers who identify a societal problem and envision an AI solution, or is it the technical teams who translate that vision into a functional system? Both play crucial roles, and their responsibilities must be distinguished. The policy-makers are accountable for ensuring that the problem statement is well-defined, that the intended beneficiaries are clearly identified, and that the proposed AI solution aligns with public values and legal requirements. They are also accountable for initiating the process of ethical impact assessments. The technical teams, on the other hand, are accountable for the fidelity of the AI's design and implementation to the stated objectives, for proactively identifying potential biases within the data and algorithms, and for adhering to security and privacy protocols throughout the development process. This might involve data scientists being accountable for the rigor of their model selection and training, and software engineers being accountable for the secure coding practices and the robustness of the deployment architecture.

Moving into the deployment phase, accountability becomes even more complex due to the integration of AI into existing operational environments. Who is accountable for the successful integration of the AI system into public services? Who is responsible for ensuring that end-users are adequately trained and understand the capabilities and limitations of the AI? And critically, who is accountable for the ongoing monitoring of the AI's performance in real-world conditions? A public agency implementing an AI-powered system for processing citizen applications, for instance, must assign clear responsibility for the change management process, user training, and the establishment of feedback mechanisms. The agency must hold specific individuals or teams accountable for validating the AI's outputs against human benchmarks, for monitoring for performance degradation or drift, and for ensuring that any unexpected or anomalous behavior is immediately flagged and investigated. This might involve appointing a dedicated AI system administrator or a cross-functional oversight committee.

The role of technology providers in this ecosystem also demands precise definition. When a government agency procures an AI solution from a private vendor, the contractual agreements must explicitly outline the vendor's responsibilities. This includes not only the functional performance of the AI but also its compliance with relevant regulations, its security posture, and its commitment to transparency and explainability where feasible. For example, the vendor might be accountable for delivering an AI model that meets pre-defined accuracy metrics, for providing regular software updates to address security vulnerabilities, and for cooperating with the agency's internal auditors in their review of the AI's performance. Furthermore, the contract should specify the vendor's responsibilities in the event of AI-induced harm or failure. This could range from providing technical support to investigate incidents, to offering remediation, or even to indemnifying the agency in certain circumstances, depending on the nature of the failure and the contractual terms.

External auditors and oversight bodies play a vital role in bolstering accountability. Their mandate must be clearly defined, including the scope of their reviews, the frequency of their assessments, and their authority to recommend or mandate corrective actions. These independent entities serve as a critical check on the internal processes of the agency and the performance of the technology provider. They are accountable for conducting thorough and impartial evaluations, for identifying potential risks and compliance issues, and for reporting their findings to relevant stakeholders, including legislative bodies or the public, where appropriate. For an AI system used in resource allocation, for instance, an external auditor might be tasked with verifying that the AI's decision-making process does not perpetuate systemic biases, and that its outputs are equitable and align with policy objectives. Their accountability lies in the integrity and rigor of their independent oversight.

To operationalize these defined roles and responsibilities, public sector organizations should consider implementing structured accountability matrices, such as a RACI (Responsible, Accountable, Consulted, Informed) chart, tailored for AI systems. A "Responsible" party is one who does the work. An "Accountable" party is one who owns the work and has the authority to approve it. "Consulted" parties provide input, and "Informed" parties are kept up-to-date. Applying this to an AI project might mean that a data science team is "Responsible" for developing a predictive model, while the Chief Data Officer or a designated AI ethics officer is "Accountable" for its ethical implications and performance. The legal department might be "Consulted" on data privacy aspects, and the public relations team might be "Informed" about potential public-facing implications.

Moreover, accountability must extend to the ongoing maintenance and evolution of AI systems. AI models are not static; they require continuous monitoring, updating, and retraining to remain effective and safe. Who is accountable for the timely retraining of a model that begins to exhibit performance degradation due to changes in the underlying data distribution? Who is responsible for deciding when an AI system has become obsolete or poses an unacceptable risk, and needs to be retired? These are critical questions that require clear ownership. For instance, a government agency utilizing an AI system for fraud detection would need to assign responsibility for the proactive identification of concept drift – the phenomenon where the statistical properties of the target variable change over time, rendering the model less accurate. This responsibility might fall to a dedicated AI operations team, which would be accountable for monitoring model performance metrics, initiating retraining cycles, and managing the deployment of updated models.

The establishment of robust feedback loops is intrinsically linked to accountability. When individuals or communities are negatively impacted by an AI system, they need clear channels through which to report their experiences and seek redress. The agency or official accountable for receiving and responding to these concerns must be clearly identified. This process should not be an afterthought but a critical component of the AI system's design and operational plan. For example, if an AI-powered public transport scheduling system consistently fails to serve certain neighborhoods adequately, there must be a clearly designated point of contact within the transit authority who is accountable for investigating these complaints, addressing the underlying causes, and implementing necessary improvements. The public's ability to provide feedback and receive a substantive response is a tangible manifestation of accountability.

Furthermore, the principles of explainability and transparency, while technically challenging for some AI models, are crucial enablers of accountability. Even if a full step-by-step explanation of an AI's decision is not possible, efforts must be made to provide a rationale that is understandable to relevant stakeholders. Who is accountable for developing and communicating these explanations? This responsibility might lie with the technical teams, in collaboration with domain experts and communication specialists. The aim is to ensure that decisions made by AI systems, particularly those that have significant impacts on individuals' lives, are not entirely opaque. For an AI system used in determining eligibility for social services, the agency must be accountable for providing applicants with a clear understanding of the key factors that influenced the AI's decision, even if the underlying algorithm is complex.

The concept of "accountability by design" is also paramount. This means that the structures for accountability are not bolted on as an afterthought but are integral to the design and development process itself. From the earliest stages of conceptualization, discussions must revolve around who will be accountable for what, and how that accountability will be verified. This requires a shift in mindset, moving away from simply asking "Can we build this?" to "Should we build this, and if so, who is responsible for ensuring it is built and used ethically and safely?" This proactive approach helps to prevent the emergence of accountability gaps by embedding oversight and responsibility into the very fabric of the AI system. For example, when designing an AI system to assist in loan application processing, the design team should proactively consider how to ensure accountability for potential discriminatory outcomes, perhaps by building in fairness metrics and audit trails from the outset.

Finally, the legal and regulatory landscape plays a significant role in shaping the contours of accountability. While this chapter focuses on defining internal roles, it is important to acknowledge that these definitions are often informed by and must align with external legal obligations. Agencies and officials must be aware of the specific legal frameworks governing AI use within their jurisdictions and ensure that their internal accountability structures are sufficient to meet these external requirements. This might involve collaborating closely with legal counsel to interpret relevant statutes and regulations and to ensure that contracts with technology providers adequately address legal liabilities. Ultimately, the goal is to create a robust ecosystem of accountability where every participant understands their role, accepts their responsibilities, and can be held answerable for the outcomes of AI systems deployed in the public interest. This comprehensive approach to defining roles and responsibilities is not merely an administrative formality; it is a cornerstone of building trustworthy and ethically sound AI systems that serve the public good.

The preceding discussion has underscored the critical imperative of clearly defining roles and responsibilities across the entire AI lifecycle within the public sector. This foundational work is essential for fostering a culture of accountability. However, establishing these lines of responsibility, while vital, is only one part of the equation. Even with the most meticulously crafted accountability frameworks, AI systems can, and sometimes will, fall short of expectations, lead to unintended consequences, or cause harm to individuals or communities. It is in these moments that the true test of an accountability regime lies: its capacity to provide effective mechanisms for grievance and redress.

Without robust pathways for citizens to voice concerns, challenge decisions, and seek remedies when they believe an AI system has acted unfairly or caused them harm, the promise of responsible AI in public service remains hollow. The absence of such mechanisms can lead to a sense of powerlessness, erode public trust, and perpetuate systemic inequalities. Therefore, a crucial component of any comprehensive AI accountability strategy is the deliberate and proactive establishment of accessible, transparent, and equitable grievance and redress procedures.

These mechanisms must be designed not as an afterthought, but as an integral part of the AI system's deployment and ongoing operation.

The first principle in establishing effective grievance mechanisms is accessibility. Information about how to file a complaint, and to whom, must be readily available and comprehensible to all segments of the population, regardless of their technical literacy, language, or disability. This means providing multiple channels for individuals to lodge their grievances. For instance, a government agency deploying an AI system for determining eligibility for social benefits should ensure that information on how to appeal an automated decision is prominently displayed on its website, included in all official correspondence, and made available through in-person service centers and phone hotlines. The language used in these communications should be plain and free of jargon, ensuring that individuals are not deterred from seeking recourse due to a lack of understanding. Consider the potential impact on elderly citizens or those with limited digital access; relying solely on online portals would effectively exclude them from the grievance process, creating a significant accountability gap. Therefore, a multi-modal approach, encompassing digital, telephonic, and in-person options, is essential to ensure equitable access to redress.

Beyond mere accessibility of information, the channels themselves must be user-friendly. Complaint forms, whether digital or paper-based, should be straightforward, asking only for necessary information without being overly burdensome. The process should guide individuals through the steps required, offering assistance where needed. For AI systems involved in complex decision-making processes, such as those used in the justice system or for critical infrastructure management, the complaint process might need to be more nuanced, allowing for the submission of supporting evidence and detailed explanations of perceived errors or biases. The key is to minimize the friction for individuals seeking to have their concerns heard, recognizing that they may already be in a vulnerable or distressed state due to the AI's impact.

Once a grievance is lodged, a clear and transparent investigation procedure must be initiated. This procedure should outline the steps that will be taken to review the complaint, who will be involved in the investigation, and the typical timelines for resolution. Transparency in this process is paramount for building trust. Individuals should be kept informed about the progress of their complaint, even if it is simply to acknowledge receipt and provide an estimated timeframe for a substantive response. The investigation itself should be conducted by individuals or a team that is independent of the direct operators of the AI system in question, or at least has the capacity to impartially assess the situation. In situations where an AI system is developed and operated by a third-party vendor, the contract should specify the vendor's cooperation in the grievance investigation process, including providing access to relevant data and logs, subject to privacy and security considerations.

The composition of the investigative team is also critical. For AI systems with significant societal implications, such as those used in law enforcement or public health, the investigation team might benefit from a multi-disciplinary approach, including technical experts, domain specialists, ethicists, and potentially, representatives from the affected community. This ensures that the investigation considers not only the technical aspects of the AI's performance but also its broader societal impact and ethical implications. For example, if a complaint arises regarding an AI system used for predictive policing that is perceived to disproportionately target certain communities, the investigation should involve not only an analysis of the algorithm and data but also an understanding of community concerns and potential systemic biases.

Timeliness is another non-negotiable aspect of effective grievance redress. A protracted or delayed response can exacerbate the harm caused by an AI system and lead to significant frustration. While complex investigations may require time, reasonable deadlines should be set and communicated. The public sector should strive for prompt acknowledgement of complaints, with a defined period for conducting the investigation and issuing a decision. For urgent matters, such as an AI system incorrectly denying essential services, expedited review processes should be in place. The goal is to provide a resolution in a timeframe that is meaningful to the individual affected, demonstrating that their concerns are taken seriously. This might involve establishing service level agreements (SLAs) for grievance resolution, similar to those used for other public service complaints.

The outcome of a grievance investigation should be a clear and reasoned decision. This decision should explain the findings of the investigation, whether the complaint was upheld or dismissed, and the rationale behind that conclusion. If the grievance is upheld, appropriate remedies must be offered. These remedies can vary widely depending on the nature of the harm. They could include: a reversal of an AI-driven decision, financial compensation for damages incurred, the implementation of corrective measures to prevent recurrence, or an apology. For instance, if an AI system incorrectly flagged an individual for fraud, leading to a suspension of benefits, the remedy would involve rectifying the error, restoring the benefits, and potentially providing compensation for any financial hardship caused.

In cases where the AI system's decision is upheld, it is still crucial to provide a clear explanation. This explanation should not only reiterate the basis for the decision but also guide the individual on any further avenues for appeal or recourse, if available. The process of seeking redress should not be a dead end after the initial investigation. There should be a clear escalation path, allowing individuals to seek review of the initial decision by a higher authority or an independent review panel. This layered approach ensures that decisions are not arbitrary and that there are checks and balances within the grievance and redress system itself.

A critical consideration for public sector AI is the potential for systemic issues. A single grievance might be an isolated incident, but a pattern of similar complaints can indicate a fundamental flaw in the AI system's design, data, or deployment. Therefore, grievance mechanisms should be integrated with broader AI monitoring and auditing functions. Data on grievances, along with the outcomes of investigations and implemented remedies, should be systematically collected, analyzed, and used to inform improvements to AI systems and related policies. This feedback loop is essential for continuous learning and for ensuring that the public sector's use of AI evolves responsibly. Analyzing the types of grievances received can highlight areas where AI systems are underperforming, exhibiting bias, or failing to meet ethical standards, prompting necessary adjustments or even decommissioning of the system.

Furthermore, the establishment of an independent ombudsman or a dedicated AI ethics review board could serve as a powerful mechanism for grievance and redress. Such an entity would operate independently of the specific agencies deploying AI systems, providing an impartial avenue for citizens to seek resolution when internal complaint processes have been exhausted or are perceived as inadequate. This external oversight can lend significant credibility to the public sector's commitment to accountability and provide a vital check on the power of AI. The ombudsman's office would be empowered to investigate complaints, mediate disputes, and make recommendations for systemic improvements. Their reports, made public, would offer a valuable insight into the challenges and successes of AI deployment in public service.

The nature of AI means that its impacts can sometimes be diffuse and difficult to attribute to a single cause. This can complicate the process of assigning responsibility and providing redress. For instance, an AI system that subtly influences hiring decisions over time might contribute to a lack of diversity in an organization without any single, identifiable discriminatory act. In such cases, the grievance and redress mechanisms need to be sufficiently sophisticated to address these emergent and complex harms. This might involve moving beyond individual complaints to collective redress mechanisms, or the establishment of impact funds to compensate groups of individuals who have been demonstrably disadvantaged. The focus shifts from pinpointing a specific faulty decision to addressing a broader pattern of adverse impact.

Public sector organizations must also consider the role of public communication in their grievance and redress strategies. Transparency about the existence and effectiveness of these mechanisms builds public confidence. Sharing anonymized data on the types of grievances received, the resolutions provided, and lessons learned can demonstrate a commitment to continuous improvement and accountability. This open dialogue is vital for maintaining a healthy relationship between the government and the governed, especially as AI plays an increasingly significant role in public services. For example, publishing an annual report detailing AI-related complaints and their resolution can serve as a valuable accountability tool.

In conclusion, the establishment of robust grievance and redress mechanisms is not merely an optional add-on to an AI accountability framework; it is a fundamental requirement for ensuring that AI is deployed in the public sector in a manner that is just, equitable, and trustworthy. These mechanisms must be designed with the end-user in mind, prioritizing accessibility, transparency, fairness, and timeliness. By providing clear avenues for citizens to voice their concerns and seek remedies, public sector organizations can not only address individual harms but also foster a culture of continuous improvement and accountability that underpins the responsible and ethical use of artificial intelligence for the benefit of all. This commitment to providing recourse is a tangible demonstration that the public sector takes seriously its obligation to protect the rights and well-being of the citizens it serves, even when navigating the complexities of advanced technologies.

The preceding discussion has illuminated the vital necessity of robust grievance and redress mechanisms for public sector AI systems, ensuring that individuals have avenues to challenge adverse outcomes and seek remedies. However, the very effectiveness of these redress pathways, and indeed the overall trustworthiness of AI deployed in public service, hinges upon a parallel and equally critical set of safeguards: independent auditing and oversight. While grievance procedures address harms after they have occurred, auditing and oversight proactively work to prevent such harms by ensuring AI systems are designed, developed, and deployed responsibly and ethically, adhering to established standards and regulations.

Establishing clear lines of accountability is not merely about assigning blame or offering recourse when things go wrong. It is fundamentally about building systems and processes that are inherently sound, reliable, and aligned with public values. This is where the concept of independent auditing and oversight becomes paramount. Just as financial institutions are subject to rigorous audits to ensure fiscal probity and prevent fraud, AI systems, particularly those impacting citizens' lives, require a similar level of scrutiny. This scrutiny must extend beyond mere technical functionality to encompass the ethical implications, fairness, transparency, and adherence to legal and policy frameworks. Without such mechanisms, the potential for AI to embed and even amplify existing societal inequalities, or to introduce new forms of discrimination and injustice, becomes significantly heightened, undermining the very purpose of public service.

The challenge of auditing AI is multifaceted, stemming from the inherent complexity and evolving nature of these technologies. Unlike traditional software, AI systems, especially those employing machine learning, can exhibit emergent behaviors that are not explicitly programmed. Their decision-making processes, particularly in deep learning models, can be opaque, making it difficult to fully understand *why* a particular outcome was reached. This "black box" problem necessitates specialized auditing techniques that go beyond simply checking code for bugs. Audits must delve into the data used for training, the algorithms themselves, the validation processes, and the ongoing performance of the system in real-world deployments.

Several models for AI oversight and auditing can be envisioned within the public sector, each with its strengths and potential limitations. A foundational approach involves **internal review boards or committees**. These bodies, composed of individuals from various departments within a public agency, could be tasked with reviewing AI project proposals, setting ethical guidelines, and monitoring deployed systems. Their internal nature allows for a deep understanding of the agency's mission and operational context. However, their effectiveness can be hampered by potential conflicts of interest, a lack of specialized expertise, or insufficient authority to challenge deeply entrenched practices or politically sensitive AI deployments. To mitigate these issues, internal review boards should ideally include diverse perspectives, such as ethicists, legal counsel, data scientists, and representatives from departments that directly interact with the public. Clear mandates, robust reporting structures, and a degree of independence from project development teams are also crucial. For instance, a city's planning department might establish an internal AI ethics committee to review proposals for AI-powered traffic management systems, ensuring that such systems do not disproportionately impact certain neighborhoods or exacerbate existing congestion issues in underserved areas. This committee would examine the data sources, the fairness metrics used in the algorithms, and the projected impact on different demographic groups before a system is even piloted.

A more robust layer of oversight involves **external, independent auditing bodies**. These could take several forms, ranging from specialized government agencies to contracted third-party auditors. The key characteristic is their independence from the agencies deploying the AI systems. These auditors bring an objective perspective and can employ a wider array of specialized tools and methodologies. For example, a national digital agency or a dedicated AI oversight commission could be established to set standards, conduct periodic audits of critical public sector AI systems, and investigate significant incidents. These bodies would need statutory powers to access information, demand explanations, and enforce compliance. Such an approach mirrors the oversight mechanisms found in regulated industries like finance or healthcare, where external bodies ensure adherence to safety, security, and fairness standards. Consider the development of an AI system for the criminal justice system aimed at assisting in sentencing or parole decisions. An independent external audit would be essential to scrutinize the system for biases in the data used to train it – perhaps historical sentencing data that reflects systemic racial disparities. The audit would verify that the system's outputs do not perpetuate or amplify these biases and that it adheres to principles of due process and fairness, even if its recommendations are not binding.

Another model for oversight is the
**establishment of ombudsman-like roles or dedicated AI ethics commissioners**. These figures, operating independently, would serve as a point of contact for citizens and civil society to raise concerns about AI systems and to conduct investigations into alleged misuses or harms. While not conducting routine technical audits, they would play a critical role in oversight by investigating specific complaints, identifying systemic issues, and advocating for policy changes. Their investigative powers, coupled with their independence, can provide a powerful check on the unchecked deployment of AI. Imagine a situation where an AI system used for allocating social housing is perceived by community groups to be unfairly prioritizing certain applicants. While internal grievance procedures might exist, an AI ombudsman could be approached to conduct an independent inquiry into the system's design, data, and operational outcomes, potentially uncovering broader issues of bias or lack of transparency that individual complaints might not reveal.

The specific
**types of audits** required for public sector AI systems are diverse and must be tailored to the system's purpose, the data it uses, and the potential risks associated with its deployment. At a minimum, audits should focus on several key areas:

Firstly,
**data integrity and provenance audits** are fundamental. AI systems are only as good as the data they are trained on. Audits must verify that the

data used is accurate, representative, and free from known biases that could lead to discriminatory outcomes. This involves tracing the origin of the data, understanding how it was collected and processed, and assessing its suitability for the intended application. For instance, an AI system designed to predict public health needs based on historical demographic and health data must be audited to ensure that the data accurately reflects the current population, is not skewed by historical underreporting in certain communities, and does not inadvertently encode past discriminatory healthcare practices. The audit would scrutinize data collection methodologies, identify any missing data points, and assess the imputation strategies used to fill gaps.

Secondly,
**algorithmic fairness and bias audits** are crucial. These audits aim to detect and measure unfair discrimination against protected groups, such as those based on race, gender, age, or disability. This involves employing a range of fairness metrics, which can include demographic parity (ensuring outcomes are similar across groups), equalized odds (ensuring prediction accuracy is similar across groups), or counterfactual fairness (ensuring an individual's outcome would not change if their protected attribute were different). These audits require specialized statistical techniques and tools. For example, if a government agency uses an AI system to assess risk in loan applications or welfare benefits, an audit would rigorously test whether the system disproportionately rejects applicants from minority groups or assigns them higher risk scores, even when other factors are equal. The audit would not just identify bias but also explore its root causes, whether in the data or the model design, and recommend mitigation strategies.

Thirdly,
**performance and robustness audits** are necessary to ensure the AI system functions as intended and can withstand unexpected inputs or adversarial attacks. This includes testing the system's accuracy,

reliability, and stability under various conditions, including edge cases and novel scenarios. For AI systems used in safety-critical applications, such as autonomous vehicles or medical diagnosis, the standards for robustness are exceptionally high. An audit might involve simulated environments to test how an AI-powered emergency response system performs under different disaster scenarios, or how a system for identifying fraudulent tax claims handles slightly altered or intentionally misleading submissions. The goal is to ensure that the AI system does not fail catastrophically or produce unreliable results when faced with real-world complexities.

Fourthly,
**privacy and data protection audits** are indispensable. Public sector AI systems often handle sensitive personal data. Audits must verify that the system complies with all relevant data protection regulations, such as GDPR or similar national laws. This involves assessing how data is collected, stored, used, and shared, ensuring that appropriate anonymization or pseudonymization techniques are employed, and that access controls are robust. For an AI system used to personalize educational content for students, a privacy audit would ensure that student data is used only for its intended educational purpose, is not shared with third parties without consent, and is securely stored, protecting against potential data breaches that could reveal sensitive information about a student's learning progress or personal circumstances.

Fifthly,
**explainability and transparency audits** are vital for building public trust and enabling effective oversight. While deep learning models can be complex, efforts should be made to ensure that their decision-making processes can be understood, at least to a degree that allows for meaningful review and accountability. Audits would assess whether the system provides clear explanations for its outputs, particularly for

decisions that significantly impact individuals. This could involve evaluating the use of techniques like LIME (Local Interpretable Model-agnostic Explanations) or SHAP (SHapley Additive exPlanations) to generate insights into model behavior. For an AI system used by a social services agency to recommend interventions for at-risk families, an audit would examine whether the system can articulate *why* a particular family is flagged, citing specific contributing factors that are understandable to human caseworkers and, where appropriate, to the families themselves, thereby ensuring decisions are not arbitrary.

Finally,
**ethical compliance audits** are essential. These audits go beyond legal compliance to assess whether the AI system aligns with broader ethical principles and societal values. This might involve evaluating whether the system promotes fairness, equity, human dignity, and autonomy, and whether it avoids creating new societal harms or exacerbating existing ones. These audits can be more qualitative and may involve stakeholder consultations. For an AI system designed to assist in public recruitment processes, an ethical compliance audit would assess whether it actively promotes diversity and inclusion, rather than simply optimizing for a narrow set of performance indicators that might inadvertently favor certain demographics or educational backgrounds. It would consider whether the system's use could lead to a reduction in human judgment and empathy in hiring decisions.

The implementation of these auditing and oversight mechanisms is not a one-time event but an ongoing process. AI systems are dynamic; they learn, adapt, and evolve. Therefore, audits should not be confined to the pre-deployment phase but must be conducted regularly throughout the system's lifecycle. Continuous monitoring, periodic re-audits, and mechanisms for rapid reassessment in response to significant changes or incidents are all critical components of effective oversight. Furthermore, the findings of these audits must be acted upon. A report detailing biases or vulnerabilities is of little value if it does not lead to concrete improvements, policy changes, or, in extreme cases, the decommissioning of a flawed system. This necessitates clear protocols for addressing audit findings, assigning responsibility for remediation, and tracking progress.

The public sector's commitment to responsible AI deployment can be demonstrably strengthened through transparency regarding its auditing and oversight processes. While proprietary algorithms and sensitive data may require protection, the methodologies used for auditing, the standards applied, and the general findings of audits (appropriately anonymized and aggregated) could be made public. This transparency can foster public trust and allow for informed public discourse about the use of AI in government. Publishing reports on regular AI audits, for example, could reveal patterns of bias that need to be addressed systemically, or showcase successful implementations that adhere to high ethical standards.

In conclusion, while establishing clear lines of accountability and providing robust grievance mechanisms are fundamental to responsible AI in the public sector, they are incomplete without a comprehensive system of independent auditing and oversight. These mechanisms act as crucial checks and balances, proactively ensuring that AI systems are developed and deployed in a manner that is fair, transparent, secure, and aligned with public values. By embracing diverse models of oversight, conducting rigorous and varied audits, and committing to continuous monitoring and transparency, public sector organizations can build and maintain the trust of the citizens they serve, ensuring that AI serves as a tool for public good rather than a source of unintended harm or injustice. This commitment to scrutiny and accountability is not merely a procedural requirement; it is an ethical imperative that underpins the very legitimacy of government in the age of artificial intelligence.

The integration of artificial intelligence into public sector operations offers profound opportunities for efficiency, improved service delivery, and evidence-based policymaking. However, the very power and potential of these systems also necessitate careful consideration of how accountability is maintained, especially when AI outputs influence critical decisions affecting citizens' lives. As explored in the preceding sections, robust grievance mechanisms, independent auditing, and comprehensive oversight are indispensable pillars for building trust and ensuring responsible AI deployment. Yet, these structures, while vital, are ultimately complemented by a fundamental requirement: the persistent role of human judgment and the indispensable capacity for human intervention. Accountability, at its most profound level, is not an abstract concept delegated entirely to algorithms; it remains, and must remain, a human endeavor.

The argument for maintaining human involvement in AI-driven decision-making, particularly in high-stakes public sector applications, is rooted in both ethical principles and practical necessity. AI systems, no matter how sophisticated, operate based on historical data, programmed parameters, and statistical probabilities. They lack the nuanced understanding of human context, the capacity for empathy, and the moral reasoning that are integral to sound judgment in complex societal situations. When AI is deployed in domains such as criminal justice, social welfare, healthcare, or resource allocation, its recommendations or classifications can have life-altering consequences. In such scenarios, the notion that an algorithm alone should bear the ultimate responsibility for a decision is not only untenable but also deeply problematic. Instead, accountability is most effectively solidified when a human agent retains the authority and the responsibility to review, validate, and, if necessary, override the AI's output.

Consider, for example, an AI system designed to assist judges in determining bail recommendations or parole eligibility. Such systems often analyze vast datasets to predict the likelihood of recidivism. While they can identify statistical correlations and patterns that might be invisible to human analysts, they cannot fully grasp the unique circumstances of an individual case, the potential for rehabilitation beyond statistical measures, or the broader societal implications of a decision. A judge, endowed with legal expertise and a deep understanding of justice principles, must be able to consider the AI's recommendation as one piece of information among many. If the AI flags an individual as high-risk based on factors that may be proxies for socioeconomic disadvantage or historical discrimination, the judge must possess the agency to question this assessment and apply their own judgment. This human intervention ensures that the decision is not merely an outcome of data processing but a considered judgment that respects individual circumstances and upholds the principles of justice. The ultimate accountability for the bail or parole decision rests squarely with the judge, not the algorithm that informed their consideration.

Similarly, in social welfare services, AI might be employed to identify families or individuals who are at high risk of requiring intervention or who are eligible for specific support programs. While AI can efficiently sift through demographic data, service utilization patterns, and other indicators to flag potential cases, human caseworkers are essential for a holistic assessment. They can engage directly with families, understand the subtleties of their situations, build trust, and assess needs in a way that algorithms cannot. A caseworker might observe signs of resilience or emergent strengths that an AI, focused on statistical risk factors, might overlook. The AI's output should serve as an alert, a prompt for a human to investigate further, rather than an automatic trigger for action.

The caseworker, empowered to exercise professional judgment, remains accountable for the support plan developed and implemented, ensuring that it is tailored to the family's actual needs and circumstances, not just dictated by algorithmic predictions. This preserves the human element of care and strengthens accountability by ensuring that decisions are made by individuals who can be held responsible for their actions and judgments.

The principle of "human-in-the-loop" is therefore not simply a desirable feature; it is a foundational requirement for ensuring meaningful accountability in public sector AI. This means designing systems and establishing protocols that explicitly delegate ultimate decision-making authority to human operators, supervisors, or designated officials. It implies providing these individuals with adequate training, clear guidance, and the necessary tools to effectively interpret AI outputs and to make informed decisions about whether to accept, reject, or modify them. Moreover, it requires cultivating an organizational culture that values and supports human judgment, rather than one that blindly defers to technological pronouncements. When AI systems are implemented, the focus should be on how they can augment human capabilities, provide insights, and streamline processes, but never on how they can entirely replace human judgment in critical decision-making contexts.

The importance of human intervention is further amplified in situations where AI systems may exhibit unforeseen errors or biases. Despite rigorous auditing and testing, no AI system is infallible. Data can become outdated, algorithms can encounter edge cases they were not trained for, and emergent societal dynamics can render previous patterns obsolete. In such instances, a human operator, attuned to the nuances of the real-world context, is far more likely to detect anomalies or deviations from expected outcomes than an automated system. For example, an AI system used by a transportation department to manage traffic flow might, due to an unexpected event or a data anomaly, begin to reroute traffic in a way that causes significant disruption or even safety hazards. A human supervisor monitoring the system would be expected to notice the unusual patterns, investigate the cause, and override the AI's instructions to restore optimal and safe traffic management. Without this human intervention, the AI's erroneous directives could persist, leading to escalating problems. The supervisor, in this scenario, is accountable for ensuring that the AI's operation remains safe and effective, and for taking corrective action when it does not.

Moreover, the concept of accountability is intrinsically linked to human agency and moral responsibility. Algorithms do not possess consciousness, intentions, or the capacity for moral deliberation. They execute code based on inputs and parameters. To attribute accountability to an algorithm in the human sense is a category error. True accountability requires an entity that can understand the ethical implications of its actions, that can feel remorse, and that can learn from its mistakes in a way that reflects moral growth. This is fundamentally a human capacity. Therefore, when an AI system contributes to a harmful outcome, the locus of accountability must reside with the human individuals or institutions that designed, deployed, or oversaw its use. This includes the policymakers who authorized its implementation, the developers who built it, the data scientists who trained it, and the public officials who ultimately relied on its outputs. The human-in-the-loop principle ensures that there is always a human agent who can be questioned, held responsible, and who has the authority to make amends or prevent future harms.

This is not to suggest that AI has no role in decision support. On the contrary, its ability to process information and identify patterns at scale can be invaluable. However, the AI's role should be understood as that of a highly sophisticated assistant or advisor, not an autonomous decision-maker in sensitive public sector contexts. The language used to describe AI's function is important; it should be framed in terms of "support," "assistance," "recommendation," or "identification of potential issues," rather than "decision-making" or "judgment." This subtle but crucial distinction helps to reinforce the understanding that the ultimate responsibility remains with human actors. For instance, an AI flagging potential tax fraud should be presented to a human auditor as a case requiring further investigation, accompanied by the reasons for the flag, rather than as a definitive determination of guilt. The auditor then applies their expertise to verify the suspicion, conduct interviews, and make the final determination, thereby retaining accountability for the outcome.

The implementation of effective human oversight and intervention mechanisms requires careful consideration of several practical aspects. First, appropriate training for human operators is paramount. Staff who interact with AI systems must understand how these systems work, their limitations, the types of data they use, and the potential sources of error or bias. They need to be trained to critically evaluate AI outputs, to recognize when an output might be erroneous or questionable, and to know how to use the override functions correctly. This training should not be a one-off event but an ongoing process, as AI systems and their applications evolve.

Second, clear protocols and guidelines must be established for when and how human intervention should occur. These protocols should specify the circumstances under which an AI recommendation must be reviewed, the criteria for overriding an AI decision, and the steps to be taken when an override occurs. For example, in a healthcare setting, an AI might recommend a particular treatment plan based on a patient's medical history and diagnostic data. However, the attending physician must have the authority to deviate from this recommendation if, in their professional judgment, it is not in the patient's best interest, perhaps due to unstated patient preferences, unique physiological responses, or the availability of alternative treatments not considered by the AI. The protocol should ensure that such deviations are documented and reviewed.

Third, the design of user interfaces for AI systems is critical. These interfaces should be intuitive and user-friendly, providing clear visualizations of AI outputs and the underlying rationale, where possible. They should also make it straightforward for human operators to review, question, and override AI recommendations. The process of overriding an AI decision should not be overly cumbersome, as this could discourage its use, thereby undermining the very principle of human control. However, it should also be deliberate and well-documented, to ensure that such actions are not taken lightly.

Fourth, there must be a mechanism for feedback and continuous improvement. When human operators intervene and override an AI, this information should be captured and used to refine the AI system. This feedback loop is essential for learning from errors, identifying blind spots in the AI's capabilities, and continuously enhancing its accuracy and fairness. This iterative process ensures that the AI evolves in a way that is aligned with human judgment and societal values, rather than becoming a static, potentially flawed, element within public service delivery.

Finally, organizational policies must explicitly uphold the principle that human accountability is non-negotiable. This means that even in situations where an AI system was demonstrably the source of an error, the ultimate responsibility will lie with the humans who were tasked with overseeing its operation. This clarity of responsibility is crucial for fostering a culture of diligence and care in the implementation and use of AI. It encourages public officials to be thorough in their oversight, to question AI outputs critically, and to always prioritize the well-being and rights of citizens over the perceived efficiency of automated processes.

In conclusion, while the preceding discussions have emphasized the critical importance of robust grievance mechanisms, independent auditing, and comprehensive oversight in establishing accountability for public sector AI, this final consideration brings the focus back to the indispensable role of human judgment and intervention. True accountability in the age of AI is not about absolving humans of responsibility by delegating it to machines. Instead, it is about augmenting human capabilities with AI while ensuring that humans remain firmly in control, capable of applying their unique capacities for reason, empathy, and moral judgment. By maintaining humans in the loop, providing them with the necessary tools and training, and establishing clear protocols for intervention, public sector organizations can ensure that AI serves as a tool to enhance, rather than diminish, the accountability that is fundamental to legitimate and trustworthy governance. The ultimate responsibility for decisions that impact citizens' lives must always reside with accountable human agents, who can be called upon to explain, justify, and ultimately bear the weight of their judgments.

# Chapter 9: Continuous Monitoring and Iterative Improvement

The evolution of artificial intelligence systems within the public sector is not a singular event, but rather an ongoing journey. Unlike static software programs that operate consistently within predefined boundaries, AI systems are inherently dynamic. They are born from data, trained on specific patterns, and deployed into environments that are themselves in constant flux. This inherent dynamism means that an AI system that performs optimally today might exhibit degraded performance, unexpected biases, or simply become irrelevant tomorrow. This necessitates a fundamental shift in how we approach AI implementation, moving away from a model of "set it and forget it" towards one of continuous vigilance, adaptation, and proactive management.

The reasons for this dynamic nature are multifaceted. Firstly, the data upon which AI systems are trained is rarely a perfect reflection of reality, and even if it were at the moment of training, reality itself changes. For example, an AI system trained to predict traffic congestion patterns might have been developed using data from a period before a major new highway was opened, before a significant shift in commuting habits due to remote work policies, or before the introduction of new public transportation options. As these external factors evolve, the historical data that trained the AI becomes a less accurate predictor of current or future conditions. The patterns the AI learned to recognize may no longer hold true, leading to increasingly inaccurate predictions and suboptimal routing decisions. This "data drift" is a pervasive challenge, requiring ongoing monitoring of incoming data streams to ensure they remain representative of the operational environment.

Secondly, user behavior and interactions with AI systems can also change over time, influencing system performance. Consider an AI-powered chatbot designed to assist citizens with navigating government services. Initially, users might interact with the chatbot in predictable ways, asking common questions based on existing documentation. However, as users become more familiar with the chatbot, or as new services are introduced, their queries might become more complex, nuanced, or even exploit loopholes in the chatbot's understanding. If the AI is not continuously learning from these interactions, it can become less effective at understanding and responding to user needs. The very success of an AI can, paradoxically, lead to its obsolescence if it cannot adapt to the evolving ways in which humans choose to interact with it. This underscores the importance of capturing and analyzing user interaction logs to identify emerging patterns and areas where the AI's understanding falters.

Furthermore, the operational environment itself can introduce dynamic changes. This could include shifts in regulatory frameworks, the introduction of new technologies that interact with the AI, or even subtle changes in the physical environment. For instance, an AI system used for environmental monitoring might be trained to detect specific pollutant signatures. However, new industrial processes or the introduction of novel materials could lead to the emission of pollutants with signatures that the AI has never encountered. Without mechanisms to detect these new signatures and adapt its models, the AI would fail to identify these emerging environmental hazards, rendering it ineffective. Similarly, an AI deployed in a smart city infrastructure might need to adapt to the introduction of new sensor networks or changes in communication protocols.

The implication of this dynamic nature is that the initial deployment of an AI system is not an endpoint, but a starting point. The process of ensuring AI's continued effectiveness, fairness, and relevance is an iterative one, demanding a robust framework for continuous monitoring and adaptation. This involves establishing mechanisms to detect performance degradation, identify emerging biases, and proactively update or retrain AI models. It requires a shift in mindset from treating AI as a finished product to viewing it as a living system that requires ongoing care and attention, much like any other critical infrastructure or service.

One of the primary aspects of managing this dynamic nature is establishing comprehensive monitoring systems. These systems should go beyond simply checking if the AI is "online" and instead focus on a range of performance indicators. For an AI used in fraud detection, this would involve tracking not only the number of fraudulent transactions identified but also the false positive and false negative rates. A gradual increase in false positives, for example, could indicate that the AI is becoming overly sensitive or is flagging legitimate transactions based on outdated patterns. Conversely, an increase in false negatives would suggest that new fraud techniques are emerging that the AI is not detecting. Similarly, for an AI assisting in loan application processing, monitoring metrics like approval rates, rejection rates, and crucially, the demographic breakdown of these outcomes, becomes essential. A drift in these metrics, particularly a widening disparity across different demographic groups, would be a strong indicator of emerging bias, even if the initial system was designed to be fair.

These monitoring systems need to be equipped to detect "concept drift" and "data drift." Data drift refers to changes in the statistical properties of the input data over time. This could mean that the average value of a certain feature changes, or the distribution of values within a feature shifts. Concept drift, on the other hand, refers to changes in the relationship between the input variables and the target variable that the AI is trying to predict. For example, a traditional indicator of financial distress might no longer be a reliable predictor of bankruptcy as economic conditions or market behaviors evolve. Identifying these drifts is crucial because they are often precursors to performance degradation. Advanced monitoring tools can use statistical tests to detect these shifts automatically, flagging them for human review and intervention.

Beyond automated monitoring, there is a critical need for human oversight and interpretation. While automated systems can flag anomalies, human judgment is often required to understand the root cause of performance drift. For instance, if an AI system used for predicting demand for public services begins to underpredict needs in a particular neighborhood, an automated system might flag this. However, a human analyst would be best placed to investigate whether this is due to a genuine change in demand, an influx of a new demographic with different service utilization patterns, or a data anomaly in the reporting from that specific neighborhood. This human element ensures that the response to performance drift is appropriate and informed by contextual understanding that an algorithm alone cannot possess.

The response to detected performance degradation or emerging biases must be a clearly defined iterative improvement process. This typically involves a cycle of detection, diagnosis, correction, and redeployment. Once an issue is detected through monitoring, the next step is to diagnose its cause. This might involve examining the training data, analyzing recent user interactions, or reviewing changes in the operational environment. Following diagnosis, a correction strategy is devised. This could range from simple recalibration of thresholds to more complex retraining of the AI model using updated data, or even a fundamental redesign of certain aspects of the system.

Retraining an AI model is a significant undertaking. It requires access to representative, up-to-date data. The process of data collection, cleaning, and annotation for retraining can be resource-intensive. Furthermore, retraining a model might inadvertently introduce new issues or regressions in performance on aspects that were previously working well. Therefore, rigorous testing and validation are essential before any updated model is redeployed into the live operational environment. This validation process should include not only checks against historical benchmark datasets but also, where feasible, A/B testing in a controlled live environment to compare the performance of the old and new models.

The iterative improvement cycle also highlights the importance of maintaining version control and audit trails for AI models. Just as software developers track changes to code, public sector organizations deploying AI must maintain detailed records of different model versions, the data they were trained on, the parameters used, and the performance metrics observed. This is crucial for accountability, for debugging issues, and for understanding the evolution of the system over time. If a particular model version leads to unforeseen negative consequences, the ability to roll back to a previous, stable version is invaluable.

Moreover, the concept of "continuous learning" in AI, while promising, also introduces complexities. Some AI systems are designed to learn and adapt in real-time as they operate. While this can enable rapid adaptation to changing conditions, it also poses significant challenges for governance and accountability. If an AI is continuously modifying its own behavior, it can become difficult to pinpoint exactly why it made a particular decision at a specific point in time. This makes traditional auditing methods challenging. Therefore, even in systems designed for continuous learning, there is a need for safeguards, such as periodically freezing the model to perform in-depth audits, or implementing mechanisms that allow for controlled learning with human review before updates are integrated.

The dynamic nature of AI also has profound implications for fairness and equity. An AI system that was initially unbiased can develop biases over time as the data it encounters or the environment it operates within changes. For example, an AI system used for resource allocation in urban planning might be trained on historical data reflecting past discriminatory practices in service provision. If the system is not regularly audited for fairness, it can perpetuate and even amplify these historical inequities. New patterns of discrimination might emerge that were not present in the original training data but become evident as the AI interacts with the contemporary world. Therefore, continuous monitoring for fairness metrics, disaggregated by relevant demographic groups, is not an optional add-on but a core requirement for responsible AI deployment in the public sector. This includes monitoring for disparate impact, where a seemingly neutral policy or system has a disproportionately negative effect on certain groups.

The commitment to continuous monitoring and iterative improvement also necessitates investment in the necessary skills and infrastructure. Public sector organizations need to cultivate internal expertise in data science, AI ethics, and MLOps (Machine Learning Operations). This involves not only hiring skilled personnel but also providing ongoing training and professional development opportunities. Furthermore, the technological infrastructure must support the demands of continuous data collection, model retraining, and robust monitoring. This might involve investing in cloud computing resources, specialized AI platforms, and data governance tools.

In essence, the recognition of AI systems as dynamic entities shifts the paradigm from a one-off deployment to a lifecycle management approach. This lifecycle encompasses initial design and development, rigorous testing and validation, careful deployment, and crucially, ongoing monitoring, evaluation, and adaptation. Each phase is interconnected, and the success of the entire lifecycle hinges on a commitment to continuous improvement. This is not merely a technical challenge; it is an ethical imperative. By embracing the dynamic nature of AI and implementing robust processes for continuous monitoring and iterative improvement, public sector organizations can better ensure that their AI systems remain beneficial, equitable, and accountable for the long term, serving the public good effectively in an ever-changing world.

Establishing a robust monitoring framework is paramount for ensuring that artificial intelligence systems deployed within the public sector continue to serve their intended purpose effectively, ethically, and equitably over time. This framework is not a static checklist but a dynamic, integrated system designed to capture a holistic view of an AI's performance in its real-world operational environment. It moves beyond mere technical functionality to encompass the broader societal and ethical implications of the AI's deployment. The goal is to cultivate a continuous feedback loop that informs ongoing improvements, mitigates risks, and ultimately upholds public trust.

At its core, a comprehensive monitoring framework must systematically track a multifaceted array of indicators. These indicators can be broadly categorized into system performance, fairness and equity, user experience, and operational efficiency. Each category provides a crucial lens through which to evaluate the AI's effectiveness and identify areas for intervention. Without a structured approach to collecting and analyzing data across these dimensions, public sector organizations risk deploying AI systems that may appear functional on the surface but are subtly failing to meet their objectives, or worse, are actively causing harm.

System performance metrics form the foundational layer of any monitoring framework. These are the quantitative indicators that directly reflect how well the AI is achieving its designed objectives. For an AI system tasked with predicting the likelihood of a student dropping out of a vocational training program, key performance indicators (KPIs) might include accuracy, precision, recall, and F1-score. Accuracy, while a common metric, can be misleading in imbalanced datasets; therefore, understanding the trade-offs between precision (minimizing false positives – wrongly identifying a student as at risk) and recall (minimizing false negatives – failing to identify a student who is actually at risk) becomes critical. A high false negative rate, for instance, might mean that students who need intervention are being missed, directly undermining the program's goals. Similarly, for an AI used to optimize public transportation routes, metrics such as on-time performance, passenger wait times, and route adherence would be essential. Deviations from established benchmarks in these metrics could signal a need to reassess the AI's underlying model, the data it's processing, or the operational context itself. It's also vital to track the stability of these metrics over time. A sudden drop in accuracy or a significant increase in wait times, even if still within acceptable historical ranges, should trigger an investigation. This involves establishing baseline performance levels during initial deployment and setting thresholds for acceptable deviation. When these thresholds are breached, automated alerts should be generated, prompting an immediate review by the responsible team.

Beyond raw performance, the monitoring framework must embed rigorous checks for fairness and equity. This is arguably one of the most critical and ethically charged aspects of AI monitoring in the public sector. An AI system, even if technically accurate, can perpetuate or even amplify existing societal biases if not carefully monitored. This requires moving beyond aggregate performance metrics to disaggregated analyses across various demographic groups. For an AI used in screening job applications for public service roles, this means tracking selection rates, rejection rates, and the quality of candidates recommended, broken down by gender, ethnicity, age, disability status, and other protected characteristics. If the AI disproportionately rejects candidates from a particular demographic group, even if it's based on seemingly neutral criteria learned from historical data, it signals a serious fairness issue. Similarly, an AI used for allocating social housing or emergency relief funds must be monitored to ensure equitable distribution, preventing any demographic group from being systematically disadvantaged. This involves defining clear fairness criteria upfront, such as equal opportunity (ensuring that individuals with similar qualifications have an equal chance of a positive outcome) or demographic parity (ensuring that the proportion of positive outcomes is the same across different groups). The framework must include mechanisms to quantify these fairness metrics and generate alerts when they fall below acceptable standards. This often necessitates dedicated tools and expertise in algorithmic fairness, moving beyond standard machine learning evaluation libraries. The process of defining and measuring fairness is itself complex and may require iterative refinement as understanding of the system's impact evolves.

User feedback and engagement constitute another vital pillar of the monitoring framework. AI systems in the public sector are often designed to interact directly with citizens, service providers, or internal stakeholders. Understanding how these users perceive and interact with the AI is invaluable for identifying usability issues, unexpected behaviors, and areas where the AI might be falling short of user expectations. This can be gathered through various channels: direct feedback mechanisms within the AI interface (e.g., "Was this helpful?" buttons, satisfaction surveys), analysis of user support requests and complaints related to the AI's output, and qualitative research methods like user interviews and focus groups. For a public-facing AI chatbot assisting with navigating government services, tracking the frequency of users abandoning the conversation, repeatedly asking the same questions, or escalating to human agents can indicate that the AI is not effectively understanding or addressing their needs. Similarly, if frontline staff using an AI-powered decision support tool express frustration or distrust in its recommendations, it signals a need for deeper investigation. This feedback isn't just about identifying bugs; it's about understanding the nuanced ways in which the AI integrates into human workflows and whether it is genuinely augmenting human capabilities or creating new burdens. Aggregating and analyzing this qualitative and quantitative user feedback allows for the identification of patterns that might not be apparent from technical performance metrics alone.

The operational costs associated with deploying and maintaining AI systems must also be a key component of the monitoring framework. While the promise of AI is often efficiency and cost savings, these systems can incur significant expenditures in terms of computational resources, data storage, specialized personnel, licensing fees for third-party tools, and ongoing model retraining. A robust framework will track these costs against the expected benefits and budget allocations. For example, an AI system that requires extensive GPU hours for inference could become prohibitively expensive if its usage scales unexpectedly or if its computational efficiency is lower than anticipated. Similarly, the cost of data labeling for continuous retraining, or the salaries of specialized MLOps engineers, need to be factored into the total cost of ownership. By monitoring these operational expenditures, public sector organizations can ensure that the AI remains a cost-effective solution and identify any potential budgetary overruns or inefficiencies. This cost monitoring should ideally be linked to performance metrics; for instance, if performance is degrading but costs are increasing, it indicates a particularly concerning trend. Furthermore, understanding the cost breakdown can help in prioritizing areas for optimization, such as investing in more efficient algorithms or cloud resource management strategies.

To effectively implement such a comprehensive monitoring framework, several foundational elements are required. First,
**clear ownership and accountability** must be established. Who is responsible for monitoring specific metrics? Who is empowered to act when alerts are triggered? This requires defining roles and responsibilities across different teams, potentially including data scientists, AI ethicists, operational managers, and legal counsel. Without clear lines of accountability, monitoring efforts can become fragmented and ineffective, with critical issues falling through the cracks.

Second,
**the right tools and infrastructure** are essential. This involves investing in platforms that can collect, store, process, and visualize data from various sources. This might include specialized MLOps platforms, data warehousing solutions, business intelligence tools, and potentially custom-built dashboards. The infrastructure must be capable of handling the volume and velocity of data generated by AI systems and enabling timely analysis and reporting. Automating data collection and alert generation is crucial to avoid overwhelming human oversight.

Third,
**defining clear, actionable thresholds and alerts** is critical. Simply collecting data is insufficient; the framework must specify what constitutes a deviation from expected performance or fairness. These thresholds should be informed by regulatory requirements, organizational policies, and expert judgment. When a threshold is breached, automated alerts should be generated and routed to the appropriate individuals or teams, along with sufficient context to understand the nature and potential impact of the issue. The alert system should be designed to be informative without being overly noisy, distinguishing between minor anomalies and critical failures.

Fourth,
**regular review and adaptation of the framework itself** is necessary. The AI landscape and the operational environments in which these systems function are constantly evolving. Therefore, the monitoring framework cannot remain static. Periodically, the framework should be reviewed to ensure that the metrics being tracked are still relevant, that the thresholds are appropriate, and that the tools and processes are effective. This review process should incorporate lessons learned from past incidents and evolving best practices in AI governance and ethics. For instance, as new research emerges on identifying specific types of algorithmic bias, the fairness metrics and monitoring techniques within the framework should be updated accordingly.

Finally, the framework must be integrated with the broader
**AI governance and risk management processes** of the public sector organization. Monitoring is not an isolated activity but a continuous feedback loop that informs decision-making, policy updates, and future AI development. The insights gained from monitoring should feed directly into the organization's understanding of AI risks, its strategies for risk mitigation, and its overall approach to responsible AI

deployment. This ensures that monitoring is not just a compliance exercise but a strategic enabler of trustworthy AI.

In summary, developing a comprehensive monitoring framework for AI systems in the public sector is a complex but indispensable undertaking. It requires a systematic approach to tracking system performance, ensuring fairness and equity, understanding user experiences, and managing operational costs. By establishing clear ownership, leveraging appropriate tools, defining actionable alerts, and committing to regular review, public sector organizations can create a resilient system that safeguards the integrity, efficacy, and ethical deployment of AI, ultimately fostering greater public trust and delivering on the promise of technology for the public good. This deliberate and ongoing process of observation and analysis is what transforms an AI system from a deployed technology into a continuously improving public service.

The preceding discussion has underscored the critical importance of a robust monitoring framework for AI systems in the public sector, emphasizing its role in maintaining performance, fairness, and public trust. However, the mere act of observation and data collection, while foundational, is only one half of the equation for ensuring long-term AI efficacy and ethical deployment. The true power of monitoring lies in its ability to fuel a cycle of continuous improvement. This is where the concept of feedback loops becomes not just beneficial, but absolutely indispensable. A sophisticated monitoring framework, by its very nature, generates a wealth of data that, if properly harnessed, can drive iterative refinement of AI systems. This proactive and adaptive approach allows public sector organizations to move beyond a static understanding of AI performance and into a dynamic realm where systems evolve and improve in response to their real-world interactions and impacts.

Building these effective feedback loops requires a deliberate and structured approach to integrating insights gleaned from ongoing monitoring and direct user input directly into the AI system's improvement cycle. It's about creating a circulatory system for knowledge, where data flows from observation to analysis, and then directly back into the refinement of the AI itself. This encompasses the establishment of clear mechanisms for collecting, meticulously analyzing, and, crucially, acting upon data that highlights system errors, instances of user dissatisfaction, or any emerging fairness concerns that might have been missed during initial deployment or initial monitoring phases. The overarching objective is to leverage this vital information to refine algorithms, update or augment training data, and judiciously adjust deployment strategies to ensure the AI continues to serve the public good effectively and equitably.

The genesis of a strong feedback loop often begins with the categorization of incoming data. Monitoring systems, as discussed, will generate a continuous stream of performance metrics, fairness indicators, and user feedback. The initial step in leveraging this data for improvement is to ensure it is properly categorized and triaged. For instance, performance degradations – a drop in accuracy, an increase in latency, or a surge in false positives/negatives – need to be distinguished from user-reported issues that might stem from usability problems or misinterpretations of the AI's output. Similarly, fairness alerts, such as a widening disparity in outcomes across demographic groups, demand a different investigative pathway than a technical glitch. Establishing clear protocols for how each type of information is flagged, prioritized, and routed is the first practical step in transforming raw data into actionable intelligence.

Consider an AI system deployed by a local government to assist citizens in identifying and applying for social services. The monitoring system might flag an increase in the number of users abandoning the application

process midway. Technically, the AI's underlying predictive model might still be performing within acceptable parameters. However, this user behavior data, when fed back into the improvement cycle, signals a potential problem with the user interface, the clarity of the questions being asked, or perhaps the perceived complexity of the eligibility criteria. A feedback loop would then trigger an investigation not into the AI's core logic, but into the user experience. This might involve a UX designer analyzing the application flow, conducting usability testing with a sample of citizens, and gathering direct feedback on points of confusion. The insights gained would then inform iterative design changes to the application interface or the accompanying explanatory text. This isn't about retraining the AI model itself, but about improving the
*way* the AI system is presented and interacted with, a crucial aspect of its overall effectiveness and public acceptance.

Another critical channel for feedback lies in the direct reporting of errors or inaccuracies. Public sector AI systems, particularly those involved in decision support or information provision, can sometimes produce outputs that are factually incorrect or contextually inappropriate. For example, an AI assisting social workers in identifying potential child welfare risks might, on occasion, generate a high-risk alert based on incomplete or misinterpreted information. When such an error occurs, and is subsequently identified – either through internal audits, cross-referencing with human expert judgment, or through a citizen reporting the issue – this incident becomes a vital piece of feedback. The feedback loop here would involve a thorough post-mortem analysis of the specific case. What data points led to the erroneous conclusion? Was the data input incomplete? Was there ambiguity in the data that the AI misinterpreted? Was the algorithm's weighting of certain factors inappropriate for this specific context? The answers to these questions are not merely diagnostic; they are prescriptive. They inform the necessary adjustments. This might involve refining the data ingestion process to ensure more complete inputs, enhancing data validation checks, or even recalibrating the weights within the AI model to better reflect the nuanced realities of the domain.

The realm of fairness and equity provides perhaps the most ethically urgent feedback loop. When monitoring reveals disparities in how the AI system treats different demographic groups, swift and decisive action is paramount. Imagine an AI used by a housing authority to prioritize applications for affordable housing. If monitoring data shows that applicants from a particular racial or socioeconomic background are consistently being ranked lower, even when their stated needs and qualifications appear similar, this is a critical fairness alert. The feedback loop must initiate a deep dive into the factors influencing these rankings. This investigation would go beyond simply identifying the disparity; it would seek to understand its root cause. Is the AI inadvertently learning proxies for protected characteristics from historical data (e.g., zip codes correlating with race, which then influence neighborhood desirability scores)? Are certain desirable but unstated factors (like "community ties" or "long-term residency") being implicitly weighted in a way that disadvantages newer or less established communities? The feedback loop's output would be a set of targeted interventions. These could range from re-engineering the feature set used by the AI to exclude or de-emphasize biased proxies, to augmenting the training data with more diverse and representative examples, to developing specific fairness-aware algorithms that actively work to equalize outcomes across groups. This iterative process might involve multiple cycles of adjustment, monitoring, and further refinement, underscoring that achieving algorithmic fairness is often an ongoing journey rather than a one-time fix.

User dissatisfaction, even when not tied to outright errors or fairness breaches, also serves as a powerful catalyst for improvement. Citizens interacting with public services expect helpfulness, clarity, and respect. If an AI system is perceived as unhelpful, bureaucratic, or dismissive, it erodes public trust and diminishes the system's utility. For instance, a public sector chatbot designed to answer queries about local bylaws might be technically accurate in its responses but fail to understand the user's underlying intent or colloquial language. This can lead to frustration, as users repeatedly rephrase their questions or are directed to irrelevant information. The feedback loop here would involve analyzing transcripts of user interactions, identifying common points of confusion or failure, and perhaps employing natural language processing (NLP) techniques to improve the AI's ability to understand varied linguistic inputs. This might also include incorporating a mechanism for users to provide explicit feedback on the helpfulness of each response, creating a direct line of sentiment data that can guide the refinement of the AI's conversational capabilities. The goal is to make the AI not just functionally correct, but also experientially positive for the user.

The integration of these feedback loops into the AI lifecycle requires specific architectural and process considerations. Firstly, robust logging and auditable trails are essential. Every piece of data that feeds into the feedback loop – performance anomalies, user complaints, fairness violations – must be meticulously logged, along with the context of its origin and the time it was generated. This ensures transparency and accountability. Secondly, dedicated teams or roles are often necessary to manage the feedback process. This might involve a combination of data scientists who can analyze technical performance and fairness metrics, UX researchers who can delve into user feedback, and domain experts who can provide crucial context about the operational environment and potential societal implications. Establishing clear workflows for how feedback is collected, reviewed, prioritized, and acted upon is critical to prevent these insights from languishing in an inbox.

Furthermore, the process of acting on feedback needs to be iterative and experimental. When adjustments are made to an AI system based on feedback, these changes should be deployed cautiously, ideally in a controlled environment or phased rollout. This allows for a new round of monitoring to assess the impact of the changes. Did the algorithmic adjustment improve fairness metrics without negatively impacting performance? Did the UI redesign reduce user abandonment rates? This cyclical approach – monitor, analyze, act, re-monitor – is the essence of continuous improvement. It acknowledges that AI systems are not static entities but dynamic tools that require ongoing attention and adaptation to remain effective and ethical.

The nature of the "action" taken in response to feedback can vary significantly. It could involve:

**Algorithmic Adjustments:** This is often the most direct response. If monitoring reveals bias, performance degradation, or suboptimal decision-making, the underlying machine learning model may need to be retrained, fine-tuned, or even replaced with a different architecture. This might involve adjusting hyperparameters, modifying the loss function to incorporate fairness constraints, or employing techniques like adversarial debiasing.
**Data Augmentation or Refinement:** The quality and representativeness of the training data are paramount. If feedback indicates that the AI is struggling with certain scenarios, or if biases are detected, it may be necessary to collect more data, label existing data more accurately, or generate synthetic data that addresses the identified gaps or imbalances. This also applies to real-time data streams; if the

input data quality degrades, the feedback loop should trigger alerts for data cleaning or validation processes.

**Feature Engineering and Selection:** The features or variables that an AI model uses to make predictions can have a profound impact on its performance and fairness. Feedback might reveal that certain features are acting as unintended proxies for protected characteristics, or that important contextual information is missing. This can lead to the removal of problematic features, the creation of new ones, or a re-evaluation of how existing features are processed.

**User Interface and Experience (UI/UX) Modifications:** As highlighted with the social services application example, not all improvements require changes to the core AI algorithm. Sometimes, the AI's effectiveness is limited by how users interact with it. Feedback can drive improvements to the user interface, the clarity of instructions, the accessibility of the system, or the way information is presented.

**Policy and Procedural Updates:** In some cases, feedback might indicate that the AI system, while technically sound, is operating within a flawed policy framework or procedure. For instance, an AI flagging potential fraud in benefit claims might be highly accurate, but if the appeal process for flagged individuals is too cumbersome or lacks due process, this feedback loop might necessitate changes in the broader operational policy rather than just the AI itself.

**Deployment Strategy Adjustments:** The way an AI system is deployed can also be subject to iterative improvement. Feedback might suggest that the AI is being used too broadly, or not in the optimal context. This could lead to decisions to restrict its application, introduce more human oversight at specific decision points, or pilot it in a more limited setting before wider deployment.

The success of these feedback loops is inextricably linked to the organizational culture. Public sector entities that embrace transparency, encourage open communication about AI performance, and view errors or user complaints as opportunities for learning rather than failures are far more likely to establish effective iterative improvement cycles. This requires a commitment from leadership to allocate resources – both human and financial – to the ongoing monitoring, analysis, and refinement of AI systems. It also necessitates a culture where frontline staff and citizens feel empowered to provide feedback without fear of reprétrocussions.

Moreover, the feedback loop must be designed to be resilient and adaptive to changing external conditions. Regulatory landscapes evolve, societal norms shift, and the data-generating environment itself can change. A feedback loop that is too rigid might fail to capture novel issues or adapt to new ethical considerations. Therefore, periodic reviews of the feedback mechanisms themselves are necessary. Are the right metrics being tracked? Are the thresholds for flagging issues still appropriate? Are the channels for feedback still accessible and effective? This meta-level of feedback, applied to the improvement process itself, ensures that the system remains relevant and effective over time.

In essence, leveraging feedback loops transforms AI deployment in the public sector from a one-off installation into an ongoing civic partnership. It acknowledges that AI systems are not infallible, nor are they static. By establishing clear pathways for data and insights to flow from the real world back into the AI's development and operation, organizations can ensure that these powerful tools remain aligned with public values, adapt to evolving needs, and ultimately contribute more effectively and equitably to the public good. This continuous dialogue between the AI, its users, its operators, and its data is the engine that drives responsible and impactful AI in the public sphere, ensuring that technology serves humanity, not the other way around.

The dynamic nature of societal needs and values presents a continuous challenge to the longevity and efficacy of AI systems deployed within the public sector. As societies evolve, so too do their expectations regarding technology, fairness, privacy, and the very definition of the "public good." An AI system that was considered cutting-edge and ethically sound at its inception might, within a few years, become outdated, misaligned, or even detrimental if it fails to adapt. This necessitates a proactive and ongoing commitment to ensuring that AI remains a beneficial tool, rather than an artifact of a bygone era. This adaptation is not merely a technical undertaking; it is deeply interwoven with the ethical fabric of public service and requires a sophisticated understanding of how societal shifts manifest and how they can be incorporated into the AI lifecycle.

One of the primary drivers of this evolution is the continuous re-evaluation of fairness and equity. What constitutes fair treatment can shift over time, influenced by public discourse, academic research, and evolving legal interpretations. For instance, early AI systems might have been designed to optimize for efficiency based on historical data that, in retrospect, contained implicit biases. As societal awareness of systemic inequalities grows, so too does the imperative to scrutinize and rectify these biases. This might involve re-examining the very definition of fairness being used by the AI – moving from a purely statistical parity to more nuanced concepts like counterfactual fairness or individual fairness. The process of adapting an AI to these evolving fairness standards involves a deep dive into its decision-making processes, potentially requiring the identification and mitigation of algorithmic proxies for protected attributes that were not previously recognized as problematic. This could lead to the retraining of models with carefully curated datasets designed to reflect contemporary understandings of equity, or the implementation of algorithmic fairness interventions that actively work to counterbalance historical disadvantages. The challenge lies in discerning genuine shifts in societal values from transient trends, and in ensuring that the AI's adaptation is robust and genuinely beneficial, rather than a superficial adjustment.

Privacy expectations also undergo significant transformations. What was once considered acceptable data collection and usage might now be viewed as intrusive or a violation of personal autonomy. Public sector AI systems often handle sensitive personal information, and their data practices must remain aligned with contemporary privacy norms and regulations. For example, an AI system initially designed to improve traffic flow by anonymizing vehicle movement data might, in a later era, face scrutiny if the anonymization techniques are deemed insufficient or if the scope of data collection is perceived as overly broad. Adapting to these changing privacy landscapes might involve implementing more robust anonymization or differential privacy techniques, limiting the types of data collected to only what is strictly necessary, or enhancing user consent mechanisms. It also requires ongoing dialogue with privacy advocates and the public to understand emerging concerns. The legal and regulatory frameworks governing data privacy are rarely static; they are often reactive to technological advancements and public opinion. Therefore, public sector AI governance must anticipate these shifts and build in the flexibility to comply with new or strengthened privacy protections, even if it means redesigning core functionalities of the system.

Beyond fairness and privacy, the very definition of the "public good" that AI is intended to serve can be redefined by society. New challenges emerge, and old problems may demand novel solutions that AI can address. Consider the rapid evolution of public health concerns or the growing urgency of climate change. An AI system initially deployed for administrative efficiency might need to be repurposed or augmented to address these emerging societal priorities. This requires a strategic foresight that anticipates future needs and ensures that the AI infrastructure is adaptable. It might involve developing new AI models trained on different data, integrating AI capabilities with other technological solutions, or even fundamentally changing the objectives and operational parameters of existing systems. The process of adapting an AI to serve evolving public goods involves continuous engagement with policymakers, community stakeholders, and domain experts to identify these new frontiers and assess how AI can be ethically and effectively deployed to meet them. This adaptability ensures that public sector AI remains relevant and contributes to addressing the most pressing issues of the day.

The regulatory landscape surrounding AI is another critical factor driving the need for adaptation. As governments and international bodies grapple with the implications of AI, new laws, guidelines, and ethical frameworks are constantly being developed. Public sector AI systems must not only comply with existing regulations but also anticipate and prepare for future ones. This might involve building in mechanisms for explainability and transparency, adhering to new standards for algorithmic auditing, or incorporating ethical guardrails that may not have been a priority at the time of initial deployment. For instance, a mandate for algorithmic impact assessments might require public sector organizations to systematically evaluate the potential societal and ethical implications of their AI systems before deployment and at regular intervals thereafter. Adapting to these regulatory shifts requires a flexible system architecture, robust documentation, and a commitment to ongoing legal and ethical review. It means viewing regulatory compliance not as a burden, but as an integral part of ensuring responsible AI deployment.

Furthermore, public trust is a fluid asset that is directly influenced by how well AI systems align with societal values. Incidents of bias, privacy breaches, or perceived unfairness can erode public confidence, making it more difficult for public sector organizations to leverage AI effectively. Conversely, AI systems that are demonstrably fair, transparent, and aligned with community expectations can foster trust and enhance public engagement. Therefore, adapting AI systems involves a continuous effort to build and maintain this trust. This means being responsive to public concerns, engaging in transparent communication about AI capabilities and limitations, and demonstrating a clear commitment to ethical principles. It also involves making mechanisms for redress and appeal readily available when AI-driven decisions have adverse impacts on individuals. The iterative improvement process, as discussed previously, plays a crucial role here, as it allows organizations to address issues that might undermine public trust before they become systemic problems.

The practical implementation of adapting AI systems to changing societal needs and values is a complex, multi-faceted process. It requires more than just technical adjustments; it necessitates a strategic and organizational commitment. Firstly, establishing cross-functional teams that include AI experts, ethicists, legal counsel, policy analysts, domain specialists, and community representatives is crucial. This diverse group can provide the breadth of perspectives needed to identify emerging societal needs and values and to assess the implications of AI deployment. Secondly, continuous horizon scanning is essential. This involves actively monitoring public discourse, academic research, policy developments, and international best practices related to AI ethics, fairness, privacy, and societal impact. This proactive approach helps to identify potential shifts before they become critical issues.

Thirdly, the AI governance framework itself must be designed for adaptability. This means that policies and procedures for AI development, deployment, and monitoring should not be static but should include provisions for regular review and updates. This might involve scheduled "value audits" or "societal alignment reviews" where AI systems are periodically assessed against current societal expectations and emerging ethical considerations. Fourthly, fostering a culture of continuous learning and improvement within public sector organizations is paramount. This involves encouraging open dialogue about the ethical challenges of AI, providing training on emerging ethical principles and best practices, and empowering staff to raise concerns and propose solutions. When AI systems are seen not as immutable artifacts but as evolving tools that require ongoing stewardship, adaptation becomes a natural part of their lifecycle.

Consider the example of an AI system used for predictive policing. Initially deployed to optimize resource allocation based on historical crime data, such a system might face significant societal re-evaluation as public understanding of systemic biases in the justice system grows. Society's values might shift from prioritizing purely proactive enforcement to emphasizing community-based policing, restorative justice, and addressing the root causes of crime. An AI system designed for the former might become misaligned with the latter. Adapting such a system would involve a thorough re-examination of its objectives. Should it be retrained to identify social determinants of crime rather than predicting crime hotspots? Should its outputs be used to guide resource allocation towards community programs rather than increased patrols? Should its use be significantly curtailed or even discontinued if it cannot be ethically repurposed? These are not purely technical questions; they are deeply ethical and societal ones that require broad consultation. The feedback loops previously discussed become indispensable here, not just for technical performance, but for sensing these societal shifts and driving the necessary strategic reorientation of the AI's purpose and function.

Another illustrative case might involve AI used in social welfare eligibility determination. As societal debates around poverty, social safety nets, and the dignity of individuals evolve, the criteria and processes that AI uses might need recalibration. For example, if there's a societal shift towards recognizing non-traditional forms of work or caregiving, an AI trained on traditional employment data might unfairly disadvantage certain populations. Adapting such a system would involve incorporating new data streams, adjusting algorithmic parameters to recognize broader definitions of economic contribution, and ensuring that the AI's decision-making process aligns with contemporary understandings of social justice and equitable support. This could also involve implementing more human-in-the-loop processes where borderline cases or complex individual circumstances are always reviewed by a human caseworker, ensuring that AI augments, rather than replaces, empathetic and nuanced decision-making.

The challenge of adapting AI to changing societal needs and values is thus an ongoing dialogue between technology and society. It is a testament to the fact that AI in the public sector is not a finished product but a continually developing entity that must remain accountable to the evolving aspirations and ethical standards of the communities it serves. This requires a commitment to long-term stewardship, recognizing that the ethical deployment of AI is a journey, not a destination, and that continuous adaptation is the key to ensuring that these powerful tools remain instruments of progress and public good. The willingness to revisit, re-evaluate, and fundamentally change AI systems in response to societal evolution is perhaps the most critical element in ensuring their enduring value and trustworthiness. It underscores that the most successful public sector AI is not the one that is deployed and forgotten, but the one that is continuously engaged with the society it is meant to serve, growing and adapting alongside it.

When an artificial intelligence system has served its purpose, or has become outdated, ineffective, or misaligned with the evolving needs and ethical standards of the public sector, its retirement becomes a critical, yet often under-addressed, phase of its lifecycle. Unlike the continuous monitoring and iterative improvement that characterize an AI's active life, the decommissioning of an AI system requires a distinct set of strategies to ensure a responsible, orderly, and ethical transition. This involves careful planning, meticulous execution, and transparent communication with all relevant stakeholders. The goal is not simply to switch off a system, but to manage its cessation in a manner that safeguards data, preserves knowledge, and minimizes disruption.

The process of managed AI retirement begins long before the actual shutdown date. It necessitates proactive planning that aligns with the AI system's initial design and deployment phases, embedding considerations for its eventual end-of-life into its core architecture and governance framework. A fundamental aspect of this planning is the identification of clear criteria for retirement. These criteria should be established upfront and regularly reviewed, encompassing factors such as technological obsolescence (e.g., when underlying hardware or software becomes unsupported), decreased performance or accuracy that cannot be rectified through retraining, a fundamental shift in public policy or societal values that renders the AI's purpose obsolete or unethical, the emergence of superior alternative solutions, or significant security vulnerabilities that cannot be patched. Without clearly defined triggers for retirement, AI systems can linger in operation long past their useful lifespan, consuming resources and potentially posing risks.

A crucial element of managed retirement is the comprehensive data archival strategy. AI systems, particularly those in the public sector, often process and generate vast amounts of sensitive data. Upon retirement, this data must be handled with the same rigor and ethical considerations applied during its active use. Archival plans must detail what data needs to be retained, for how long, and under what conditions, in accordance with legal and regulatory requirements, as well as organizational data retention policies. This might involve anonymizing or pseudonymizing datasets to protect individual privacy while preserving historical insights. The methods for long-term storage must be secure, accessible for authorized purposes (such as audits or future research), and economically viable. Consideration should also be given to the metadata associated with the data – information about the data itself, such as its origin, collection methods, and any transformations it underwent – as this is vital for understanding and interpreting the archived information in the future. Failing to plan for data archival can lead to the loss of valuable historical information or, conversely, the indefinite retention of sensitive data that becomes an unnecessary liability.

The transition to a new system, or the redistribution of the functions previously performed by the retiring AI, is another critical component of a well-managed retirement. This often involves a period of parallel operation, where both the old and new systems run concurrently, allowing for comparison and validation of the replacement's performance. The objective is to ensure a seamless handover, minimizing any disruption to public services or internal operations. This transition phase may also involve migrating relevant data, user interfaces, and operational workflows from the old system to the new one. If no direct replacement is planned, the retirement strategy must clearly outline how the responsibilities of the AI will be absorbed by other systems or human processes, and what the implications of this change will be for service delivery and operational efficiency. Thorough testing and validation of the new system or processes before the complete decommissioning of the old one are paramount to prevent unforeseen issues.

Communication with stakeholders is perhaps the most sensitive aspect of AI retirement. This includes internal staff, affected citizens or communities, oversight bodies, and potentially vendors. Transparency about the decision to retire an AI system, the reasons behind it, and the timeline for decommissioning is essential. For internal teams, clear communication ensures that they are prepared for the changes, understand new workflows, and are aware of where to direct any residual queries. For citizens or community groups who may have interacted with or been impacted by the AI, communication should explain what the changes mean for them, how their data will be handled, and what recourse, if any, they have regarding past decisions made by the AI. This might involve public notices, updated website information, or direct outreach to specific groups. For oversight bodies and regulators, providing detailed information about the retirement plan demonstrates due diligence and adherence to governance protocols. The communication strategy should be tailored to the specific audience, addressing their likely concerns and providing them with the necessary information in an accessible format.

Furthermore, the retirement process must include a comprehensive review and documentation of the AI system's performance, its limitations, and any lessons learned during its operational life. This post-mortem analysis is invaluable for informing the design and deployment of future AI systems. It should capture not only technical performance metrics but also ethical considerations, societal impacts, and stakeholder feedback. Documenting the rationale for decisions made during the AI's lifecycle, including its eventual retirement, creates an institutional memory that can prevent the repetition of mistakes and foster continuous learning within the organization. This documentation should be readily accessible to relevant personnel and integrated into organizational knowledge management systems.

The decommissioning of an AI system also presents an opportunity to reassess the broader AI strategy of the public sector organization. Is the retiring AI part of a larger suite of systems? What are the implications of its removal or replacement on other AI initiatives? This moment can prompt a strategic evaluation of the organization's overall approach to AI, ensuring that future investments align with evolving public needs, ethical priorities, and technological advancements. It might reveal gaps in expertise, governance structures, or technological infrastructure that need to be addressed.

Consider the hypothetical scenario of an AI system used for optimizing public transport routes. Over its operational life, this AI has gathered extensive data on passenger flows, traffic patterns, and operational costs. As societal priorities shift towards environmental sustainability and the promotion of active transportation, the initial objective of pure efficiency might become secondary to objectives like reducing carbon emissions, improving accessibility for all demographic groups, or encouraging walking and cycling. If the AI cannot be recalibrated to accommodate these new, complex objectives – perhaps due to limitations in its original design or the data it was trained on – then its retirement becomes necessary. The managed retirement would involve:

First, a clear definition of the retirement triggers. This might include a predefined operational lifespan, a significant decline in its ability to meet new, broader service objectives, or the availability of a new, more adaptable AI solution.

Second, a detailed data archival plan. The historical passenger data, traffic logs, and route performance metrics would need to be preserved in a secure, anonymized format, adhering to privacy regulations and future research needs. This data could be invaluable for understanding long-term trends or for training future, more sophisticated models.

Third, a smooth transition strategy. If a new AI system is to replace it, this would involve a period of parallel running, rigorous testing of the new system's ability to balance efficiency with sustainability and accessibility goals, and thorough training for public transport operators and planners. If the functions are to be managed manually or by a less sophisticated system, the plan would detail the process for reallocating human resources and updating operational procedures.

Fourth, a transparent communication plan. This would inform the public transport authority staff, transport unions, city planners, and the general public about the upcoming changes, the reasons for retiring the old system, and the benefits of the new approach, including its focus on environmental and social equity goals.

Finally, a comprehensive post-operational review. This would document the performance of the retiring AI, the challenges encountered in adapting it to new societal goals, and the lessons learned for future AI development, particularly regarding the importance of designing AI systems with adaptability and evolving public values in mind from the outset.

Another example could be an AI system deployed in a social services agency to triage applications for housing assistance. Over time, societal understanding of the complexities of homelessness, the impact of gentrification, and the need for trauma-informed care may evolve. An AI system that was designed to prioritize applications based on a narrow set of criteria, such as immediate need and statistical likelihood of successful placement, might no longer align with contemporary ethical imperatives that emphasize dignity, equity, and addressing systemic causes.

In such a case, managed retirement would involve:

Clearly defined criteria for retirement, such as failure to adapt to new service mandates, evidence of persistent, unmitigated biases that run counter to current equity goals, or the development of a more holistic, human-centered approach that renders the AI's purely quantitative approach inadequate.

A robust data archival process. All application data, decision logs, and outcomes would need to be retained in a secure, anonymized format. This would allow for historical analysis of the system's impact, identification of any residual biases, and potentially inform the design of new assessment tools, ensuring compliance with data protection laws.

A careful transition plan. This might involve shifting to a more integrated case management system that combines AI-assisted preliminary assessments with mandatory human review by trained social workers. The plan would detail how existing cases are managed, how staff will be trained on new protocols, and how the effectiveness of the new, hybrid system will be measured against contemporary ethical and service delivery standards.

Proactive and empathetic stakeholder communication. This would be crucial for informing both agency staff and individuals seeking assistance. Explanations would need to articulate how the new approach aims to be more equitable, provide better support, and respect the dignity of applicants. Mechanisms for feedback and redress would need to be clearly communicated.

A thorough post-retirement evaluation. This would capture the lessons learned from the AI's operational period and its eventual retirement, focusing on the ethical challenges encountered, the effectiveness of the transition, and the evolving definition of 'fairness' and 'support' within the context of social services. This review would contribute to the organization's repository of knowledge for future technology implementations.

The practice of managed AI retirement is not merely a technical necessity; it is an ethical imperative. It reflects a commitment to responsible stewardship of public resources and a dedication to ensuring that technology serves the public good not just in its prime, but throughout its entire lifecycle. By embracing thoughtful, proactive retirement strategies, public sector organizations can ensure that their AI systems, even in their absence, continue to contribute to a more ethical, efficient, and responsive government. This disciplined approach to decommissioning reinforces public trust and demonstrates a mature understanding of the responsibilities that accompany the deployment of powerful AI technologies.

# Chapter 10: Data Governance and Privacy in AI

The bedrock upon which any artificial intelligence system is built, especially within the intricate landscape of the public sector, is the integrity of its data. This isn't merely a technical prerequisite; it is a fundamental ethical and operational pillar that underpins the entire functionality, fairness, and trustworthiness of AI. In the realm of public service, where decisions made by AI can profoundly impact the lives and livelihoods of citizens, the quality and governance of data are not abstract concepts but tangible determinants of justice, equity, and accountability. Data, in this context, transcends its role as mere input; it is a critical asset, a reservoir of information that must be curated, protected, and managed with an unwavering commitment to excellence and responsibility. The act of building trust with the public is intrinsically linked to the transparency and robustness of these data practices. Citizens grant governments the privilege of access to their personal information with the implicit understanding that this information will be handled with the utmost care, used for legitimate public purposes, and protected from misuse or compromise. When AI systems are deployed, this expectation of responsible data stewardship is amplified, as these systems can process and analyze vast quantities of sensitive data at speeds and scales previously unimaginable. Therefore, the foundation of any AI initiative in the public sector must be a comprehensive and ethically grounded approach to data integrity and trust.

Data integrity, in essence, refers to the accuracy, completeness, consistency, and validity of data throughout its lifecycle. For AI systems, compromised data integrity can lead to a cascade of negative consequences. Imagine an AI system designed to allocate limited public housing resources. If the data it relies on for assessing applicant needs is incomplete (missing crucial income details), inaccurate (outdated employment status), or inconsistent (differing formats for addresses), the AI's recommendations will inevitably be flawed. This could result in deserving individuals being denied essential housing, while others who are less in need, but whose data is erroneously presented, might receive preferential treatment. Such outcomes erode public confidence and undermine the very purpose of public service, which is to serve all citizens equitably. The AI, in this scenario, becomes an amplifier of existing societal inequities, not a tool for their mitigation, all because its foundational data was not sound.

Ensuring data integrity begins at the source – with data collection. In the public sector, data is gathered from a myriad of touchpoints: citizen applications, sensor networks, administrative records, public surveys, and more. Each of these collection points presents unique challenges and opportunities for maintaining integrity. For instance, when collecting data through online forms, robust validation checks are essential. These checks act as gatekeepers, preventing the entry of erroneous or malformed data. This could include ensuring that fields requiring numerical input only contain numbers, that email addresses follow a standard format, and that mandatory fields are not left blank. Beyond simple format checks, more sophisticated validation might involve cross-referencing data with existing databases to confirm accuracy or identify duplicates. For example, if an applicant provides a national identification number, the system could, with appropriate legal and privacy safeguards, cross-reference this with a national registry to confirm its validity. Similarly, when data is collected through physical means, such as during a census or a public health survey, rigorous training for enumerators on data recording procedures, clear definitions of terms, and quality control mechanisms during data entry are paramount. The principle is to embed integrity checks at every step, minimizing the introduction of errors before they can propagate through the system.

Beyond initial collection, data integrity must be maintained as data flows through various stages of processing, storage, and analysis. Data pipelines, the intricate pathways through which data moves, are often complex, involving transformations, aggregations, and integrations from multiple sources. Each transformation step presents a risk of introducing errors or altering the data in unintended ways. Therefore, meticulous documentation of all data transformation processes is crucial. This documentation, often referred to as a data lineage or data provenance record, should clearly outline what operations were performed on the data, by whom, when, and why. This allows for traceability – if an anomaly is detected in the AI's output, one can trace the data back through its transformations to pinpoint where the integrity may have been compromised. Think of it as a digital audit trail for every piece of data. Automated checks at various points in the pipeline can also help maintain consistency. For example, if a dataset is supposed to contain only entries from a specific geographical region, an automated check could flag any records falling outside that defined area.

The sheer volume and velocity of data, characteristic of many public sector applications, further complicate the challenge of maintaining integrity. Big data environments can make it difficult to manually inspect every data point. This is where robust data governance frameworks become indispensable. A strong governance framework establishes clear policies, procedures, and responsibilities for data management. It defines who owns the data, who is responsible for its quality, and what standards must be adhered to. This includes establishing data quality metrics and regularly monitoring them. These metrics might include fields like "percentage of complete records," "data entry error rate," or "record consistency score." By establishing these benchmarks, public sector organizations can quantify their data integrity and identify areas requiring improvement. Furthermore, governance frameworks often dictate the use of standardized data dictionaries and metadata repositories. A data dictionary provides clear definitions for each data element, ensuring that everyone in the organization understands what a particular data point represents and how it should be interpreted. Metadata, or "data about data," captures information about the data's origin, format, ownership, and quality, providing essential context for its use and interpretation.

Trust, the other critical component of this subsection, is not merely a byproduct of data integrity; it is an active pursuit that requires deliberate and ongoing effort. For the public sector, earning and maintaining citizen trust in AI systems means being transparent about how data is used, how decisions are made, and how individuals can exercise control over their information. This transparency extends to acknowledging the limitations of AI and the inherent imperfections in data. It is far more conducive to trust to admit that a dataset might contain biases or that an AI's prediction is probabilistic, rather than to present AI outputs as infallible truths.

One of the most significant challenges to data trust in AI, particularly in the public sector, stems from the potential for bias within datasets. AI systems learn from the data they are trained on. If this data reflects historical societal biases – whether related to race, gender, socioeconomic status, or any other characteristic – the AI will likely perpetuate and even amplify these biases in its decision-making. For example, an AI used for predictive policing, trained on historical arrest data that disproportionately features certain communities due to biased policing practices, might erroneously flag individuals from those same communities as higher risk, leading to discriminatory surveillance and interventions. This is not a hypothetical concern; numerous real-world examples have demonstrated how biased data can lead to unfair outcomes in areas like loan applications, hiring processes, and even criminal justice.

Addressing data bias requires a multi-faceted approach that begins with a deep understanding of the data's origins and the context in which it was collected. Critical examination of historical data is essential. This involves asking difficult questions: Who collected this data? For what purpose? What were the prevailing societal conditions and biases at the time of collection? Were certain groups underrepresented or overrepresented? Techniques for bias detection and mitigation are an active area of research and practice. These can include statistical methods to identify disparities in data representation or outcomes, and algorithmic approaches to "de-bias" datasets or adjust AI models to compensate for known biases. However, it is crucial to recognize that algorithmic solutions are not a panacea. They must be complemented by human oversight and a commitment to ethical principles.

Transparency in data usage is another cornerstone of trust. Citizens have a right to know how their data is being collected, used, and protected, especially when AI is involved. This means moving beyond vague privacy policies filled with legal jargon. Public sector organizations should strive for clear, accessible explanations of their AI systems' data practices. This might involve publishing "AI fact sheets" that detail the types of data used, the purpose of the AI, the potential risks and benefits, and the safeguards in place. For AI systems that make direct decisions affecting individuals, such as those related to benefit eligibility or permit applications, providing individuals with explanations for those decisions, even if facilitated by AI, is essential. This "explainability" of AI, while technically challenging for some complex models, is crucial for building user trust and enabling recourse when decisions seem unfair or incorrect.

The concept of data ownership and individual control is also central to building trust. While the public sector collects data for collective benefit, individuals should have a degree of agency over their personal information. This might involve mechanisms for data access, correction, and, where appropriate, deletion. Regulations like the General Data Protection Regulation (GDPR) in Europe and similar frameworks emerging globally are establishing clear rights for individuals concerning their data. Public sector AI initiatives must be designed to comply with these regulations and, more broadly, to respect the principles of data minimization (collecting only what is necessary) and purpose limitation (using data only for the stated purposes).

Furthermore, establishing independent oversight and accountability mechanisms is vital for reinforcing trust. This can involve internal ethics review boards, external audits, and mechanisms for public feedback and grievance redressal. When AI systems are deployed, it is important to have clear lines of accountability. If an AI system makes a harmful or discriminatory decision, who is responsible? Is it the data scientists who built the model, the officials who deployed it, or the agency itself? A well-defined accountability framework helps ensure that corrective actions are taken and that lessons are learned. Regular audits of AI systems, focusing on data integrity, bias detection, and adherence to ethical guidelines, can provide an objective assessment of performance and identify areas for improvement. These audits should be conducted by individuals or bodies with the necessary expertise and independence.

The public sector's commitment to data integrity and trust is not a static achievement but an ongoing journey. It requires continuous investment in data management infrastructure, the development of skilled data professionals, and a culture that prioritizes ethical considerations. It means acknowledging that AI is not a magical solution but a powerful tool whose effectiveness and fairness are inextricably linked to the quality of the data it consumes and the governance structures that surround it. By diligently focusing on data integrity and fostering transparent, trustworthy data practices, public sector organizations can lay a solid foundation for AI systems that serve the public good, enhance efficiency, and ultimately, earn the confidence of the citizens they are designed to support. This meticulous attention to the data itself is what distinguishes responsible AI deployment from simply adopting new technology; it is about ensuring that technology serves humanity, not the other way around.

The ethical deployment of artificial intelligence within the public sector hinges on a robust and principled approach to data handling. This is not merely a matter of compliance; it is a fundamental imperative for ensuring fairness, accountability, and the very trustworthiness of AI systems that impact citizens' lives. At the heart of responsible data handling lie several core principles, each designed to safeguard individual rights and promote the responsible use of information throughout the entire lifecycle of an AI system.

One of the foundational tenets of responsible data handling is **data minimization**. This principle dictates that organizations should collect and retain only the data that is strictly necessary for a clearly defined purpose. In the context of public sector AI, where vast datasets are often tempting to collect "just in case," data minimization acts as a crucial safeguard against overreach and potential misuse. For instance, an AI system designed to optimize public transportation routes might require data on passenger volume and travel times. However, it does not necessarily need to collect individual passenger identities, their precise origins and destinations beyond what is aggregated for route analysis, or any personal demographic information that is not directly relevant to the stated purpose of route optimization. By adhering to data minimization, the risk of a data breach exposing sensitive personal information is reduced, and the potential for data to be used for unintended or profiling purposes is significantly curtailed. This proactive approach ensures that the data collected serves the public good without unnecessarily infringing upon individual privacy.

Closely intertwined with data minimization is the principle of **purpose limitation**. Once data is collected for a specific, legitimate purpose, it should not be further processed or used for any other incompatible purpose without explicit consent or legal justification. This principle acts as a firewall, preventing the "scope creep" that can afflict data usage over time. Consider an AI system developed to analyze energy consumption patterns in public buildings to identify areas for efficiency improvements. If, at a later stage, the organization contemplates using this same data to predict the occupancy of specific public spaces for security reasons, this would likely constitute a secondary purpose that is incompatible with the original intent. Without a clear legal basis or renewed consent from affected individuals (if applicable), such repurposing would violate the principle of purpose limitation. This ensures that the trust placed in an organization to handle data for one purpose is not eroded by its subsequent use for another, potentially less transparent or more intrusive, objective.

The imperative of **accuracy** is paramount in any responsible data handling framework. Data must be accurate, complete, and kept up to date to ensure that the AI systems relying on it make sound and fair decisions. Inaccurate data can lead to erroneous outputs, discriminatory outcomes, and a profound loss of public trust. For example, an AI system used to assess eligibility for social welfare programs must be fed with accurate and current information regarding an applicant's income, employment status, and household composition. If the data is outdated or contains errors, an individual might be wrongly denied essential support or receive an incorrect amount, causing significant hardship. Public sector bodies must therefore implement rigorous data validation processes, establish clear procedures for data correction, and regularly audit their datasets for accuracy. This might involve cross-referencing data from multiple reliable sources, implementing automated data quality checks, and providing clear channels for individuals to report and correct inaccuracies in their own data.

Ensuring the
**security** of data is a non-negotiable aspect of responsible handling, especially in the public sector where sensitive information is routinely managed. This encompasses implementing robust technical and organizational measures to protect data from unauthorized access, disclosure, alteration, or destruction. For AI systems, this security imperative is amplified due to the potentially vast quantities of data they process and the sophisticated nature of cyber threats. Measures can include strong encryption for data both in transit and at rest, strict access controls based on the principle of least privilege, regular security audits and penetration testing, and comprehensive incident response plans. A data breach in a public sector AI system could have far-reaching consequences, from financial loss and reputational damage to identity theft and the compromise of national security. Therefore, a proactive and layered security strategy is essential, treating data protection as an ongoing commitment rather than a one-time task.

Adherence to
**legal and regulatory frameworks** is the bedrock upon which responsible data handling is built. Numerous laws and regulations globally govern how personal data can be collected, processed, and stored. Prominent among these are the General Data Protection Regulation (GDPR) in the European Union, the California Consumer Privacy Act (CCPA) and its successor, the California Privacy Rights Act (CPRA), in the United States, and various national data protection laws in

other jurisdictions. These regulations often establish rights for individuals, such as the right to access their data, the right to rectification, the right to erasure, and the right to data portability. Public sector organizations deploying AI must ensure that their data handling practices are fully compliant with all applicable data protection laws. This requires a thorough understanding of these legal requirements, embedding compliance into the design of AI systems (privacy by design), and conducting regular legal reviews of data processing activities. For instance, if an AI system is designed to process the personal data of EU citizens, it must comply with the GDPR's requirements regarding lawful bases for processing, data subject rights, and cross-border data transfers. Failure to comply can result in significant fines, legal challenges, and a severe erosion of public trust.

Beyond legal mandates, embracing
**best practices** further strengthens responsible data handling. This involves adopting industry-standard methodologies and ethical guidelines that may go above and beyond minimum legal requirements. For example, employing techniques for anonymization and pseudonymization where appropriate can significantly reduce privacy risks. Anonymization involves irreversibly removing identifying information so that the data subject cannot be identified, while pseudonymization replaces direct identifiers with artificial ones, allowing for re-identification only under specific, controlled circumstances. Such techniques are invaluable when training AI models or sharing data for research purposes, ensuring that the data can still be useful without exposing individuals' identities. Furthermore, maintaining detailed records of data processing activities, often referred to as a "record of processing activities," is a best practice that supports accountability and transparency. This record should document what data is collected, why, how it is processed, who has access, and how it is secured, providing a clear audit trail.

The concept of

**data governance** provides the overarching structure for implementing these principles. It establishes the policies, processes, roles, and responsibilities necessary for managing data assets effectively and ethically. In the context of public sector AI, a robust data governance framework ensures that decisions regarding data collection, usage, sharing, and retention are made transparently and accountably. This includes defining data ownership, establishing data quality standards, and creating mechanisms for dispute resolution and oversight. Without strong data governance, the individual principles of responsible data handling can become fragmented and inconsistently applied, leaving the organization vulnerable to errors, misuse, and breaches of trust. Effective data governance ensures that data is treated as a strategic asset that is managed with care, integrity, and a deep sense of responsibility towards the citizens it represents.

Implementing these principles is not a static, one-time endeavor but an ongoing commitment that requires continuous vigilance and adaptation. As AI technologies evolve and new data sources emerge, the challenges and complexities of responsible data handling will also change. Public sector organizations must therefore foster a culture of continuous learning and improvement, regularly reviewing and updating their data handling policies and practices to remain aligned with ethical standards, legal requirements, and the evolving expectations of the public. This commitment to responsible data handling is not just about mitigating risks; it is about proactively building trust and ensuring that AI serves as a force for good, enhancing public services and empowering citizens in a fair and equitable manner.

When delving deeper into the practical implementation of these principles, consider the lifecycle of data within an AI project. It begins with

**data collection**. As previously touched upon, data minimization and purpose limitation are critical at this initial stage. For instance, if a local government agency is developing an AI to predict areas prone to illegal dumping, it might initially consider collecting detailed location data from public sensors. However, a strict application of data minimization might reveal that aggregated data on dumping incidents, combined with broader geographical information (such as zoning, proximity to waste facilities, and socio-economic indicators), is sufficient for predictive modeling. Collecting granular, real-time location data from sensors in public spaces could raise significant privacy concerns for individuals present in those areas, even if they are not directly involved in dumping. Therefore, the agency would need to justify any collection of more detailed data, ensuring it is strictly necessary for the defined purpose and that appropriate privacy safeguards are in place. This often involves conducting Data Protection Impact Assessments (DPIAs) or privacy impact assessments (PIAs) to systematically evaluate and mitigate privacy risks before data collection commences.

Following collection, data enters the

**storage and processing** phase. Here, accuracy and security become paramount. For a public health AI system analyzing disease outbreak patterns, the accuracy of diagnostic codes, patient demographics, and geographical information is vital for generating reliable predictions and informing public health interventions. Inaccurate data could lead to misallocation of resources, ineffective containment strategies, or even unnecessary public alarm. Security measures are equally critical. Storing sensitive health data requires robust encryption, access controls, and regular security audits to prevent unauthorized access that could lead to breaches of patient confidentiality, a severe violation of trust and legal requirements like HIPAA in the United States. Furthermore, the processing of this data must adhere strictly to purpose limitation. If the

AI was developed to predict outbreaks, using that same data to track individuals' movements without explicit consent or a clear legal basis would be a violation. Pseudonymization techniques can be employed here, where patient identifiers are replaced with a code, allowing analysis to proceed while significantly reducing the risk of direct identification.

**Data sharing and dissemination** present another critical juncture for responsible handling. Public sector organizations may need to share data with other government bodies, research institutions, or even the public to foster innovation and collaboration. However, this must be done with utmost care. The principle of purpose limitation remains in effect; data shared for one purpose should not be used for another without proper authorization. Accuracy of shared data is also crucial; disseminating inaccurate information can have wide-ranging negative consequences. Security during transfer is vital, employing secure file transfer protocols and encryption. When sharing data with external parties, rigorous data sharing agreements should be established, clearly outlining the permissible uses of the data, security requirements, and accountability provisions. For instance, if a city's transit authority shares anonymized ridership data with urban planners for infrastructure development, the agreement should explicitly state that the data is for planning purposes only and that re-identification of individuals is strictly prohibited. The effectiveness of anonymization in such cases must be thoroughly tested to ensure it withstands potential re-identification attacks, a consideration that falls under both accuracy and security.

Finally,
**data retention and deletion** conclude the data lifecycle. Responsible data handling dictates that data should not be retained indefinitely. **Data minimization** extends to retention periods; data should only be kept for as long as it is necessary for the purpose for which it was collected, or as

required by law. Once the retention period expires, data must be securely deleted or anonymized. For example, records of citizen complaints handled by an AI system might be retained for a specified period to allow for appeals or audits, but they should be permanently deleted thereafter to prevent the indefinite accumulation of potentially sensitive personal information. Establishing clear data retention policies and automated deletion processes is essential to ensure compliance and to minimize the long-term risks associated with data storage. The secure deletion of data is as important as its secure storage; simply deleting files from a user interface does not necessarily erase the data from all storage media, requiring specialized tools and procedures for true data destruction.

The practical application of these principles often requires a combination of technical solutions, robust policies, and a strong ethical culture. For example, implementing
**accuracy** checks can involve automated data validation rules, cross-referencing with authoritative external datasets, and human review processes. For **security**, this translates to using strong authentication mechanisms, regular vulnerability assessments, and employee training on cybersecurity best practices. **Data minimization** can be achieved through careful system design that only requests and stores essential fields, and by employing techniques like aggregation and generalization to reduce the granularity of data where full detail is not required. **Purpose limitation** is reinforced through access controls and audit trails that track how data is used, flagging any deviations from the intended purpose.

Crucially, public sector organizations must recognize that these principles are not merely technical checklists but ethical commitments. They require ongoing dialogue, transparency, and a willingness to be accountable to the public. This means not only complying with regulations but also actively seeking to build trust through clear communication about data practices, providing individuals with meaningful control over their data, and establishing mechanisms for redress when things go wrong. The development and deployment of AI in the public sector offer immense potential for improving services and addressing societal challenges. However, realizing this potential responsibly depends entirely on a foundation of ethical and secure data handling practices.

The principles of data minimization, purpose limitation, accuracy, and security, when integrated within a robust legal and governance framework, form the cornerstone of responsible AI deployment in the public sector. These are not abstract ideals but practical mandates that shape how data is collected, stored, processed, shared, and ultimately, how it contributes to public value or poses risks to individuals. Each principle requires dedicated attention and continuous effort to ensure that AI systems serve the public good without compromising fundamental rights and freedoms. By embedding these principles into the very fabric of AI development and deployment, public sector entities can foster the trust necessary for these transformative technologies to truly benefit society.

Safeguarding individual privacy is not an optional add-on for public sector AI systems; it is an indispensable requirement that underpins public trust and the ethical legitimacy of these technologies. As artificial intelligence becomes increasingly integrated into public services, from healthcare and social welfare to urban planning and law enforcement, the potential for inadvertent or deliberate misuse of personal data escalates. Consequently, understanding and rigorously applying data privacy-preserving techniques is paramount. These methods are designed to enable the extraction of valuable insights from data without compromising the identity or sensitive information of the individuals from whom the data originates.

One of the most foundational approaches to protecting privacy is **anonymization**. The core idea behind anonymization is to remove or obscure personally identifiable information (PII) from a dataset so that individuals cannot be reasonably identified. This is a crucial step before data is used for training AI models, for statistical analysis, or for sharing with third parties for research or public benefit. True anonymization aims to make re-identification impossible, even when combined with other available data. For instance, if a public health agency wants to analyze patterns of flu outbreaks across different regions, it would collect data including dates, locations, and reported symptoms. To anonymize this data, all direct identifiers like names, addresses, and social security numbers would be removed. Beyond these obvious identifiers, more subtle quasi-identifiers also need careful consideration. These are pieces of information that, when combined, can indirectly identify an individual. Examples include date of birth, gender, zip code, or occupation. Techniques like generalization (e.g., replacing a precise age with an age range), suppression (removing specific records or attributes entirely), and aggregation (combining data into summary statistics) are often employed to reduce the risk of re-identification from quasi-identifiers.

However, achieving perfect anonymization is a complex challenge, and the effectiveness of anonymization techniques can be debated and is highly context-dependent. What might seem anonymized at first glance could, with enough external information and sophisticated analytical capabilities, potentially be de-anonymized. This is why it is often more accurate to speak of "risk of re-identification" rather than absolute anonymity. For example, a dataset containing a unique combination of a person's zip code, date of birth, and gender might seem innocuous. Yet, if an attacker has access to public voter registration records or other readily available databases, they might be able to cross-reference this information and pinpoint the individual. This underscores the need for a layered approach to privacy protection and a thorough understanding of the potential attack vectors.

A closely related technique, often employed as an intermediate step or an alternative to full anonymization, is **pseudonymization**. Unlike anonymization, which aims to irreversibly remove identifiers, pseudonymization replaces direct identifiers with artificial identifiers, or pseudonyms. This means that while the data is no longer directly identifiable, it is still possible to re-identify the individual if the mapping between the pseudonym and the original identifier is retained. This mapping is typically stored separately and under strict security controls. Pseudonymization is particularly useful when ongoing analysis or interaction with the data is required, or when the data needs to be linked across different datasets for the same individual without direct identification. For instance, a public sector organization might use pseudonymization for patient records in a medical research study. Each patient would be assigned a unique pseudonym, allowing researchers to track their progress, analyze their treatment outcomes, and link their data to other relevant datasets (e.g., genetic information, lifestyle data) without exposing their real names. The key benefit here is that the pseudonymized data can be handled by a wider range of personnel or even shared more broadly, with the original identifiers kept secure and accessible only by a limited number of authorized individuals. This provides a strong balance between data utility and privacy protection.

The effectiveness of pseudonymization hinges on the security of the mapping process and the pseudonym itself. If the pseudonym is too simple or predictable, or if the mapping mechanism is compromised, the privacy benefits diminish. Furthermore, the principle of purpose limitation remains critical. Even with pseudonymized data, the original purpose for which it was collected must be respected. Using pseudonymized data for a completely unrelated purpose without appropriate justification or consent would still be an ethical and potentially legal violation.

Moving beyond simple removal or substitution of identifiers, more advanced techniques aim to enable data analysis while offering stronger privacy guarantees.
**Differential privacy** is a mathematically rigorous framework that provides a quantifiable guarantee of privacy. In essence, a differentially private algorithm is designed such that the output of the algorithm (e.g., a statistical query or a machine learning model) is nearly indistinguishable whether or not any single individual's data was included in the input dataset. This is typically achieved by adding carefully calibrated random noise to the results of queries or to the model parameters during training. The amount of noise is determined by a parameter, epsilon (ε), which controls the trade-off between privacy and utility. A smaller epsilon means stronger privacy but potentially less accurate results, while a larger epsilon offers higher utility but weaker privacy.

The power of differential privacy lies in its mathematical proof. It doesn't rely on assumptions about what an attacker might know or be able to do; it provides a privacy guarantee that holds regardless of external information. This makes it particularly attractive for public sector applications where data might be highly sensitive and the potential for sophisticated adversaries is a concern. For example, a government agency might use differential privacy to release aggregate statistics about employment or income levels in different regions. By adding a controlled amount of noise to the aggregate figures, they can publish useful data for economic analysis and policy-making while ensuring that the statistics do not reveal information about any specific individual's employment status or income. Similarly, in the context of training AI models, differential privacy can be applied to the training process itself, ensuring that the resulting model does not memorize or inadvertently reveal sensitive information from the training data.

However, implementing differential privacy can be complex, and there are trade-offs to consider. The addition of noise, while protecting privacy, can reduce the accuracy and utility of the data. For highly granular analyses or when dealing with small datasets, the noise required for strong differential privacy might render the results unusable. Therefore, careful calibration of the privacy parameter ($\varepsilon$) is essential to strike the right balance for the specific application. Furthermore, not all AI tasks are equally amenable to differential privacy. For instance, highly complex deep learning models trained on massive datasets might require significant computational resources and careful engineering to achieve differential privacy without sacrificing too much performance.

Another innovative approach that addresses privacy concerns, particularly in the context of distributed data, is **federated learning**. Traditional machine learning often requires centralizing vast amounts of data on a single server for training. This poses significant privacy risks, especially when dealing with sensitive data from numerous sources. Federated learning offers a solution by allowing AI models to be trained collaboratively across multiple decentralized edge devices or servers holding local data samples, without exchanging the data itself. In a typical federated learning scenario, a global model is initialized and sent to participating devices. Each device then trains the model locally on its own data. After local training, the devices send only the model updates (e.g., gradients or parameter changes) back to a central server. The central server aggregates these updates to improve the global model.

This approach significantly enhances privacy because the raw data never leaves the user's device or the local server. For example, a public health initiative aiming to build an AI model for early disease detection could involve hospitals in different regions. Instead of each hospital sending its patient data to a central repository, a federated learning approach would allow each hospital to train a local model on its own patient records. Only the aggregated, anonymized model updates would be shared with a central entity for global model improvement. This preserves patient confidentiality while enabling the collective development of a more robust and accurate diagnostic tool.

Federated learning is particularly well-suited for scenarios where data is inherently distributed and sensitive, such as in healthcare, finance, or mobile device applications. It respects data sovereignty and can help organizations comply with data localization regulations. However, while it protects raw data from being shared, it's important to note that the model updates themselves can sometimes leak information about the underlying data, especially if the updates are sent without additional privacy protections. Therefore, federated learning is often combined with other privacy-enhancing technologies, such as differential privacy (applied to the model updates) or secure aggregation techniques, to further strengthen its privacy guarantees. Secure aggregation protocols ensure that the central server can only access the aggregated updates, not individual updates from specific devices, thus preventing the reconstruction of information about individual data points.

Beyond these specific techniques, a broader concept of **privacy-preserving machine learning (PPML)** encompasses a range of methods and principles aimed at minimizing privacy risks throughout the AI lifecycle. This includes strategies like:

**Homomorphic Encryption:** This advanced cryptographic technique allows computations to be performed on encrypted data without decrypting it first. The results of these computations, when decrypted, are the same as if the computations had been performed on the original plaintext data. While computationally intensive and still largely in the realm of research for complex AI models, homomorphic encryption offers the potential for highly secure data analysis where data remains encrypted throughout its entire journey, from collection to processing. Imagine a scenario where a public sector entity needs to analyze sensitive citizen data held by multiple private organizations. Using homomorphic encryption, the analysis could be performed on the encrypted data held by each organization, with only the final encrypted result being decrypted by the authorized entity.

**Secure Multi-Party Computation (SMPC):** This cryptographic approach enables multiple parties to jointly compute a function over their inputs while keeping those inputs private. Each party only learns the final output of the computation, not the other parties' inputs. SMPC can be used for tasks like securely calculating aggregate statistics or training models on data held by different organizations without any party revealing its own data.

**Trusted Execution Environments (TEEs):** TEEs are secure, isolated areas within a processor that can be used to process sensitive data. Data and applications running within a TEE are protected from the rest of the system, even from the operating system or system administrators. This can provide a secure environment for running AI models or performing data processing tasks on sensitive data.

The selection and implementation of these privacy-preserving techniques should not be a one-size-fits-all approach. The choice of method depends heavily on the specific AI application, the nature and sensitivity of the data, the regulatory environment, the required level of utility, and the acceptable level of privacy risk. A thorough **Data Protection Impact Assessment (DPIA)** or Privacy Impact Assessment (PIA) is crucial for identifying potential privacy risks associated with an AI system and for determining the most appropriate privacy-preserving techniques to mitigate those risks. A DPIA should be conducted proactively, ideally during the design phase of an AI project, and revisited periodically as the system evolves.

Furthermore, these technical solutions must be integrated within a comprehensive **data governance framework** and supported by strong organizational policies and practices. This includes clear guidelines on data handling, access controls, data retention, and incident response. Training and awareness programs for all personnel involved in the AI lifecycle are also essential to ensure that privacy is considered a shared responsibility. Transparency with citizens about how their data is being collected, used, and protected is also a critical component of building and maintaining trust. This might involve clear privacy notices, accessible explanations of AI functionalities, and mechanisms for individuals to exercise their data rights, such as the right to access, rectify, or erase their data, where applicable and legally mandated.

In conclusion, the responsible development and deployment of AI in the public sector are inextricably linked to the effective protection of individual privacy. Anonymization, pseudonymization, differential privacy, federated learning, and other advanced PPML techniques offer powerful tools to achieve this. However, their successful application requires a deep understanding of their strengths and limitations, a careful consideration of the specific context, and their integration within robust governance structures and ethical frameworks. By prioritizing privacy through these rigorous techniques and a commitment to transparency, public sector organizations can harness the transformative potential of AI while upholding the fundamental rights and trust of the citizens they serve.

### Transparency in Data Usage and Algorithmic Decision-Making

The integration of artificial intelligence into public services, while promising efficiency and innovation, also introduces a critical imperative: transparency. Citizens have a fundamental right to understand how their data is being utilized and how decisions that affect them are being made by automated systems. This subsection delves into the crucial aspects of transparency in data usage and algorithmic decision-making, exploring methods to foster clarity, build trust, and empower individuals.

### Communicating Data Usage Policies Effectively

The foundation of transparent data usage lies in clearly articulating how personal information is collected, processed, and employed to power AI systems. This goes beyond mere legal compliance; it is about fostering an environment of informed consent and public accountability. In the public sector, where data often relates to fundamental rights and essential services, the communication of data usage policies must be exceptionally clear, accessible, and proactively shared.

Traditional privacy policies, often dense legal documents, are frequently overlooked or misunderstood by the very individuals they are meant to inform. For AI systems, the complexity of data flows and model training introduces new layers of opacity. Therefore, public sector organizations must adopt innovative approaches to communicate their data practices. This can involve:

**Layered Notices:** Presenting information in digestible layers. A short, high-level summary can be provided at the point of data collection, with options to "learn more" leading to more detailed explanations. This allows individuals to engage with the level of detail they are comfortable with. For instance, when a citizen interacts with a new online service powered by AI, a brief pop-up might state: "Your information helps us personalize your experience and improve our services using AI. Learn more about how your data is used."

**Visualizations and Infographics:** Translating complex data processes into easily understandable visual formats. Flowcharts illustrating data pathways, pie charts showing the proportion of data used for different AI functions, or simple diagrams explaining machine learning concepts can demystify the technology. For example, a local government website could use an infographic to show how anonymized traffic data is used by an AI to optimize traffic light timings for smoother commutes.

**Plain Language Explanations:** Avoiding jargon and technical terminology. Instead of "training data ingestion for model refinement," use "using information to teach our system to work better." This principle extends to explaining the types of data collected, the specific purposes for which it is used (e.g., "to predict demand for public transport," "to identify potential infrastructure maintenance needs"), and the safeguards in place to protect privacy.

**Interactive Tools:** Developing online portals or dashboards where citizens can see summaries of how data is being used within specific services. This could include information on the types of AI systems deployed and the general categories of data they process. For example, a public health agency might offer a dashboard explaining how aggregated, anonymized health data contributes to public health research and epidemic forecasting, without revealing individual health records.

**Regular Updates and Audits:** Data usage policies and practices are not static. As AI systems evolve and new data sources are integrated, policies must be updated and communicated. Public sector organizations should also consider making summaries of their data usage audits publicly available, demonstrating a commitment to accountability.

The goal is to move beyond a perfunctory acknowledgment of data usage to a genuine understanding. When citizens feel informed about how their data fuels public services, they are more likely to trust the systems and participate willingly.

## The Imperative of Explainable AI (XAI)

Beyond how data is collected and used, transparency is equally critical in understanding
*how* AI systems arrive at their decisions. This is the domain of Explainable AI (XAI), a set of techniques and principles aimed at making AI's decision-making processes understandable to humans. In the public

sector, where AI can influence critical outcomes such as loan applications, eligibility for social benefits, or even risk assessments in the justice system, the ability to explain a decision is not just desirable; it is often a legal and ethical necessity.

The "black box" nature of many advanced AI models, particularly deep learning networks, poses a significant challenge to transparency. These models can achieve high accuracy but are notoriously difficult to interpret, making it hard to ascertain why a particular output was generated. This lack of interpretability can lead to:

**Erosion of Trust:** If citizens cannot understand why a decision was made, they are more likely to distrust the system and the institution using it. This can manifest as appeals, complaints, and a general reluctance to engage with public services.
**Difficulty in Debugging and Auditing:** When an AI system produces an incorrect or biased outcome, it is challenging to identify the root cause without understanding its internal logic. This hinders efforts to improve the system, correct errors, and ensure fairness.
**Challenges in Upholding Rights:** In many legal and regulatory frameworks, individuals have a right to an explanation for decisions that affect them. Opaque AI systems can make it impossible to fulfill this right.

XAI seeks to address these issues through various approaches, often categorized by the type of explanation they provide:

**Local Explanations:** These explain individual predictions or decisions. For instance, if an AI system denies a citizen's application for a permit, a local explanation would detail the specific factors from the application that led to the denial. Techniques like LIME (Local Interpretable Model-agnostic Explanations) and SHAP (SHapley Additive exPlanations) are popular for generating these explanations. They work by approximating the complex model's behavior around a specific data point with a simpler, interpretable model. For example, a SHAP explanation might highlight that "the denial was primarily due to the applicant's incomplete documentation (contributing 60% to the decision) and a recent violation report (contributing 30%), while the applicant's residency status had a minor impact."

**Global Explanations:** These aim to provide an understanding of the overall behavior of the AI model. This might involve identifying the most important features the model uses across all its predictions, or visualizing decision boundaries. A global explanation for a traffic management AI, for instance, might reveal that real-time traffic density and weather conditions are the most influential factors in its route optimization algorithms.

**Model-Specific vs. Model-Agnostic Explanations:** Some XAI techniques are designed for specific types of models (e.g., decision trees are inherently interpretable). Others, known as model-agnostic methods, can be applied to any AI model, regardless of its underlying architecture. This flexibility is crucial for public sector organizations that may use a variety of AI systems.

**Surrogate Models:** A common XAI technique involves training a simpler, interpretable model (like a decision tree or linear regression) to mimic the behavior of the complex "black box" model. While this surrogate model might not perfectly replicate the original, it can provide insights into the key drivers of its decisions.

### Implementing Transparency in Practice

Integrating transparency into public sector AI systems requires a multi-faceted strategy that combines technical solutions with organizational policies and citizen engagement.

### 1. Transparency in Data Usage:

**Data Inventories and Catalogs:** Maintaining comprehensive inventories of all data assets, including their origin, characteristics, and intended use. These catalogs should be accessible internally and, where appropriate, externally. For public-facing services, a simplified version can inform citizens about what data is being collected and why.
**Purpose Limitation:** Strictly adhering to the principle that data collected for a specific purpose should not be used for unrelated purposes without explicit consent or legal basis. Any re-purposing of data for AI training must be transparently communicated.
**Data Minimization:** Collecting only the data that is strictly necessary for the AI system's intended function. This reduces privacy risks and makes communication about data usage more straightforward.
**Clear Consent Mechanisms:** For AI systems that rely on individual consent for data processing, ensuring that consent is freely given, specific, informed, and unambiguous. This involves providing clear explanations of what consent entails.

**Data Sharing Agreements:** When data is shared with third parties for AI development or operation, ensuring that these agreements include robust transparency clauses and outline data usage limitations.

**2. Transparency in Algorithmic Decision-Making:**

**Explainability by Design:** Incorporating XAI considerations from the very inception of an AI project. This involves selecting models that are inherently more interpretable or planning for the application of XAI techniques during the development phase.
**Decision Thresholds and Criteria:** Clearly defining the thresholds and criteria that an AI system uses to make decisions. For instance, if an AI system is used to flag applications for further human review, the criteria for flagging should be documented and understandable.
**Human Oversight and Appeal Mechanisms:** Ensuring that AI-driven decisions are not final and can be reviewed or overridden by human personnel. Providing clear pathways for individuals to appeal AI-generated decisions is crucial. This human element acts as a critical backstop, allowing for the correction of AI errors and the application of nuanced judgment.
**Impact Assessments:** Conducting regular Algorithmic Impact Assessments (AIAs) or similar evaluations to identify potential biases, fairness issues, and the need for greater explainability in AI systems used in public services. These assessments should be transparent in their findings and recommendations.
**Public Registries of AI Systems:** In some contexts, maintaining public registries of AI systems used by government agencies can enhance accountability. These registries could detail the purpose of the AI, the data it uses, and the general mechanisms for accountability and oversight.

## Fostering Trust Through Transparency

Transparency in data usage and algorithmic decision-making is not merely a technical challenge; it is a fundamental aspect of democratic governance and the ethical deployment of AI. By embracing clear communication, employing explainable AI techniques, and embedding transparency into the fabric of their AI governance, public sector organizations can build and maintain the trust of the citizens they serve. This trust is the bedrock upon which the successful and equitable integration of AI into public life must be built. Without it, the potential benefits of AI risk being overshadowed by public apprehension and a breakdown in the relationship between government and the governed. The journey towards transparent AI in the public sector is an ongoing one, requiring continuous learning, adaptation, and a steadfast commitment to the principles of openness and accountability.

To further elaborate on the practical implications of transparency, consider the example of an AI system used by a municipal housing authority to prioritize applications for affordable housing. The system might take into account factors like income level, family size, current housing situation, and duration of residency.

### Transparency in Data Usage for the Housing AI:

**Initial Communication:** Upon applying for housing, applicants would be presented with a clear, concise notice stating: "Your application details, including income, family size, and residency information, are used to help our AI system assess your eligibility and prioritize applications for affordable housing. We use this data solely for this purpose and protect it with robust security measures. You can learn more about our data practices [link to detailed policy]."

**Detailed Policy:** The linked policy would break down:

**What data is collected:** Specific data fields (e.g., proof of income, rental history, household composition).

**Why it's collected:** To accurately assess needs and allocate resources fairly according to established criteria.

**How it's used:** "This data trains and informs an AI algorithm designed to objectively rank applications based on predefined community housing priorities. The AI does not make the final decision; it provides a prioritized list for human review."

**Who has access:** Limited staff involved in housing allocation and AI system maintenance.

**Data retention:** How long application data is stored.

**Security measures:** Encryption, access controls, and regular audits.

**Transparency in Algorithmic Decision-Making for the Housing AI:**

**Explanation of Prioritization:** If an applicant is placed lower on the priority list, they should receive an explanation. This could be an automated notification stating: "Your application has been prioritized based on factors including income level, current housing circumstances, and length of residency, as per the municipal housing allocation guidelines. The AI system identified [specific factors, e.g., 'a lower score for current housing stability' or 'a higher score for number of dependents'] as significant contributors to your current ranking. You have the right to appeal this assessment."

**Feature Importance:** The housing authority could also provide general information about the AI's decision drivers. For example, a public report might state that "the AI system has found that a combination of verifiable low income and demonstrated long-term residency in the city are the most significant factors influencing priority rankings for affordable housing."

**Human Review and Appeal:** Crucially, the AI's output is a recommendation. A human caseworker reviews the AI's prioritized list and the individual applications. Applicants can request a review of their assessment, during which the caseworker can explain the AI's recommendation and consider any extenuating circumstances the AI might not have captured. This human intervention is vital for fairness and empathy.

### Challenges and Mitigation Strategies

Achieving meaningful transparency is not without its challenges. The inherent complexity of AI, the potential for malicious actors seeking to exploit explanations, and the administrative burden of developing and maintaining transparent systems are significant hurdles.

**Balancing Transparency and Intellectual Property/Security:** Public sector organizations may be hesitant to reveal the inner workings of their AI systems due to concerns about proprietary algorithms or potential security vulnerabilities. However, a balance can be struck by focusing on explaining how *decisions are made and* what factors are considered, rather than revealing the exact code or proprietary weights. This can involve providing high-level model architectures, key feature importances, and simplified decision trees that represent the model's logic without disclosing trade secrets.

**The "Paradox of Explainability":** Sometimes, the most accurate AI models are the least explainable. Conversely, highly interpretable models might sacrifice predictive power. XAI techniques are continuously evolving to bridge this gap. Public sector bodies need to be aware of these trade-offs and make informed decisions based on the criticality of the application. For low-stakes applications, simpler, more interpretable models might be preferable. For high-stakes decisions, sophisticated XAI methods are essential.

**Ensuring Explainability is Understandable:** Providing explanations that are technically accurate but incomprehensible to the average citizen is counterproductive. XAI outputs need to be translated into plain language, using analogies and examples relevant to the individual's context. This often requires collaboration between AI engineers, ethicists, and communication specialists.

**Proactive vs. Reactive Transparency:** True transparency is proactive. It involves embedding explainability and clear communication into the AI development lifecycle, rather than attempting to retrofit it after a system is deployed or a problem arises. Regular AI impact assessments and pre-deployment ethical reviews are critical for this proactive approach.

Ultimately, the pursuit of transparency in data usage and algorithmic decision-making is an ongoing commitment to ethical AI and good governance. It requires a cultural shift within public sector organizations, a willingness to engage with citizens, and a sustained investment in the tools and practices that demystify AI and build lasting public trust. It is through this commitment that public sector AI can truly serve the public interest.

## Establishing Data Governance Policies for AI

The proliferation of Artificial Intelligence (AI) within public sector operations necessitates a robust and comprehensive approach to data governance. As AI systems ingest, process, and derive insights from vast quantities of data, it becomes paramount to establish clear, actionable policies that govern every stage of the data lifecycle. These policies are not merely bureaucratic requirements; they form the bedrock of trust, ensuring data integrity, security, privacy, and compliance with legal and ethical standards. Without a well-defined data governance framework, public sector AI initiatives risk compromised data quality, unauthorized access, privacy breaches, and ultimately, a loss of public confidence. This section outlines the essential components of establishing data governance policies specifically tailored for AI in public services, focusing on ownership, access, retention, auditability, and incident response.

### Defining Data Ownership and Stewardship

A foundational element of any data governance policy is the clear designation of data ownership and stewardship. In the context of public sector AI, this can be a complex undertaking, as data often originates from multiple sources and serves various departmental functions. Ownership refers to the ultimate accountability for a given dataset, including its accuracy, security, and compliance. Stewardship, on the other hand, pertains to the day-to-day management and custodianship of the data, ensuring it is used in accordance with defined policies.

For AI systems, it is crucial to identify which entity or department is the definitive owner of the data used for training, operation, and analysis.

This designation should consider the source of the data and its primary intended use. For instance, if a municipal tax department provides anonymized property data for an AI model designed to predict infrastructure maintenance needs, the tax department may retain primary ownership, while the public works department acts as the steward for the AI application. This division of responsibility ensures that there is always a clear point of accountability. Policies must clearly articulate these roles, defining responsibilities such as data quality assurance, metadata management, and adherence to privacy regulations. Establishing a data governance council or committee, composed of representatives from relevant departments, can facilitate these decisions and provide ongoing oversight. This council would be tasked with resolving disputes over ownership, approving new data sources for AI use, and ensuring consistent application of governance principles across the organization. The policy should also address the ownership of data generated
*by* AI systems, such as predictions, classifications, or insights. This derived data, while a product of AI, still requires clear ownership to manage its subsequent use, storage, and potential for bias.

## Implementing Robust Access Controls and Permissions

Once data ownership is established, stringent access controls become critical. AI systems, by their nature, often require broad access to large datasets. However, unrestricted access poses significant security and privacy risks. Data governance policies must therefore define granular permissions based on the principle of least privilege, ensuring that individuals and systems only have access to the data necessary for their specific roles and functions.

This involves categorizing data based on its sensitivity – for example, personally identifiable information (PII), sensitive personal data, or de-identified datasets. Policies should detail the criteria for granting access to each category, the approval workflows required, and the duration of such access. For AI models, access might be granted to specific features or aggregated forms of data rather than raw PII, depending on the model's requirements and the sensitivity of the information. Role-based access control (RBAC) is a standard practice, assigning permissions to defined roles (e.g., AI data scientist, system administrator, privacy officer) rather than individual users. This simplifies management and ensures consistency. Furthermore, policies should mandate regular reviews of access permissions to ensure they remain appropriate and to revoke access when it is no longer needed. For example, when an AI project concludes or an employee transitions to a different role, their access to the associated datasets must be promptly rescinded. Multi-factor authentication should be a mandatory requirement for accessing any system containing sensitive data used by AI. The governance policy should also specify protocols for handling emergency access requests, ensuring that such access is time-limited, logged, and immediately reviewed.

### Establishing Clear Data Retention and Disposal Policies

The duration for which data is stored has significant implications for privacy, security, and resource management. AI models often benefit from historical data, but retaining data indefinitely increases the attack surface and the potential impact of a data breach. Conversely, premature disposal of data can hinder future analysis, model retraining, or regulatory compliance. Data governance policies for AI must therefore establish clear data retention schedules based on legal requirements, operational needs, and privacy considerations.

These policies should differentiate retention periods for raw data, processed data, and AI model outputs. For instance, raw PII used for initial model training might have a shorter retention period than anonymized or aggregated data that continues to be valuable for ongoing analysis or model monitoring. The policies should specify the rationale behind each retention period, linking it to relevant legislation (e.g., GDPR, CCPA, or specific public sector data protection laws) or business requirements. Crucially, policies must also outline secure data disposal procedures. Simply deleting data may not be sufficient; secure wiping or physical destruction of media may be necessary to prevent data recovery. The process of data disposal must be documented, providing auditable proof that data has been disposed of according to policy. For AI models themselves, policies might address the retention of model versions, training datasets, and performance logs, balancing the need for reproducibility and auditability with the risk of retaining outdated or potentially biased models. Consideration should also be given to data minimization principles during retention; if certain features are no longer used or are found to be redundant, they should be identified for early disposal within the retention schedule.

### Mandating Comprehensive Audit Trails and Monitoring

Audit trails are indispensable for accountability, security, and troubleshooting in AI deployments. They provide a historical record of who accessed what data, when, and for what purpose. For AI systems, these trails are crucial for tracking data lineage, monitoring model behavior, detecting unauthorized access, and investigating incidents. Data governance policies must mandate the implementation of comprehensive audit logging and ongoing monitoring mechanisms.

These policies should specify what events must be logged, including data access, data modification, model training activities, model predictions, and changes to system configurations. The level of detail required for these logs should be defined to ensure they are sufficiently informative for retrospective analysis. For example, an audit log entry for data access might include the user ID, the timestamp, the specific data or dataset accessed, and the operation performed (e.g., read, write, delete). For AI model predictions, logs might capture the input data, the model version used, and the output generated. Policies should also stipulate how these audit logs are stored, protected from tampering, and retained for a specified period, aligning with data retention policies. Regular review and analysis of audit logs are essential. This can be facilitated by automated monitoring tools that flag suspicious activities, such as unusual access patterns, repeated failed login attempts, or access to unusually large volumes of data. The governance policy should define the procedures for responding to such alerts, including escalation protocols and investigation steps. Furthermore, audit trails are vital for demonstrating compliance with regulations and for supporting internal or external audits of AI systems. They provide the evidence necessary to verify that data is being used appropriately and that security controls are effective.

## Developing Robust Incident Response Plans

Despite best efforts, data breaches and security incidents can occur. A well-defined incident response plan is a critical component of data governance, ensuring that public sector organizations can react quickly and effectively to mitigate the damage caused by such events. For AI systems, incidents can range from data breaches to model manipulation or the generation of harmful or biased outputs.

The data governance policy must mandate the creation and regular testing of an AI-specific incident response plan. This plan should outline the steps to be taken in the event of a security incident, including:

**Identification:** How an incident is detected and confirmed. This includes monitoring audit logs, security alerts, and user reports.
**Containment:** Actions to prevent the incident from spreading, such as isolating affected systems, revoking compromised credentials, or disabling specific AI functionalities.
**Eradication:** Steps to remove the cause of the incident, which might involve patching vulnerabilities, removing malware, or retraining compromised AI models.
**Recovery:** Restoring affected systems and data to their operational state, ensuring data integrity and system functionality.
**Post-Incident Analysis:** Conducting a thorough review of the incident to understand its root cause, identify lessons learned, and update security measures and policies accordingly.

The incident response plan should clearly define roles and responsibilities, including the establishment of an incident response team with representatives from IT security, legal, communications, and the relevant AI development or operations teams. Communication protocols are also vital, detailing how stakeholders, including affected individuals and regulatory bodies, will be informed in accordance with legal obligations and ethical considerations. For AI-related incidents, the plan should include procedures for assessing the impact of a compromised AI model, potentially involving re-evaluating its outputs and performing forensic analysis of model behavior. This could also involve the decision to temporarily halt the operation of the AI system until the incident is fully resolved and the system's integrity is re-established. The governance policy should ensure that this plan is regularly reviewed, updated, and tested through simulations or tabletop exercises to ensure its effectiveness.

## Ensuring Data Quality and Integrity for AI

The performance and reliability of any AI system are directly dependent on the quality and integrity of the data it uses. "Garbage in, garbage out" is a fundamental truth in AI development. Therefore, data governance policies must incorporate stringent measures to ensure data quality and integrity throughout the AI lifecycle.

This begins with defining data quality standards, including accuracy, completeness, consistency, timeliness, and validity. Policies should mandate data profiling and validation processes to identify and rectify data errors before they are fed into AI models. For AI systems in the public sector, where decisions can have significant societal impacts, these standards must be particularly high. Data lineage is also a crucial aspect of data integrity. Policies should require the tracking of data from its source through all transformations and uses within AI systems. This end-to-end data lineage provides transparency and allows for the tracing of errors or biases back to their origin. Mechanisms for data cleansing and enrichment should be clearly defined, specifying the tools and techniques to be employed and the approval processes for changes to data. Furthermore, policies should address data validation at multiple stages: during data ingestion, before model training, and during ongoing model operation. Regular data quality audits should be conducted to assess compliance with defined standards and to identify areas for improvement. The governance framework should also empower data stewards to flag and resolve data quality issues, ensuring a proactive approach to maintaining data integrity.

**Upholding Privacy and Compliance**

At the heart of data governance for AI in the public sector lies the unwavering commitment to privacy and regulatory compliance. AI systems often process sensitive personal information, making adherence to data protection laws and ethical principles non-negotiable. The data governance policy must integrate privacy-by-design and by-default principles into all AI initiatives.

This involves conducting Data Protection Impact Assessments (DPIAs) or equivalent privacy risk assessments
*before* deploying AI systems that process personal data. These assessments should identify potential privacy risks associated with the AI's functionality, data inputs, and outputs, and outline mitigation strategies. Policies should clearly articulate the legal bases for processing personal data for AI purposes, such as consent, legitimate interest, or legal obligation, and ensure that these are meticulously documented and adhered to. Mechanisms for handling data subject rights, such as the right to access, rectification, erasure, and objection, must be integrated into AI workflows. For AI-generated decisions, policies should ensure that individuals can exercise these rights effectively, potentially requiring human review processes. Furthermore, data anonymization and pseudonymization techniques should be applied wherever feasible to reduce privacy risks. The governance policy should specify the acceptable methods and validation processes for these techniques. Continuous monitoring of regulatory changes and updates to data protection laws is also essential. The policy should designate

responsible parties for staying abreast of these developments and for ensuring that AI systems and their associated data governance practices remain compliant. Training and awareness programs for all personnel involved in AI development and deployment are critical to embed a culture of privacy and compliance throughout the organization.

In essence, establishing comprehensive data governance policies for AI in public services is not a one-time task but an ongoing process of definition, implementation, monitoring, and refinement. It requires a holistic approach that considers the entire data lifecycle, from acquisition and processing to retention and disposal, all underpinned by a steadfast commitment to security, privacy, and public trust. By meticulously crafting and diligently enforcing these policies, public sector organizations can harness the transformative power of AI responsibly, ensuring that these technologies serve the public good with integrity and accountability.

## Chapter 11: Ensuring Transparency and Explainability

The increasing reliance on Artificial Intelligence (AI) within public sector operations, while promising significant advancements in efficiency and service delivery, concurrently introduces a complex challenge: the 'black box' problem. This phenomenon refers to AI systems whose internal mechanisms and the intricate pathways leading to their outputs are largely inscrutable, even to the very individuals who design, deploy, and manage them. Unlike traditional algorithms, where logic might be traceable through lines of code, many modern AI models, particularly those based on deep learning architectures, develop emergent properties and complex, non-linear relationships within their parameters that defy simple explanation. This opacity is not merely an academic curiosity; in the context of public services, where decisions profoundly impact citizens' lives, it poses a significant ethical and practical hurdle.

The core of the 'black box' issue lies in the very nature of how sophisticated AI models learn. These systems, often trained on vast datasets, identify patterns and correlations that are too complex for humans to explicitly program or even fully comprehend. For instance, a deep neural network used for credit risk assessment might learn that a specific combination of subtle linguistic cues in a loan application, combined with an unusual pattern in a user's browsing history, strongly correlates with a higher risk of default. While the model may achieve high predictive accuracy, explaining *why* that specific combination of factors leads to a negative assessment becomes exceedingly difficult. The model does not reason in a human-understandable, step-by-step logical progression; instead, it operates through a complex web of interconnected nodes and weighted connections, resulting in a decision without a readily available, human-interpretable justification. This is akin to asking a highly skilled artisan to explain precisely how their intuition allows them to create a masterpiece, a process that often transcends explicit, codified knowledge.

The implications of this inscrutability are particularly acute in public service delivery. When AI systems are deployed to make decisions that affect individuals' access to benefits, opportunities, or even legal judgments, the ability to understand and challenge those decisions is a fundamental tenet of democratic governance and individual rights. Imagine an AI system used to allocate scarce public housing, determine eligibility for social welfare programs, or even assist in judicial sentencing. If such a system denies an application or recommends a particular sentence, the affected individual has a right to know *why*. A response of "the algorithm decided" is fundamentally insufficient. It erodes accountability, as no human can be held directly responsible for a decision made by an opaque system. It also undermines due process, as individuals cannot effectively appeal or contest a decision if they do not understand the basis upon which it was made. The very legitimacy of public institutions relies on transparency and the ability for citizens to trust that decisions affecting them are fair, equitable, and justifiable.

Furthermore, the 'black box' nature of AI exacerbates existing biases. AI models learn from the data they are fed. If this data reflects historical societal biases – such as racial discrimination in policing or gender disparities in employment – the AI system will inevitably learn and perpetuate these biases, often in ways that are difficult to detect or correct. For example, an AI used for hiring might inadvertently learn to favor candidates with certain demographic characteristics if those characteristics were disproportionately represented in successful past hires due to systemic bias, not merit. Without transparency into how the AI arrives at its conclusions, identifying and mitigating these embedded biases becomes a formidable, if not impossible, task. This can lead to discriminatory outcomes that are not only unethical but also illegal, further eroding public trust and disproportionately harming already marginalized communities. The public sector has a unique responsibility to ensure equitable outcomes, making the 'black box' problem a critical barrier to achieving this goal.

The challenge extends beyond understanding individual decisions to the broader accountability of the systems themselves. If an AI system makes a harmful error, who is responsible? Is it the data scientists who trained it, the engineers who deployed it, the policymakers who authorized its use, or the data itself? The opaqueness of the AI's decision-making process obscures the causal chain of events leading to the error, making it difficult to assign blame or implement effective corrective measures. This lack of clarity can lead to a diffusion of responsibility, where no single entity feels accountable for the system's performance or its consequences. In the public sector, where the stakes are exceptionally high and public scrutiny is intense, this ambiguity is unacceptable. There must be clear lines of accountability for all technologies used, and this requires a degree of insight into their operation.

Consider the example of an AI-powered diagnostic tool in a public health system. If the tool misdiagnoses a patient, leading to delayed or incorrect treatment, the ramifications are severe. While the tool might achieve a high overall accuracy rate across thousands of cases, the mechanism by which it arrived at the incorrect diagnosis in a specific instance is crucial. Was it due to an unusual presentation of symptoms that the model was not trained to recognize? Was it a data anomaly in that particular patient's record? Or was it a systemic flaw in the model's architecture or training data? Without the ability to delve into the AI's reasoning, healthcare professionals and patients are left without a clear understanding of the error's origin, hindering both immediate corrective action and future improvements to the system. This lack of insight can also lead to a reluctance among medical professionals to fully trust and adopt AI tools, even when they offer potential benefits, for fear of the unknown consequences of their opaque decision-making.

The 'black box' problem also poses a significant challenge for ongoing monitoring and auditing. Public sector systems are typically subject to rigorous auditing and oversight mechanisms to ensure compliance with regulations, ethical standards, and operational efficiency. When an AI system is a 'black box,' it becomes exceedingly difficult to perform meaningful audits. Auditors may be able to verify that the system produces certain outputs, but they cannot verify the integrity of the process by which those outputs were generated. This makes it challenging to detect potential fraud, manipulation, or unintended consequences. For example, if an AI system is used to manage financial transactions or detect fraudulent claims, the ability to audit its decision-making process is essential for ensuring that it is operating as intended and not being exploited. The lack of transparency can therefore create blind spots in oversight, leaving public institutions vulnerable.

Moreover, the 'black box' nature of AI can stifle innovation and improvement. Developers and domain experts often learn from understanding how a system works. If the inner workings of an AI are impenetrable, it becomes harder to identify areas for optimization, to discover novel approaches, or to adapt the system to new or evolving circumstances. Imagine trying to improve a complex machine when you can only observe its inputs and outputs, but never see the gears, levers, and circuits inside. This limitation can lead to stagnation, where AI systems, once deployed, become difficult to refine or upgrade, potentially falling behind best practices or failing to adapt to changing societal needs. For public services that require agility and responsiveness, this inflexibility is a significant drawback.

The imperative to address the 'black box' problem is thus not merely a technical pursuit but a fundamental requirement for ethical AI deployment in the public sector. It touches upon core democratic values, citizen rights, and the very trust that underpins the relationship between government and the governed. While complete transparency may be technically infeasible for some highly complex AI models, the goal must be to achieve a sufficient level of explainability and interpretability. This means developing methods and tools that allow us to understand, to a meaningful degree, how and why an AI system makes the decisions it does. This is crucial for ensuring that these powerful technologies serve the public good, upholding fairness, accountability, and the fundamental rights of all citizens. The subsequent sections will delve into the various strategies and techniques employed to peel back the layers of the 'black box,' striving for a future where AI in public services is not only intelligent but also understandable and trustworthy.

The journey toward trustworthy AI in the public sector necessitates a nuanced understanding of what "transparency" truly entails. It is not a monolithic concept but rather a spectrum, encompassing various degrees of insight into an AI system's operation. Recognizing these different levels is paramount, as different stakeholders require different types and depths of understanding. Just as a citizen seeking to understand why their benefit claim was denied needs a different explanation than a data scientist tasked with auditing the system's performance, so too must our approach to transparency be tailored.

At the most fundamental level, we can speak of *goal and purpose transparency*. This refers to a clear articulation of what the AI system is designed to achieve, the specific public service it aims to improve, and the overarching objectives it is intended to meet. This level of transparency is crucial for public buy-in and for setting expectations. Citizens need to know *why* an AI system is being deployed, what problems it is meant to solve, and what outcomes are anticipated. For instance, if a municipality implements an AI system to optimize waste collection routes, the public should be informed that the goal is to reduce fuel consumption, improve efficiency, and potentially lower costs, rather than to spy on citizens or collect extraneous data. This high-level understanding establishes a baseline of trust and allows for informed public discourse about the adoption of such technologies. Regulatory bodies also benefit from this level of clarity, as it informs their mandate and enables them to assess whether the AI's purpose aligns with public interest and legal frameworks.

Moving a step further, we encounter *data transparency*. This level focuses on the information used to train and operate the AI system. It involves disclosing the types of data collected, the sources from which it is derived, and how it is processed. For a public sector AI system, understanding the data is critical for identifying potential biases and ensuring fairness. For example, if an AI system is being developed to predict areas at high risk of requiring emergency services, transparency about the training data would involve detailing whether historical crime statistics, demographic information, socioeconomic indicators, or geographical data were used. This allows for scrutiny to determine if the data itself is representative and free from historical discrimination. If, for instance, the historical data over-represents certain neighborhoods due to biased policing patterns, the AI could inadvertently learn to perpetuate these biases, leading to disproportionate resource allocation. Data transparency also encompasses information about data governance, privacy protections, and any anonymization or aggregation techniques employed. This reassures individuals that their personal information is handled responsibly and ethically. Regulators will scrutinize this level of transparency to ensure compliance with data protection laws and to assess the risk of data misuse. Developers, too, need access to this information to understand the limitations and potential pitfalls of the data they are working with.

The next layer of transparency delves into *algorithmic transparency*. This concerns the underlying models and methodologies employed by the AI system. While not always requiring the disclosure of proprietary code or intricate mathematical formulas, it involves explaining the general approach taken by the AI. For instance, is it a rule-based system, a machine learning model, a deep learning network, or a hybrid? What are the key parameters and how are they tuned? For a predictive policing algorithm, algorithmic transparency might involve explaining whether it uses statistical correlations with historical crime data, real-time sensor inputs, or a combination of factors. It could also entail disclosing the general architectural design of the model, such as the number of layers in a neural network or the type of algorithm used for classification or regression. This level of insight helps domain experts and technical auditors to understand the fundamental logic that drives the AI's operations, even if the precise decision-making process remains complex. Citizens might not need to understand the mathematical intricacies, but they should be able to grasp the general principles. For example, knowing that an AI assisting in loan application processing uses a "gradient boosting model" is less important to a citizen than understanding that it evaluates factors like credit history, income, and debt-to-income ratio.

A more granular, and often more challenging, form of transparency is *process transparency*. This examines the step-by-step operations of the AI system. It seeks to illuminate how the AI processes inputs, makes intermediate calculations, and arrives at its final output. This is where the 'black box' problem becomes most acute, particularly with complex deep learning models. For public sector applications, understanding the process is crucial for debugging, auditing, and ensuring fairness in individual decisions. Consider an AI system used to screen job applications for public service roles. Process transparency would involve detailing how the AI analyzes resumes, scores qualifications, and filters candidates. While a full line-by-line explanation might be impossible for a neural network, stakeholders should be able to understand the sequence of operations: how keywords are identified, how experience levels are weighted, and how the system generates a shortlist. For regulatory bodies, this level of detail is essential for verifying compliance with anti-discrimination laws. For developers, it is vital for identifying and rectifying errors or inefficiencies in the AI's workflow.

Finally, at the most detailed end of the spectrum, lies *decision transparency* or *explainability*. This focuses on providing a justification for individual decisions made by the AI. When an AI system influences a decision that directly affects an individual – such as approving or denying a permit, allocating a social service, or recommending a course of action for law enforcement – the ability to explain *why* that specific decision was made is paramount. This is the bedrock of due process and the right to an explanation. For a citizen whose application for affordable housing has been rejected by an AI, a decision transparency explanation would go beyond stating the outcome. It would ideally articulate the key factors that led to the rejection. For instance, it might explain that the application was denied because the applicant's income exceeded the eligibility threshold for that particular housing program, or that a required document was missing, or that the applicant did not meet the residency requirements as stipulated in policy X. This explanation should be presented in a clear, accessible language, avoiding technical jargon.

This tiered approach to transparency is not merely an academic exercise; it has profound practical implications for different groups of people who interact with or are affected by AI systems in the public sector.

**Citizens and the Public:** For the general public, the most relevant levels of transparency are **goal and purpose transparency** and **decision transparency**. Citizens need to understand *why* an AI is being used in their community and what benefits it is expected to bring. When an AI system makes a decision that directly impacts them, they have a fundamental right to know the reason behind it. This empowers them to challenge erroneous or unfair outcomes and fosters trust in public institutions. The explanations provided to citizens must be cou laconic and easy to comprehend, focusing on the actionable reasons and policy implications rather than technical details. For example, if an AI system is used to flag potential tax fraud, a citizen who is audited should be informed that the AI identified unusual patterns in their filings consistent with known fraud indicators, and that their specific return was flagged due to the presence of these indicators, rather than a blanket statement of suspicion.

**Regulators and Oversight Bodies:** These stakeholders require a more comprehensive understanding, spanning **goal and purpose transparency**, **data transparency**, **algorithmic transparency**, and **process transparency**. Regulators must ensure that AI systems comply with laws, regulations, and ethical guidelines. They need to scrutinize the data used to identify and mitigate biases, assess the appropriateness of the algorithms employed, and understand the operational processes to detect potential manipulation or unintended consequences. For instance, a financial regulatory body overseeing an AI used for anti-money laundering might need to examine the data sources, the model's architecture for identifying suspicious transactions, and the rules governing its decision-making to ensure it is effective and fair. This level of transparency is essential for their auditing and enforcement functions.

**Public Sector Employees and Decision-Makers:** These individuals are often the direct users or overseers of AI systems. They benefit from all levels of transparency, but particularly from **goal and purpose transparency**, **algorithmic transparency**, and **process transparency**. Understanding the AI's objectives helps them integrate it effectively into their workflows. Grasping the algorithmic approach and operational process allows them to interpret the AI's outputs critically, to identify situations where the AI might be faltering, and to exercise their own judgment and discretion when necessary. A social worker using an AI to help prioritize cases might need to understand not only the AI's goal (e.g., identifying high-risk families) but also how it weighs different risk factors (algorithmic transparency) and the steps it takes to flag a case (process transparency), so they can corroborate the AI's assessment with their own professional expertise. They also play a critical role in providing feedback for system improvement, which requires a solid understanding of its mechanics.

**AI Developers and Technical Teams:** While they build and maintain the systems, even developers often require heightened transparency, especially when working with complex, pre-trained models or collaborating across teams. They need deep insights into **data transparency**, **algorithmic transparency**, and **process transparency**, and often desire granular detail on **decision transparency** for debugging and refinement. Understanding the data is crucial for identifying potential data drift or quality issues. Familiarity with the algorithm allows them to optimize performance and identify architectural flaws. Detailed knowledge of the process is essential for troubleshooting errors. When debugging a deep learning model used in medical imaging analysis, for example, developers might need to trace the activation patterns of specific neurons to understand why a false positive occurred, thus requiring a very deep level of "decision transparency" at a microscopic level of the model's operation.

Tailoring explanations to these diverse needs is where the practical application of transparency principles becomes an art. A "one-size-fits-all" approach is insufficient.

For
**citizens**, explanations should be framed in terms of impact and observable outcomes. Instead of discussing model parameters, focus on the criteria used and the policy implications. Visualizations can be powerful tools, illustrating how different inputs might lead to different outputs, or showing the distribution of data used for training. Storytelling, where appropriate, can help demystify complex processes by relating them to relatable scenarios. For instance, an explanation of an AI that helps optimize traffic light timings could use a simple analogy of how a smart traffic controller "learns" from real-time traffic flow to adjust signal durations, much like a conductor adjusts the tempo of an orchestra based on the music.

For
**regulators**, explanations need to be precise, verifiable, and aligned with legal and ethical frameworks. This often involves providing access to documentation, audit logs, and potentially even a sandbox environment where they can test the AI under controlled conditions. The language used must be technical enough to satisfy their oversight requirements, clearly referencing compliance with specific statutes or ethical codes. For example, when explaining an AI used in parole decisions, regulators would expect to see evidence that the AI does not rely on protected characteristics and that its decision-making process can be audited against fairness metrics.

For
**public sector employees**, explanations should be practical and actionable, integrating seamlessly with their existing knowledge and workflows. Training sessions that combine theoretical understanding with hands-on exercises can be highly effective. Explanations should highlight the AI's limitations, the scenarios where human override is necessary, and how to interpret the AI's confidence scores or uncertainty measures. For an AI assisting in identifying potential social service needs, employees should understand not just which families are flagged, but also the degree of certainty with which the AI flags them, and the contextual information that supports the flag, enabling them to make informed judgments.

For
**developers**, transparency involves providing access to detailed documentation, code repositories, detailed error logs, and potentially tools for model introspection. The language here is highly technical, focusing on mathematical formulations, architectural diagrams, and performance metrics. Collaborative platforms and regular technical reviews are also crucial for ensuring shared understanding and facilitating effective problem-solving.

Ultimately, achieving meaningful transparency in public sector AI is an ongoing process. It requires a commitment to open communication, a willingness to adapt and evolve our methods as AI technologies advance, and a deep respect for the diverse needs of all stakeholders. By understanding and implementing these different levels of transparency, we can move closer to building AI systems that are not only effective but also equitable, accountable, and worthy of public trust. This layered approach ensures that while the inner workings of highly complex AI may remain intricate, the core principles, goals, and critical decision-making rationales are accessible and comprehensible to those who need to understand them most.

Achieving explainable AI (XAI) is not a singular feat but a collection of methodologies and techniques designed to shed light on the often opaque decision-making processes of artificial intelligence systems. These methods aim to provide insights into *why* an AI model arrives at a particular prediction or decision, moving beyond simply stating the outcome. In the context of public sector applications, where accountability, fairness, and public trust are paramount, the adoption of robust XAI techniques is not merely beneficial but essential. The challenge lies in making complex AI, particularly sophisticated machine learning models like deep neural networks, understandable to a range of stakeholders, from technical experts to the general public. This subsection will delve into some of the prominent practical methodologies employed to achieve this crucial explainability.

One of the most widely adopted and versatile techniques for achieving model interpretability is
**Local Interpretable Model-agnostic Explanations (LIME)**. The beauty of LIME lies in its name: "local," "interpretable," "model-agnostic," and "explanations." "Local" signifies that LIME explains individual predictions, not the entire model globally. It achieves this by approximating the complex model's behavior in the vicinity of a specific data point with a simpler, interpretable model (like a linear model or a decision tree). The "interpretable" aspect refers to the fact that these simpler models are inherently easy for humans to understand. Crucially, LIME is "model-agnostic," meaning it can be applied to any machine learning model, regardless of its internal structure – be it a neural network, a support vector machine, or a random forest. This makes it an incredibly powerful tool for explaining predictions from existing or legacy systems where direct access to the model's internal workings might be limited or proprietary.

The process of LIME involves perturbing the input data around the instance to be explained, generating a dataset of these perturbed samples, and then getting predictions from the original complex model for these samples. LIME then weighs these perturbed samples based on their proximity to the original instance and trains a simple, interpretable model on this weighted data. The explanation is then derived from this local, interpretable model. For example, consider an AI system used by a local council to assess the likelihood of a citizen defaulting on their property tax payments. If the AI flags a particular individual as high-risk, LIME could be used to explain this specific prediction. LIME would generate variations of the individual's financial profile (e.g., slightly altering income figures, changing debt-to-income ratios within a plausible range) and observe how the AI's risk score changes for these variations. It would then construct a local linear model that best approximates the AI's behavior for this individual. The explanation would then highlight the features that most significantly influenced the high-risk prediction for *that specific individual*, such as a recent increase in outstanding debt or a decline in declared income, presented in an easily digestible format. This allows the tax official to understand the primary drivers behind the AI's assessment for that citizen, facilitating a more informed and transparent review.

Another highly influential XAI technique is **SHapley Additive exPlanations (SHAP)**. Developed by drawing on cooperative game theory, SHAP provides a unified framework for interpreting predictions from any machine learning model. SHAP values represent the marginal contribution of each feature to the prediction of a specific instance, averaged over all possible combinations of feature values. The core idea is to fairly distribute the "payout" (the prediction difference from a baseline or average prediction) among the "players" (the features). This attribution provides a rigorous and theoretically sound way to understand feature importance at both a local (individual prediction) and global (overall model behavior) level.

SHAP values are additive, meaning the sum of the SHAP values for all features, plus the expected value of the model, equals the actual prediction for that instance. This property is crucial for ensuring that explanations are consistent and understandable. For instance, in a public sector context, imagine an AI system designed to predict the likelihood of a public housing application being approved. A citizen whose application is rejected might receive an explanation based on their SHAP values. The explanation would detail how much each factor (e.g., income level, number of dependents, length of residency, credit score) contributed positively or negatively to the probability of rejection, relative to the average approval probability. Features that push the prediction towards rejection will have positive SHAP values (meaning they increased the likelihood of rejection), while features that push it towards approval will have negative SHAP values. This granular insight allows the applicant to understand precisely which aspects of their application were most influential in the outcome. Furthermore, aggregated SHAP values across many applications can reveal global feature importances, helping policymakers understand which demographic or economic factors are most predictive in the housing allocation process, which can inform policy adjustments. For instance, if SHAP analysis consistently shows that a specific employment type is a strong negative predictor, policymakers might investigate whether this reflects a genuine risk factor or an unintended bias in the data.

Beyond these model-agnostic techniques, there are approaches that focus on inherently interpretable models or methods that aim to make complex models more transparent.

**Rule-based systems**, for example, represent a foundational approach to AI that is inherently interpretable. These systems operate on a set of predefined "if-then" rules. The logic is explicit and easy to follow: if condition A and condition B are met, then action C is taken. In public administration, rule-based systems have long been used for tasks like eligibility determination for social benefits or processing standard administrative procedures. For instance, a rule-based system for determining eligibility for a specific welfare program might have rules like: "IF applicant*income* < *threshold*X AND applicant*household*size > Y THEN eligible = TRUE." The transparency here is absolute – the rules are visible, and the path to a decision is a direct logical deduction. While they may lack the predictive power of complex machine learning models for intricate tasks, their interpretability makes them ideal for scenarios where compliance with explicit regulations and clear justification for decisions are paramount.

However, many complex AI models, especially deep learning networks, are not inherently interpretable. For such "black box" models, researchers have developed techniques to extract interpretable information. One such approach is
**decision tree extraction**. This involves training a decision tree to mimic the behavior of the complex model, at least within a specific region of the input space or for a subset of classes. While a single decision tree might not perfectly replicate a deep neural network, it can provide a valuable, human-understandable approximation. For example, if a complex AI is used to predict which public infrastructure projects are most likely to experience cost overruns, a decision tree extracted from this AI could provide a set of simplified rules that highlight key indicators of overruns, such as a project's complexity, its initial budget, and the experience of the lead contractor. This simplified view can be far more accessible to project managers and oversight committees than the intricate weights and biases of the original neural network.

Another family of techniques focuses on

**feature visualization** and **saliency maps**, particularly relevant for deep learning models, especially those used in image analysis. Saliency maps highlight the pixels or regions in an input image that are most important for the model's decision. For instance, if an AI system is used by urban planning departments to identify areas prone to flooding from satellite imagery, a saliency map could visually indicate which parts of the terrain, water bodies, or existing infrastructure the AI focused on to make its prediction. This allows an expert to quickly assess if the AI is looking at relevant features or if it is being distracted by irrelevant patterns. Similarly, feature visualization techniques can reveal what abstract patterns or features different layers of a neural network have learned to detect. This can offer insights into the model's internal representations, helping developers understand what the model "sees" or "understands" about the data.

**Counterfactual explanations** offer another powerful avenue for explainability. A counterfactual explanation describes the smallest change to the input features that would alter the prediction to a desired outcome. For example, if an AI system denies a loan application, a counterfactual explanation might state: "Your loan was denied because your debt-to-income ratio was 45%. If your debt-to-income ratio had been 35% or lower, your loan would have been approved." These explanations are highly actionable for individuals, as they clearly outline what needs to change to achieve a different outcome. In public sector services, this could be applied to decisions like social service eligibility, permit approvals, or even educational program recommendations. For a citizen applying for a grant, a counterfactual explanation could detail the specific criteria they missed and what evidence they would need to provide to satisfy those criteria. This moves beyond just stating the reason for denial to providing a roadmap for achieving success.

The implementation of these XAI methods is not without its challenges.

One primary concern is the trade-off between interpretability and accuracy. Often, the most accurate models are the most complex and least interpretable. Conversely, highly interpretable models, like simple linear regressions or decision trees, might not capture the nuances of complex real-world data as effectively. Therefore, a careful balance must be struck, often employing XAI techniques as a complement to complex models rather than a replacement. The choice of XAI method also depends heavily on the specific application, the type of data, and the target audience for the explanation. A citizen requires a different type of explanation than a data scientist auditing the system.

Furthermore, the very concept of an "explanation" can be subjective. What constitutes a satisfactory explanation for one person may be insufficient for another. The fidelity of the explanation is also critical – the explanation must accurately reflect the model's behavior. A misleading explanation can be worse than no explanation at all, eroding trust rather than building it. For public sector AI, this means that XAI methods must be rigorously validated to ensure they are not only understandable but also truthful representations of the AI's reasoning. The temporal aspect also matters; AI systems, especially those that learn and adapt over time, may require ongoing explanations as their behavior evolves.

In practice, integrating XAI into public sector AI deployments requires a holistic approach. It involves selecting appropriate XAI techniques based on the specific use case and stakeholders, developing clear communication strategies for presenting explanations, and establishing robust processes for monitoring and validating the explanations themselves. It is also an iterative process, where feedback from users and stakeholders can inform improvements to both the AI system and its explanatory mechanisms. By embracing these practical methods, public sector organizations can move towards AI systems that are not only powerful and efficient but also transparent, accountable, and ultimately, more trustworthy.

The preceding sections have explored the technical intricacies of achieving explainable AI (XAI), detailing methods like LIME, SHAP, rule-based systems, decision tree extraction, feature visualization, and counterfactual explanations. While these techniques provide the foundational tools to peer inside the "black box" of AI, their true value in the public sector is only realized through effective communication. The most sophisticated explanations are rendered impotent if they cannot be understood by the very citizens and stakeholders they are meant to inform. This subsection shifts focus from *how* we explain AI decisions to *how we communicate* those explanations, ensuring that transparency translates into genuine public trust and understanding.

Communicating AI decisions to the public is not merely about translating technical jargon into simpler terms; it is about building a bridge of comprehension and trust. In the public sector, where AI is increasingly deployed for services ranging from social benefit allocation and criminal justice risk assessment to infrastructure management and public health advisories, the impact of AI decisions on individuals' lives can be profound. Therefore, the communication strategy must be as robust and carefully considered as the XAI methodology itself. This involves a multi-faceted approach that acknowledges the diverse needs and understanding levels of the public, the importance of context, and the necessity of providing clear avenues for recourse and further engagement.

At its core, effective communication of AI decisions hinges on clarity and accessibility. This means moving beyond technical outputs and presenting information in a format that is readily understandable to someone without a background in data science or artificial intelligence. For instance, when an AI system determines eligibility for a public service, simply stating a probability score or a list of feature weights is insufficient. Instead, the explanation should be framed in terms of tangible outcomes and relatable factors. If an AI flags an applicant for a housing subsidy as "high risk" based on their financial data, the communication to the applicant should articulate this in plain language: "Our system has identified certain factors in your financial profile that suggest a higher likelihood of difficulty in managing the subsidy payments. These factors include your current debt-to-income ratio and recent changes in your employment stability. We need to discuss these further to ensure the subsidy is appropriate for your circumstances." This approach humanizes the decision and invites dialogue, rather than presenting an inscrutable verdict.

The principle of providing explanations at the right level of detail is also paramount. A "one-size-fits-all" approach to communication is unlikely to succeed. Different individuals will require different depths of information. For a citizen whose application for a permit is denied by an AI, a clear, concise explanation highlighting the primary reasons and offering a path to appeal might be all that is needed. They need to know *why* and *what to do next*. For a community group advocating for a specific policy change, a more aggregated view of how AI systems are making decisions across a broader demographic or geographical area might be more relevant. This could involve visualizing trends or highlighting systemic factors that the AI consistently identifies. For elected officials or oversight bodies, detailed reports that include information on model performance, fairness metrics, and the specific XAI techniques used might be necessary. Therefore, communication strategies must be adaptable, offering layered explanations that cater to varying levels of interest and expertise.

Context is another vital element in communicating AI decisions. An AI decision rarely occurs in a vacuum. It is part of a larger process, often involving human review, regulatory frameworks, and societal values. Explaining an AI decision effectively requires situating it within this broader context. For example, if an AI is used to help judges assess the risk of recidivism for defendants, it is crucial to explain that the AI's output is a recommendation or an input to a human decision-maker, not the final arbiter. The explanation should clarify the AI's role, its limitations, and how it is intended to augment, rather than replace, human judgment. This transparency about the AI's integration into existing workflows helps manage public expectations and prevents misinterpretations about the autonomy and authority of the AI system. Communicating that human oversight is in place for critical decisions, and detailing how that oversight functions, builds a critical layer of trust.

Furthermore, the communication must be proactive, not just reactive. When AI systems are introduced into public services, an initial, broad-scale communication effort is essential. This could take the form of public information campaigns, dedicated sections on government websites, or community forums. These efforts should aim to demystify AI in general terms, explain the specific purposes for which it will be used, and outline the principles of transparency and accountability that will guide its deployment. By proactively informing the public, organizations can preempt suspicion and foster a more receptive environment for AI adoption. Such upfront communication sets a tone of openness and a commitment to public engagement, which are foundational to building long-term trust.

A key aspect of proactive communication is explaining the

*purpose* and *benefits* of using AI. Citizens are more likely to accept AI-driven decisions if they understand how these systems are intended to improve public services, increase efficiency, ensure fairness, or enhance safety. For instance, an AI system used for traffic management might be explained as a tool to optimize traffic light timings, thereby reducing congestion, shortening commute times, and lowering emissions. The explanation should highlight how the AI analyzes real-time data to achieve these benefits, emphasizing the positive impact on the community. When the public sees AI as a tool that serves their interests, they are more inclined to accept its use and the decisions it helps to make.

Providing clear pathways for appeal and redress is non-negotiable when communicating AI decisions. When an AI-driven decision negatively impacts an individual, they must have a clear understanding of how to challenge that decision. This includes readily accessible information on the appeals process, the documentation required, and the timeframe for review. Crucially, the appeals process should involve human review, offering an opportunity for individuals to present their case and for human judgment to override or correct an AI's potentially flawed or biased decision. The communication surrounding appeals must be simple and direct, avoiding bureaucratic jargon. For example, a notice of a denial should clearly state: "If you disagree with this decision, you have the right to appeal. You can submit a written appeal within 30 days to [address or online portal]. Please include any additional information you believe is relevant to your case. Your appeal will be reviewed by a member of our team." This transparency in the appeals process reinforces accountability and demonstrates that the system is designed to correct errors.

The language used in public communications is a critical lever. It must be precise yet accessible, avoiding technical jargon and euphemisms.

Instead of using terms like "algorithmic bias" without explanation, it is more effective to say "the system may unfairly favor certain groups of people due to patterns in the data it learned from." Similarly, instead of "model drift," one might explain that "the system's accuracy might decrease over time as circumstances change, and it needs to be updated." The use of analogies and real-world examples can further enhance comprehension. For instance, explaining how an AI learns by comparing it to how a human learns from experience, or how a recommendation engine works on streaming services, can make complex concepts more relatable.

Visual aids can also be powerful tools for communicating AI decisions and their underlying mechanisms. Infographics, charts, and simple diagrams can illustrate complex data flows, decision trees, or the relative impact of different factors on an AI's output. For example, a simple bar chart could show how various factors contributed to a risk assessment score, with longer bars indicating greater influence. Saliency maps, when simplified and explained, can visually highlight which parts of an image or document an AI focused on for its decision, making the process more concrete for observers. These visual representations can often convey information more effectively and efficiently than dense text.

A vital component of communication is establishing feedback loops. Public sector organizations should actively solicit feedback on their AI communication strategies and the AI systems themselves. This can be done through surveys, public consultations, and dedicated contact points for inquiries. This feedback is invaluable for identifying areas where explanations are unclear, where the AI's decisions are causing concern, or where processes need improvement. Acting on this feedback demonstrates a commitment to responsiveness and continuous improvement, further strengthening public trust. For instance, if multiple citizens report confusion about a particular explanation, the organization should review and revise that explanation.

Moreover, the "explainability" of the explanation itself is a consideration. While an explanation should be understandable to the layperson, it should also be robust enough to withstand scrutiny. This means that the underlying justification for the explanation should be verifiable. For instance, if an AI decision is explained by citing specific rules, those rules should be accessible and auditable. If it's explained by reference to feature importance, the data used to derive that importance should be available, within privacy constraints. This layered explainability ensures that while the primary explanation is accessible, deeper technical validation remains possible for those who require it, such as auditors, researchers, or watchdog groups.

The digital divide presents a significant challenge in communicating AI decisions. Not all citizens have equal access to digital platforms or the digital literacy skills to navigate them. Therefore, communication strategies must incorporate a variety of channels, including non-digital ones. This might involve providing printed materials, offering in-person consultations at community centers, or having dedicated phone lines staffed by knowledgeable personnel. Ensuring that vulnerable populations, who may be disproportionately affected by AI-driven decisions, have equitable access to clear information and redress is a fundamental aspect of ethical AI deployment.

The role of public servants and frontline staff in communicating AI decisions cannot be overstated. These individuals are often the direct point of contact for citizens interacting with AI-driven systems. They need to be adequately trained not only on how the AI systems work but also on how to communicate their outputs and implications effectively. This training should equip them with the language, empathy, and information necessary to answer questions, address concerns, and guide individuals through the appeals process. Empowering these staff members to be effective communicators is crucial for building trust at the ground level.

In essence, communicating AI decisions to the public is an ongoing process, not a one-off task. It requires a commitment to transparency, a deep understanding of the audience, and a willingness to adapt and improve. By prioritizing clear, accessible, and context-rich explanations, by providing robust avenues for appeal, and by engaging proactively with the public, governmental organizations can navigate the complexities of AI deployment in a way that fosters public trust and ensures that AI serves the public good responsibly and equitably. This dedication to transparent communication is the bedrock upon which the successful and ethical integration of AI into public services must be built.

The imperative for transparency in public sector artificial intelligence (AI) deployment inevitably encounters a significant practical hurdle: the proprietary interests of the technology vendors who develop and supply these sophisticated systems. While citizens and oversight bodies rightly demand to understand how AI influences decisions affecting their lives, vendors often safeguard their algorithms, datasets, and unique methodologies as valuable intellectual property. This creates a delicate balancing act for public sector organizations, requiring them to procure powerful AI solutions while simultaneously ensuring a sufficient level of transparency and explainability to uphold public trust and accountability. Negotiating this terrain demands careful consideration of contractual frameworks, a nuanced understanding of intellectual property law, and a strategic approach to information disclosure.

At the heart of this challenge lies the question of what constitutes "sufficient transparency" when the underlying technology is a proprietary black box. Vendors may argue that revealing their core algorithms would expose trade secrets, undermine their competitive advantage, and potentially enable malicious actors to exploit vulnerabilities. This perspective is not without merit; however, it cannot be the sole determinant of disclosure in the public sector. When AI systems are used to make decisions that impact fundamental rights, access to services, or public safety, the public's right to understand these decisions and their basis must take precedence. The goal is not to dissect every line of code or reveal every proprietary dataset, but rather to ensure that the *logic* and *impact* of the AI's operation are comprehensible and auditable.

One crucial strategy for navigating this tension begins at the procurement stage. Public sector entities must move beyond simply acquiring AI functionality and instead prioritize procurements that explicitly demand transparency and explainability as non-negotiable requirements. This involves crafting detailed Requests for Proposals (RFPs) that specify the types of explanations expected, the metrics for assessing fairness and accuracy, and the acceptable levels of algorithmic opacity. For instance, an RFP might stipulate that vendors must provide access to post-hoc explanation tools (like LIME or SHAP), demonstrate the system's performance across various demographic groups, and commit to providing detailed documentation on the AI's intended use, limitations, and data handling practices. The contractual agreements stemming from these RFPs should then codify these commitments, outlining penalties for non-compliance and establishing clear protocols for dispute resolution.

When evaluating vendor proposals, public sector organizations should look for evidence of a vendor's commitment to explainable AI (XAI) principles. This could include the vendor's existing XAI methodologies, their willingness to collaborate with the public sector on developing tailored explanations, and their transparency regarding the data used for training and validation. A vendor who actively engages in discussions about explainability, rather than dismissing it as an external constraint, is more likely to be a suitable partner. Furthermore, the procurement process can incorporate a "transparency quotient" as a significant evaluation criterion, giving greater weight to solutions that demonstrably prioritize understandable outputs and auditable processes.

In situations where a vendor's core algorithmic logic is considered an inviolable trade secret, alternative avenues for achieving transparency must be explored. One such approach involves focusing on the *inputs* and *outputs* of the AI system, as well as the decision-making *process* surrounding its deployment. While the internal workings of a proprietary algorithm might remain opaque, it is often feasible to gain significant insight by understanding what data the AI receives, how it processes that data (at a high level, without revealing proprietary formulas), and what specific recommendations or classifications it generates. Public sector organizations can contractually require vendors to provide detailed logs of inputs and outputs, along with access to the features that the AI deems most influential in its decisions. This allows for an analysis of the AI's behavior and its consistency with established policies and legal frameworks, even if the exact mathematical operations remain proprietary.

Another vital area for contractual negotiation concerns the right to audit and independent review. Public sector organizations should insist on retaining the right to audit the AI system, either internally or through an independent third party, to verify its performance, fairness, and adherence to contractual obligations. This audit might not involve a full code review but could include testing the system with carefully constructed datasets, analyzing its outputs against ground truth, and examining its behavior under various scenarios. The contract should clearly define the scope of such audits, the vendor's responsibilities in facilitating them, and the confidentiality agreements that auditors must adhere to, especially concerning proprietary information. This ensures that the AI operates as intended and that potential biases or errors are identified and rectified.

The concept of "model cards" and "datasheets for datasets," as advocated by researchers like Timnit Gebru and others, offers a structured way to document AI models and the data used to train them. Public sector organizations can mandate that vendors provide such documentation as part of their contractual obligations. A model card would typically include information about the model's intended use, performance metrics (across different subgroups), ethical considerations, limitations, and the datasets used for training, validation, and testing. Similarly, a datasheet for a dataset would detail its provenance, composition, collection process, potential biases, and intended uses. While these documents may not reveal proprietary algorithms directly, they offer a rich source of information for understanding the AI's context, potential risks, and how it is likely to perform. This level of documentation is crucial for enabling informed decision-making, risk assessment, and accountability.

When a vendor is unwilling to disclose core algorithmic details, the public sector can explore the use of "surrogate models" or "explanation models." These are simpler, more transparent AI models that are trained to mimic the behavior of the proprietary "black box" model. While the surrogate model may not perfectly replicate the proprietary model's predictions, it can often capture its essential logic and provide a good approximation that is amenable to explanation. The contract could stipulate that the vendor must collaborate in the creation and validation of such surrogate models, ensuring that they accurately reflect the proprietary system's decision-making patterns to a defined degree of fidelity. This allows for explainability without directly exposing the vendor's intellectual property.

Furthermore, it is essential to differentiate between the proprietary interest in an algorithm's
*design* and the obligation to explain its *application* in a specific context. A vendor may have a proprietary interest in the neural network architecture they developed, but they still have a responsibility to explain how that architecture, when applied to a specific public service task (e.g., processing loan applications), leads to particular outcomes. This distinction is critical. The public sector's need is not for the blueprint of the AI engine, but for a clear understanding of its operational principles and its impact within the defined use case. Contracts should be drafted to reflect this nuance, requiring vendors to provide explanations related to the *specific deployment*, rather than forcing them to reveal the underlying proprietary technology in its entirety.

Data governance and privacy considerations also play a significant role. Vendors may be reluctant to share specific datasets due to privacy concerns or their own proprietary interests in aggregated data. However, public sector organizations must ensure that the data used by AI systems is handled ethically and in compliance with relevant privacy regulations. Contractual clauses can address this by requiring vendors to demonstrate robust data anonymization techniques, secure data storage and processing, and adherence to data usage limitations. They can also stipulate that any synthetic data generated for testing or explanation purposes must be representative of the real data while preserving privacy. In some cases, instead of sharing raw data, vendors might be required to provide access to statistical summaries or differential privacy-protected outputs, offering insights without compromising sensitive information.

The role of independent research institutions and academia can be invaluable in bridging the gap between proprietary interests and the need for public scrutiny. Public sector organizations can partner with universities or research bodies to conduct independent evaluations of AI systems. These collaborations can be structured through specific research agreements that define the scope of the research, the data access protocols (often using anonymized or synthetic data), and the intellectual property arrangements for any new findings. Such partnerships can lend an additional layer of credibility and objectivity to the assessment of AI systems, especially when direct access to proprietary components is limited.

Moreover, the concept of "trusted execution environments" or "secure enclaves" is emerging as a technological solution to balance privacy and proprietary concerns. These are secure hardware-based environments where sensitive code and data can be processed without being exposed to the host system or external observers. A vendor could potentially run their proprietary algorithm within such an environment, allowing a public sector entity or an auditor to query it and receive explanations or results without ever directly accessing the algorithm's code or the raw data. While still a developing area, this technology offers a promising path for enabling analysis and verification in highly sensitive contexts.

Ultimately, the successful integration of AI into the public sector hinges on fostering a culture of trust and accountability, and this cannot be achieved without a commitment to transparency, even when dealing with proprietary technology. Public sector leaders must be diligent in their procurement processes, drawing clear boundaries around what level of explainability is essential for public good. This involves ongoing dialogue with vendors, a deep understanding of the legal and ethical implications of AI, and a willingness to invest in the expertise needed to scrutinize these complex systems. The aim is not to hinder innovation, but to ensure that innovation serves the public interest, making AI a tool for enhanced governance, not an inscrutable arbiter of public life. This requires a proactive and strategic approach to contractual negotiations, where transparency is not an afterthought, but a foundational requirement built into the very fabric of AI acquisition and deployment. The public's right to understand how decisions affecting them are made, even when enabled by complex, proprietary technology, must be vigorously defended and upheld. This delicate equilibrium between proprietary rights and public accountability is a continuous negotiation, demanding vigilance, adaptability, and a steadfast commitment to the principles of good governance in the digital age.

## Chapter 12: Building AI Literacy in Public Institutions

The escalating integration of artificial intelligence (AI) into the fabric of public services necessitates a significant evolution in the skillsets of public sector employees. It is no longer sufficient for AI implementation to be the sole domain of a select group of IT specialists or data scientists. Instead, a foundational understanding of AI principles, capabilities, and limitations must become a widespread competence, permeating through various departments and hierarchical levels. This growing need for AI competence stems from the multifaceted impact AI is having on how public services are designed, delivered, and overseen. From administrative tasks to complex decision support systems, AI is reshaping the operational landscape, and without a commensurate rise in AI literacy among the workforce, the potential benefits of these technologies risk being unrealized, or worse, misused.

Consider, for instance, the procurement of AI-powered solutions. Without a basic grasp of AI concepts, public officials responsible for acquiring these systems may struggle to articulate specific needs, effectively evaluate vendor proposals, or comprehend the technical specifications and performance metrics presented. This can lead to the selection of suboptimal technologies, the overlooking of critical risks like algorithmic bias, or the signing of contracts that do not adequately address issues of transparency and accountability. A procurement officer who understands the difference between supervised and unsupervised learning, the implications of training data on model performance, or the concept of explainability is far better equipped to engage with vendors and ensure that the acquired AI aligns with public sector goals and ethical standards. They can ask more pertinent questions, challenge unrealistic claims, and negotiate terms that safeguard public interest.

Beyond procurement, the operational deployment and management of AI systems demand a competent workforce. Imagine a scenario where an

AI algorithm is used to prioritize resource allocation for social services. While the algorithm may be sophisticated, frontline workers or middle managers who interact with its outputs need to understand its logic at a functional level. They should be able to recognize when the AI's recommendations might be questionable due to unusual data inputs, or when its predictions deviate significantly from expected outcomes. This requires more than just operating a user interface; it necessitates an understanding of the underlying assumptions the AI makes, its potential failure modes, and its inherent limitations. Without this, there's a risk of over-reliance on AI outputs, leading to flawed decisions, or conversely, a complete mistrust and underutilization of a potentially valuable tool.

Furthermore, the critical function of oversight and governance is severely hampered without widespread AI competence. Elected officials, departmental heads, and internal audit teams are tasked with ensuring that AI systems deployed in the public sector operate fairly, equitably, and in accordance with legal and ethical guidelines. If these individuals lack a fundamental understanding of AI, they cannot effectively question the systems, identify potential harms, or demand appropriate safeguards. They may be unable to discern the nuances of algorithmic bias, the challenges of data privacy in AI, or the complexities of AI explainability. This creates a significant accountability gap, where powerful technologies are deployed without adequate scrutiny, potentially leading to discriminatory outcomes, erosion of public trust, and regulatory non-compliance. A legislator who understands the basics of machine learning can ask more pointed questions about the fairness metrics of an AI used in criminal justice, or the data provenance for an AI deployed in healthcare, thereby strengthening democratic oversight.

The development of AI competence within public institutions should not be viewed as an endeavor to turn every civil servant into an AI engineer.

Rather, it is about cultivating a level of understanding that enables individuals to engage intelligently and critically with AI in their specific roles. For a policy advisor, this might mean understanding how AI can be used to model the impact of proposed legislation or identify emerging trends in public health. For a legal counsel, it might involve grasping the implications of AI-generated evidence or the legal liabilities associated with autonomous systems. For a citizen engagement officer, it could be about understanding how AI-powered chatbots or personalized service delivery platforms work and how they might affect public perception and access. This broad-based literacy ensures that AI is integrated thoughtfully, ethically, and effectively across the entire spectrum of public administration.

The challenge lies in the sheer pace of AI development. Technologies that were once considered cutting-edge are rapidly becoming commonplace, and new advancements emerge with remarkable speed. This necessitates a dynamic and continuous approach to building AI competence, moving beyond one-off training sessions. Public institutions need to establish frameworks for ongoing learning, adapting their training programs to keep pace with technological evolution. This could involve creating internal knowledge-sharing platforms, supporting employees in pursuing relevant certifications, fostering communities of practice, and embedding AI literacy into professional development pathways. The goal is to create an organizational culture where understanding and engaging with AI is seen as a continuous and essential part of professional growth, rather than a discrete, one-time task.

Moreover, the "black box" nature of some advanced AI models presents a particular challenge that increased AI competence can help to mitigate. While the previous context highlighted the difficulties of achieving transparency with proprietary systems, a workforce with a better understanding of AI principles can better navigate these challenges. They can better appreciate the trade-offs between model complexity and explainability, and advocate for solutions that provide appropriate levels of insight, even if full algorithmic disclosure is not feasible. For instance, an employee who understands the concept of feature importance can better interpret the outputs of explanation tools, even if they cannot fully dissect the underlying proprietary model. This nuanced understanding allows for more informed discussions with vendors and more effective internal assessments of AI performance and potential risks.

The integration of AI into public services also has implications for the nature of work itself. As AI automates routine tasks, public servants will increasingly focus on roles that require complex problem-solving, critical thinking, creativity, and interpersonal skills – areas where human capabilities still far surpass those of AI. However, effectively collaborating with AI systems, which will become increasingly common, requires a degree of AI competence. Public servants will need to understand how to leverage AI as a tool, how to interpret its outputs, and how to ensure that AI augments, rather than dictates, human decision-making. This symbiotic relationship between humans and AI is where the future of public service delivery lies, and AI competence is the key to unlocking this potential.

Consider the public sector's responsibility to ensure equitable access to services. AI systems, if not carefully designed and monitored, can inadvertently perpetuate or even amplify existing societal biases. For example, an AI system used for determining eligibility for housing assistance might reflect historical patterns of discrimination if trained on biased data. Without adequate AI competence among those who design, approve, and oversee such systems, these biases can go undetected, leading to unfair outcomes for vulnerable populations. A workforce trained in AI ethics and fairness metrics can proactively identify these risks, demand mitigation strategies, and ensure that AI serves to promote equity rather than undermine it. This requires understanding concepts like algorithmic fairness, bias detection techniques, and the importance of diverse and representative training data.

The adoption of AI also introduces new categories of risk that public institutions must manage. These range from cybersecurity vulnerabilities in AI systems to the potential for AI to be used for malicious purposes, such as sophisticated disinformation campaigns or automated cyberattacks. A well-trained public sector workforce can better anticipate and defend against these risks. They can understand the unique threat vectors associated with AI technologies, implement appropriate security protocols, and develop robust incident response plans. This proactive risk management is essential for maintaining public trust and ensuring the resilience of critical public services in an increasingly AI-driven world.

Furthermore, the public discourse surrounding AI is often characterized by both hype and fear, fueled by a lack of widespread understanding. This can make it challenging for public institutions to engage effectively with citizens about the use of AI in public services. A public servant who possesses AI competence can act as a bridge, translating complex technical concepts into understandable terms for the public. They can explain how AI is being used, what its benefits and limitations are, and what safeguards are in place. This enhances transparency, builds public trust, and fosters informed democratic participation in decisions about technology adoption. Effective communication about AI is itself a demonstration of competence, reassuring citizens that their government is capable of managing these powerful tools responsibly.

The development of AI competence within public institutions is not merely about upskilling; it is about fostering a cultural shift towards data-informed and technologically aware governance. It involves creating an environment where curiosity about AI is encouraged, where experimentation with new tools is supported (within ethical boundaries), and where learning from both successes and failures is a norm. This cultural transformation is essential for public institutions to remain relevant, effective, and trusted in the 21st century. Without it, they risk falling behind, unable to harness the transformative potential of AI to improve public services, enhance efficiency, and address the complex challenges facing society. The journey towards becoming an AI-literate public sector is an ongoing one, requiring sustained investment, strategic planning, and a deep commitment to empowering public servants with the knowledge and skills they need to navigate the AI era. This foundational competence is not a luxury; it is a prerequisite for effective and responsible public administration in the age of artificial intelligence.

Crafting an AI training program for the public sector is fundamentally different from designing one for private industry. While private companies might focus on driving profit margins or competitive advantage, public institutions are bound by a mandate of public service, equity, accountability, and transparency. These core values must be woven into the very fabric of any AI training initiative. The goal is not merely to impart technical knowledge, but to cultivate a mindset of responsible innovation and critical engagement. Therefore, designing effective AI training programs requires a deliberate and thoughtful approach, moving beyond generic workshops to create learning experiences that are deeply relevant, practically applicable, and ethically grounded.

The cornerstone of any effective training program is a well-defined curriculum. For public sector AI literacy, this curriculum must strike a delicate balance between breadth and depth, catering to a diverse audience with varying levels of existing technical expertise. It should begin with foundational concepts, demystifying terms like machine learning, deep learning, natural language processing, and computer vision. This foundational layer is crucial for creating a common language and ensuring that all participants, regardless of their role, grasp the basic principles of how AI operates. This might involve explaining the distinction between supervised, unsupervised, and reinforcement learning through relatable public service examples. For instance, supervised learning could be illustrated by an AI model trained to classify citizen service requests, while unsupervised learning might be demonstrated in its ability to identify patterns in public health data without pre-defined categories.

Beyond the fundamental "what," the curriculum must address the "how" and "why" of AI implementation in the public sector. This involves exploring the practical applications of AI across different government functions. For administrative tasks, training could cover AI-powered tools for document processing, automated form filling, or intelligent scheduling, emphasizing efficiency gains while also highlighting potential pitfalls like data entry errors. In service delivery, modules could focus on AI chatbots for citizen inquiries, predictive analytics for resource allocation, or personalized recommendations for public services, with a strong emphasis on accessibility and inclusivity. For policy development and oversight, training might delve into AI for data analysis, trend identification, simulation modeling, and risk assessment, always underscoring the importance of human judgment and ethical considerations in interpreting AI-generated insights.

A critical component of any public sector AI curriculum must be dedicated to the ethical dimensions of AI. This is non-negotiable. Training sessions should thoroughly explore topics such as algorithmic bias, data privacy, transparency, accountability, and fairness. Participants need to understand how biases can creep into AI systems through training data, algorithm design, or human interaction, and the profound implications this can have on equitable service delivery. Case studies are invaluable here, presenting real-world scenarios of AI systems that have either perpetuated or mitigated societal inequalities. For example, examining an AI used in the justice system that inadvertently discriminates against certain demographic groups, and contrasting it with an AI designed with fairness metrics to ensure equitable outcomes, provides powerful lessons. Discussions should also cover the legal and regulatory landscape surrounding AI, including data protection laws and emerging AI governance frameworks.

The pedagogical approach employed in these training programs is as important as the content itself. A one-size-fits-all lecture format is unlikely to be effective for a diverse public sector workforce. Instead, a blended learning approach that combines various methods is often most successful. This could include:

**Interactive Workshops:** These are essential for hands-on learning and collaborative problem-solving. Workshops can be designed around specific use cases, allowing participants to work in teams to analyze a problem, identify potential AI solutions, discuss ethical considerations, and even engage with simplified AI tools or simulations. This experiential learning fosters deeper understanding and retention. For instance, a workshop might task participants with identifying bias in a simulated dataset used for allocating public housing, prompting them to consider different fairness metrics and mitigation strategies.

**Case Study Analysis:** As mentioned earlier, real-world and hypothetical case studies are powerful tools for illustrating complex concepts and ethical dilemmas. These should be drawn from the public sector context, showcasing both successes and failures. Participants can be asked to analyze the situation, identify the AI technologies involved, assess the ethical implications, and propose solutions or best practices. This encourages critical thinking and the application of learned principles to practical scenarios.

**Guest Speakers and Expert Panels:** Inviting AI experts, ethicists, and public sector professionals who have successfully (or unsuccessfully) implemented AI can provide invaluable real-world insights. Panels can facilitate discussions on current challenges, emerging trends, and practical advice, allowing participants to ask questions and engage with diverse perspectives. This human element can make abstract concepts more relatable and inspire confidence.

**Online Learning Modules:** For foundational knowledge and self-paced learning, well-designed online modules can be highly effective. These can cover core AI concepts, terminology, and ethical principles. These modules should incorporate interactive elements such as quizzes, short videos, and knowledge checks to maintain engagement. They also offer flexibility for employees to learn at their own pace and revisit material as needed.

**"Lunch and Learn" Sessions:** These informal sessions can be used to introduce new AI developments, discuss specific AI tools, or delve into niche topics in a low-pressure environment. They are excellent for fostering a culture of continuous learning and for sparking curiosity among staff who may not have direct involvement in AI projects.

**Mentorship and Communities of Practice:** Establishing formal or informal mentorship programs where more experienced staff can guide those newer to AI can be very beneficial. Similarly, creating communities of practice – groups of employees with a shared interest in AI who meet regularly to share knowledge, discuss challenges, and collaborate – can foster organic learning and problem-solving across departments.

The challenge of engaging a diverse workforce cannot be overstated. Public institutions employ individuals with a vast spectrum of backgrounds, from administrative staff and frontline service providers to policy analysts and senior leadership. An effective AI training program must acknowledge and cater to these differences. This might involve creating tiered training paths:

**Foundational Literacy for All:** A mandatory baseline training program for all employees, covering the fundamental concepts, ethical considerations, and the general impact of AI on public services. This ensures a common understanding and fosters a general awareness across the organization.

**Role-Specific Training:** For employees whose roles involve direct interaction with AI systems or decision-making based on AI outputs, more specialized training is needed. This could focus on understanding the specific AI tools they use, interpreting their results, identifying potential errors or biases, and knowing when to escalate issues or apply human judgment. For example, a customer service representative using an AI-powered knowledge base would receive training on how to effectively query the system, evaluate the relevance of its responses, and handle situations where the AI falters.

**Technical and Leadership Tracks:** For IT professionals, data scientists, and senior leaders involved in AI procurement, development, or governance, more in-depth training is required. This might cover AI architecture, advanced machine learning techniques, AI strategy development, risk management frameworks, and the ethical implications of large-scale AI deployment. Leaders need to understand how to foster an AI-ready culture, allocate resources effectively, and ensure that AI initiatives align with organizational goals and public values.

Crucially, training programs should be designed to be practical and immediately applicable. Abstract theoretical discussions should be grounded in concrete examples relevant to the daily work of public servants. This means using datasets and scenarios that mirror those they encounter in their roles. Instead of discussing generic AI algorithms, a training session for a housing department might focus on an AI that helps predict which properties are most likely to require maintenance, and how to interpret its recommendations for proactive intervention. This direct relevance significantly enhances engagement and the perceived value of the training.

Furthermore, fostering a culture that supports continuous learning is paramount. AI is not a static field; it is constantly evolving. Training programs should not be viewed as a one-time event but as an ongoing process. This requires institutions to invest in mechanisms for staying current, such as subscribing to relevant journals, attending conferences, supporting employees in pursuing certifications, and regularly updating training materials to reflect new advancements and best practices. Creating internal knowledge-sharing platforms, establishing AI "champions" within departments, and integrating AI literacy into performance review and professional development frameworks can help institutionalize this continuous learning.

The design of AI training programs also needs to consider the human element of technology adoption. Introducing AI can sometimes be met with apprehension, fear, or resistance from employees concerned about job security or the complexity of new systems. Effective training can proactively address these concerns by emphasizing how AI can augment human capabilities, automate tedious tasks, and free up employees to focus on more meaningful and strategic work. Open communication about the goals and benefits of AI implementation, coupled with robust training and support, can help build trust and encourage buy-in. The training itself should be delivered in a way that is supportive and empowering, rather than intimidating. Trainers should be skilled not only in AI but also in adult learning principles and in managing group dynamics.

The accessibility of training materials and delivery methods is another critical design consideration. Public sector employees represent a wide range of abilities and learning preferences. Materials should be available in various formats - written, audio, visual - and delivered through accessible platforms. This ensures that employees with disabilities or different learning styles can fully participate and benefit from the training. For instance, providing transcripts for video lectures, ensuring keyboard navigability for online modules, and offering options for closed captioning are essential steps towards inclusivity.

Finally, the effectiveness of AI training programs must be continuously evaluated and iterated upon. Feedback mechanisms should be built into the training design, allowing participants to provide input on the content, delivery, and relevance of the programs. This feedback, combined with metrics on knowledge retention, application of learning in the workplace, and overall impact on AI adoption and ethical practice, should be used to refine and improve future training initiatives. This iterative approach ensures that the training remains relevant, impactful, and aligned with the evolving needs of the public institution and the dynamic landscape of AI. The ultimate aim is to build a workforce that is not only knowledgeable about AI but is also confident, critical, and capable of harnessing its potential for the public good, while rigorously safeguarding against its risks. This requires a commitment to designing and delivering training that is as dynamic, adaptable, and ethically considered as the technology itself.

Fostering a culture of responsible AI use within public institutions is not merely an add-on to effective training; it is the bedrock upon which all AI initiatives must be built and sustained. It transcends the technical aspects of AI development and deployment, delving into the deeply human and organizational elements that determine whether AI serves the public good or inadvertently causes harm. This involves actively nurturing an environment where ethical considerations are not an afterthought but are integrated into the very DNA of an organization's engagement with artificial intelligence. Such a culture empowers individuals, strengthens governance, and ultimately ensures that AI is wielded as a tool for progress and equity.

One of the most potent mechanisms for cultivating this culture is the promotion of open and honest dialogue about the ethical complexities inherent in AI. This means creating safe spaces where public servants at all levels can voice their concerns, ask probing questions, and engage in robust discussions about the potential impacts of AI technologies. These dialogues should not be confined to specialized ethics committees or IT departments; they need to permeate throughout the organization. Consider, for instance, the implementation of an AI-powered system designed to optimize the allocation of social housing. While the intent might be to improve efficiency and fairness, frontline workers and community liaisons might raise critical concerns about how the algorithm interprets various socio-economic factors, or whether it inadvertently disadvantages applicants with complex, non-standard circumstances. A culture of responsible AI use would ensure that these voices are not only heard but are actively sought out and valued, leading to adjustments in the system's design or deployment before potential inequities manifest. This can be facilitated through regular town hall meetings dedicated to AI ethics, dedicated forums on internal communication platforms, or even structured "ethical hackathons" where teams are tasked with identifying potential ethical blind spots in proposed AI applications. The key is to make these conversations accessible, ongoing, and genuinely influential in decision-making processes.

Beyond open dialogue, a crucial element is the establishment of clear and accessible channels for reporting concerns related to AI. Public servants should feel empowered and protected when they identify potential ethical breaches, biases, or unintended consequences of AI systems. This requires a robust whistleblowing or grievance mechanism that is specifically tailored to AI-related issues. Such a system must guarantee anonymity, prohibit retaliation, and ensure that reported concerns are thoroughly investigated by an impartial body. Imagine a scenario where an AI tool used for predictive policing is deployed, and a police officer notices that it consistently flags individuals from specific neighborhoods, raising suspicions of profiling. Without a secure reporting channel, this concern might go unaddressed, perpetuating systemic bias. Conversely, a well-established reporting system would allow the officer to raise this issue without fear, triggering an independent review that could lead to the algorithm's recalibration or reassessment. The transparency of the investigation process and the communication of outcomes are also vital, not only for addressing the immediate concern but also for building trust and reinforcing the organization's commitment to responsible AI.

Embedding ethical considerations directly into performance reviews and project management frameworks is another foundational step. When AI is part of an employee's responsibilities, their adherence to ethical guidelines and their proactive engagement with AI governance principles should be a recognized aspect of their performance. This signals that responsible AI use is not an optional activity but a core competency. For project managers overseeing AI initiatives, this translates into making ethical risk assessments a mandatory component of project planning, execution, and evaluation. This means explicitly including sections in project proposals that detail potential ethical challenges, proposed mitigation strategies, and metrics for assessing fairness, accountability, and transparency. Regular project reviews should include a dedicated agenda item to discuss the ethical implications of the AI system as it evolves. For example, a project team developing an AI chatbot for citizen services would be required to document its training data's potential biases, outline its privacy protection measures, and establish protocols for handling sensitive queries. Their performance could then be evaluated not just on the chatbot's efficiency or user satisfaction, but also on their diligence in addressing these ethical dimensions. This systematic integration ensures that ethical diligence is rewarded and becomes an intrinsic part of how projects are conceived and managed.

The establishment of clear ethical guidelines and frameworks specific to AI is also paramount. While general organizational values and codes of conduct provide a foundation, AI presents unique challenges that necessitate more granular guidance. These guidelines should articulate the principles that govern the development, procurement, deployment, and use of AI within the public sector. They should address issues such as data privacy, algorithmic fairness, transparency, accountability, and human oversight. For instance, a guideline might stipulate that any AI system making decisions that significantly impact individuals must include a mechanism for human review and appeal. Another might mandate the regular auditing of AI models for bias, particularly those used in high-stakes areas like healthcare or employment. These guidelines should be developed collaboratively, drawing input from a diverse range of stakeholders, including AI experts, ethicists, legal counsel, and, crucially, representatives from the communities that will be affected by the AI. Once established, these guidelines must be widely disseminated and incorporated into all relevant policies and procedures. They serve as a compass, guiding decision-making and providing a benchmark against which AI activities can be measured.

Beyond formal policies, the active leadership of senior management is indispensable in shaping a culture of responsible AI use. When leaders champion ethical AI, demonstrate a commitment to transparency, and prioritize fairness in AI deployments, it sends a powerful message throughout the organization. This can involve publicly endorsing AI ethics principles, allocating resources for ethical AI research and training, and holding teams accountable for adhering to responsible AI practices. Leaders should also be visible participants in discussions about AI ethics, showcasing their own engagement with these complex issues. For example, a mayor or agency head might issue a public statement outlining their commitment to using AI equitably, or host a series of public consultations on an upcoming AI deployment. Their willingness to engage, to ask difficult questions, and to prioritize public trust over expediency sets a crucial tone. This visible commitment from the top filters down, encouraging employees at all levels to take responsible AI use seriously.

Creating an environment that encourages continuous learning and adaptation is also fundamental. The field of AI is rapidly evolving, and so too are the ethical challenges it presents. A culture of responsible AI use must therefore be one that embraces ongoing education and reflection. This means not only providing initial training but also offering opportunities for continuous professional development, encouraging the sharing of best practices and lessons learned, and fostering a mindset of critical inquiry. This could manifest in the form of internal newsletters highlighting new AI ethics research, regular "lessons learned" sessions following AI project deployments, or even the establishment of internal "AI ethics champions" who can act as resources and facilitators within different departments. The goal is to move beyond a static set of rules to a dynamic and adaptive approach that can respond to emerging challenges and opportunities.

Furthermore, a culture of responsible AI use necessitates a commitment to transparency in AI deployments, wherever feasible and appropriate. While proprietary algorithms and sensitive data may present limitations, efforts should be made to inform the public and internal stakeholders about how AI systems are being used, what data they rely on, and what their intended outcomes are. This transparency can take many forms, from publishing plain-language explanations of AI systems on an agency's website to providing clear notifications to individuals when they are interacting with an AI-powered service. For example, if a public transit agency uses AI to optimize bus routes, they could publish information about the factors considered by the algorithm and the data used to train it, along with the expected benefits and any potential trade-offs. This proactive disclosure builds public trust and allows for greater scrutiny and feedback. It also empowers citizens by providing them with a better understanding of how decisions that affect them are being made.

The iterative nature of AI development also demands a culture that embraces evaluation and refinement. AI systems are rarely perfect upon initial deployment; they require ongoing monitoring, assessment, and adjustment. A responsible AI culture embeds this iterative process into its operational fabric. This means establishing robust mechanisms for collecting feedback on AI performance, including user feedback, operational data, and ethical impact assessments. These evaluations should be conducted regularly and systematically, with a clear process for using the findings to improve the AI system and its governance. For instance, an AI tool designed to assist in identifying potential fraudulent claims could be regularly audited to ensure it is not disproportionately flagging legitimate claims from certain demographic groups. Any identified discrepancies would trigger an investigation and subsequent adjustments to the algorithm or its operational parameters. This commitment to continuous improvement ensures that AI systems remain aligned with their intended purposes and public values, rather than becoming ossified and potentially harmful over time.

Ultimately, fostering a culture of responsible AI use is an ongoing journey, not a destination. It requires sustained commitment, continuous learning, and a willingness to adapt. It is about embedding ethical considerations into the very fabric of public sector operations, ensuring that AI is leveraged not just for efficiency or innovation, but for the betterment of society, with fairness, equity, and accountability at its core. This cultural shift is essential for public institutions to navigate the complexities of artificial intelligence with integrity and to harness its transformative potential for the public good.

**Empowering Frontline Staff and Decision-Makers**

The most sophisticated AI systems, no matter how brilliantly designed, are ultimately tools. And like any tool, their effectiveness, safety, and ethical application depend heavily on the individuals who wield them. In the context of public institutions, this means that a significant portion of the effort towards building AI literacy must be directed towards those on the front lines – the public servants who directly interact with citizens and make daily operational decisions – and the decision-makers who authorize, procure, and oversee the implementation of AI technologies. Without their informed engagement, even the most ethically sound AI can be misused, misinterpreted, or simply fail to achieve its intended beneficial outcomes.

For frontline staff, their daily work often involves navigating complex human situations, interpreting diverse needs, and applying nuanced judgment. When AI systems are introduced into these workflows, it's crucial that these staff members don't become passive recipients of AI-generated outputs. Instead, they need to be empowered with a practical understanding of what the AI is doing, why it's doing it, and importantly, when it might be wrong. This involves demystifying AI to a level that is accessible and relevant to their roles. For example, an immigration caseworker using an AI-powered tool to flag applications for further scrutiny needs to understand not just that the system has flagged an application, but also the general criteria it used. Is it looking for missing documents? Inconsistencies in dates? Or is it, perhaps, relying on patterns that might inadvertently correlate with an applicant's country of origin? A basic AI literacy for this caseworker would equip them to ask these questions, to recognize when an AI's output seems counterintuitive or potentially biased, and to know that their own judgment and human review are paramount in such instances. The AI should serve as an assistant, not an unquestioned authority. Therefore, training should focus on developing a critical awareness of AI's probabilistic nature; that it provides likelihoods and patterns, not absolute truths. This includes understanding the concept of confidence scores, the potential for false positives and false negatives, and the limitations of the data the AI was trained on. For instance, a social worker using an AI to predict child welfare risk factors must understand that the AI's prediction is based on historical data, which may contain societal biases, and that a high-risk score does not equate to a definitive conclusion. Their training should emphasize that their lived experience, understanding of the family's context, and direct observations are irreplaceable in making the final determination.

This practical literacy extends to understanding the AI's interface and how to interact with it effectively. This might involve learning how to input data accurately, how to interpret the visual cues or confidence levels provided by the system, and how to escalate issues or seek clarification when the AI's output is unclear or problematic. The goal is to foster a sense of agency, not apprehension, when engaging with these new technologies. Consider a public transit dispatcher using an AI to optimize bus routes in real-time. Their literacy would involve understanding how the AI processes traffic data, passenger demand, and vehicle availability, but also knowing what to do if the AI proposes a route that seems illogical, perhaps due to an unusual event not yet reflected in the data, or if it consistently overloads certain buses. Empowering them means giving them the mandate and the knowledge to override the AI when their on-the-ground expertise suggests a better course of action, and crucially, providing a mechanism for them to report these instances so the AI can be improved. This feedback loop is vital for the continuous refinement of AI systems, turning frontline users into active participants in the AI's evolution rather than passive observers.

For decision-makers – those who approve budgets, select vendors, and set strategic direction – AI literacy takes on a different, but equally critical, dimension. They need to understand AI at a strategic level, grasping its potential to transform services, but also its inherent risks and ethical implications. This doesn't necessarily require them to be able to code or understand complex algorithms, but rather to comprehend the fundamental concepts and to ask the right questions. When presented with a proposal for an AI-powered system, a literate decision-maker would inquire about the system's intended purpose, the data it will use, the metrics for success, and, crucially, the safeguards against bias and error. They would understand that "off-the-shelf" AI solutions might not be suitable for unique public sector contexts and that customization and thorough validation are often necessary.

A key aspect of decision-maker literacy is understanding the procurement process for AI. This means recognizing that AI systems are not commodities but complex technological and ethical packages. They need to be able to evaluate vendor claims, understand the importance of data provenance and quality, and assess the vendor's commitment to transparency, explainability, and ongoing support. For instance, when considering an AI tool for fraud detection in tax administration, a decision-maker would need to understand the potential for disparate impact on certain taxpayer groups based on the AI's training data. They would need to ask how the vendor has addressed potential biases, what mechanisms are in place for auditing the AI's performance, and what recourse taxpayers have if they believe they have been unfairly targeted by the system. The ability to articulate these concerns and demand satisfactory answers is a hallmark of AI-literate decision-making.

Furthermore, decision-makers must grasp the concept of accountability in AI. Who is responsible when an AI system makes a harmful decision? Is it the developer, the vendor, the procuring agency, or the end-user? AI literacy for decision-makers means understanding that clear lines of accountability must be established
*before* deployment. This involves understanding the role of human oversight – that AI should augment, not replace, human judgment in critical decision-making processes. A decision-maker who understands this would champion the inclusion of "human-in-the-loop" or "human-on-the-loop" mechanisms for high-stakes AI applications, ensuring that a human has the final say or the ability to intervene and correct the AI. For example, an AI system designed to help parole boards assess recidivism risk must have its recommendations subject to rigorous human review, with the parole board members empowered to consider factors beyond the AI's algorithmic assessment.

The training and development programs designed to foster this literacy must be tailored to the specific needs and roles of these different groups. For frontline staff, training should be hands-on, scenario-based, and focused on practical application and critical thinking. It should use plain language, avoid overly technical jargon, and directly address the AI tools they will encounter in their daily work. For decision-makers, training might be more strategic, focusing on policy implications, risk management, ethical frameworks, and the broader societal impact of AI. It should equip them to ask the probing questions that ensure AI is implemented responsibly and ethically.

The concept of "explainability" or "interpretability" in AI is also a crucial element for both groups. Frontline staff need to understand, at some level, why the AI is suggesting a particular course of action, especially when it contradicts their own intuition or experience. This understanding allows them to trust the AI when it's correct, and to question it when it's not. For example, an AI that helps prioritize emergency service dispatch might flag a certain call as lower priority. If the dispatcher understands that the AI is primarily weighting factors like reported severity and caller calmness, but the dispatcher has additional context from prior interactions or local knowledge suggesting a more urgent situation, they can then make an informed decision to escalate. Decision-makers, on the other hand, need to understand the implications of explainability for transparency, accountability, and fairness. They need to advocate for AI systems where the decision-making process can be understood and audited, particularly in areas affecting citizens' rights and access to services. This might involve prioritizing AI models that offer greater transparency, even if they are marginally less performant than "black box" alternatives.

Ultimately, empowering frontline staff and decision-makers is about building a shared understanding and a common language around AI. It's about ensuring that everyone involved in the public sector's engagement with AI, from the individual interacting with a citizen to the minister signing off on a procurement, has the necessary awareness to act responsibly. This is not a one-time training event, but an ongoing process of learning, adaptation, and dialogue, ensuring that public institutions can harness the power of AI while remaining grounded in their core values of public service, fairness, and accountability. The successful integration of AI into public services hinges on this human element – the educated, empowered, and ethically-minded individuals who guide its use.

The foundational imperative for any public institution embarking on the journey of AI integration lies in cultivating robust AI literacy among its policymakers and senior leadership. This cohort occupies a unique and pivotal position; they are the architects of strategy, the custodians of public trust, and the ultimate arbiters of resource allocation and regulatory direction. Their understanding of artificial intelligence—its capabilities, its limitations, and its profound societal implications—is not merely beneficial; it is indispensable for ensuring that the deployment of AI within the public sector is not only effective but also ethically sound, equitable, and aligned with the overarching societal goals and values that such institutions are sworn to uphold. Without this strategic AI literacy at the highest levels, the risk of misaligned investments, unintended negative consequences, and a failure to truly harness AI's potential for public good becomes alarmingly high.

For policymakers and senior leaders, AI literacy translates into a nuanced comprehension of the broader landscape of artificial intelligence. This does not necessitate an in-depth technical proficiency in algorithm design or machine learning model development. Rather, it demands a strategic grasp of AI's fundamental principles, its transformative potential across various public service domains, and, critically, its inherent risks and ethical challenges. When presented with proposals for AI adoption, a leader equipped with this literacy would possess the acumen to ask incisive questions that probe beyond superficial claims of innovation. Such questions might include: What specific public problem is this AI solution intended to address, and is AI genuinely the most appropriate or effective tool for that purpose? What are the underlying data sources and their provenance, and how have potential biases within that data been identified and mitigated? What are the measurable objectives and key performance indicators for this AI deployment, and how will success be objectively evaluated, particularly in terms of fairness and equity? Crucially, what mechanisms for oversight, accountability, and redress are being established to ensure responsible use and to protect citizens' rights? This level of inquiry ensures that AI is not adopted as a panacea but as a carefully considered instrument, integrated with a clear understanding of its impact.

The strategic allocation of resources is a core responsibility of policymakers and senior leadership, and AI literacy directly informs this function. Understanding the distinct costs and benefits associated with different types of AI applications—from automation of routine tasks to sophisticated predictive analytics—allows for more judicious investment decisions. A literate leader would be able to discern between genuine, high-impact AI initiatives and speculative ventures, ensuring that public funds are directed towards solutions that offer demonstrable public value and align with strategic priorities. This also extends to understanding the long-term investment required for AI, which includes not only the initial procurement but also ongoing data management, system maintenance, talent development, and continuous evaluation. For example, a government department considering an AI system to streamline the processing of social welfare claims must understand that the initial purchase price of the software is only a fraction of the total cost. They must also budget for the expert personnel needed to manage the system, ensure the quality of the data fed into it, and periodically audit its performance for fairness and accuracy. Without this holistic understanding, projects can falter due to unforeseen budget shortfalls, leading to either incomplete implementation or the deployment of underperforming systems.

Furthermore, AI literacy empowers policymakers to shape the regulatory and legal frameworks that govern AI's use within the public sector. This includes understanding the existing legal landscape, identifying gaps that require new legislation or policy guidance, and anticipating future challenges. It means grappling with complex issues such as data privacy, algorithmic transparency, algorithmic accountability, and the potential for AI to exacerbate existing societal inequalities. A well-informed leader will advocate for regulations that strike a balance between fostering innovation and safeguarding public interest. This could involve developing clear guidelines for the ethical procurement of AI systems, establishing standards for AI system testing and validation before deployment, and creating robust mechanisms for citizen recourse when AI systems lead to adverse outcomes. For instance, when a public health agency considers using AI to identify populations at higher risk for certain diseases, policymakers need to understand the ethical implications of such predictive models. They must ensure that the AI's predictions do not lead to stigmatization or discrimination against identified groups and that individuals are informed about how their data is being used. This requires proactive engagement with AI ethics experts and legal counsel to craft policies that are both forward-thinking and grounded in principles of fairness and justice.

The role of senior leadership in fostering an organizational culture that embraces responsible AI adoption cannot be overstated. This involves championing AI literacy across all levels of the institution, not just among those directly involved in AI development or deployment. Leaders must set the tone by demonstrating their own commitment to understanding AI and by encouraging open dialogue about its ethical implications. This might involve establishing internal working groups focused on AI ethics, commissioning independent reviews of AI systems, or integrating AI literacy objectives into professional development programs for all staff. A leadership team that actively promotes a culture of critical inquiry around AI encourages employees to identify potential risks and voice concerns without fear of reprisal. This proactive approach is far more effective than reactive measures taken after an AI system has caused harm. For example, a city council that publicly endorses the principles of algorithmic fairness and transparency, and actively engages community members in discussions about AI deployments that affect them, is far more likely to build public trust and ensure that AI serves the community equitably.

A critical aspect of AI literacy for policymakers and leaders is understanding the concept of "explainability" or "interpretability" in AI. While not every leader needs to understand the intricacies of specific AI models, they must grasp why understanding *how* an AI arrives at a decision is crucial, especially in public sector contexts where decisions can have significant impacts on individuals' lives. For instance, if an AI system is used to determine eligibility for certain social benefits, policymakers must understand the importance of being able to explain to an applicant why their application was denied, and whether an AI played a role in that decision. Black-box algorithms, where the decision-making process is opaque, pose significant challenges to accountability and fairness. Leaders need to advocate for AI systems where the rationale behind decisions can be understood, audited, and challenged. This understanding informs procurement decisions, pushing vendors to prioritize explainable AI (XAI) techniques, and guides the development of internal policies that mandate transparency where feasible. This is not just a technical requirement but a fundamental aspect of good governance and public accountability.

Moreover, policymakers and leaders must be cognizant of the evolving nature of AI and the continuous learning it necessitates. AI technology is not static; it is a rapidly developing field. What is state-of-the-art today may be outdated tomorrow. Therefore, AI literacy at the leadership level must be understood as an ongoing commitment, not a one-time training exercise. Institutions must invest in continuous learning opportunities for their leadership, ensuring they stay abreast of new developments, emerging ethical considerations, and best practices in AI governance. This could involve participation in relevant conferences, workshops, executive education programs, and engaging with external AI ethics bodies. By demonstrating this commitment to ongoing learning, leaders signal the importance of AI literacy throughout the organization and ensure that the institution remains agile and responsive to the challenges and opportunities presented by AI. This proactive stance is essential for maintaining public trust and for effectively navigating the complex ethical terrain of AI in the public sphere. The ability to adapt, learn, and guide the institution through the dynamic AI landscape is, therefore, a defining characteristic of AI-literate public sector leadership.

# Chapter 13: The Field Project: AI Triage in Healthcare

The relentless demand on public healthcare systems, particularly in underserved communities, presents a persistent and multifaceted challenge. Patients often face daunting wait times, from the moment they decide to seek medical attention to actually receiving it. This delay is not merely an inconvenience; it can have significant consequences for health outcomes, turning minor ailments into chronic conditions and delaying critical interventions for serious illnesses. The sheer volume of individuals seeking care, coupled with the finite resources of healthcare professionals and facilities, creates a bottleneck that strains the entire system. This strain manifests in overcrowded waiting rooms, exhausted medical staff, and, ultimately, a diminished quality of care. In many urban and rural areas alike, the accessibility of healthcare is hampered by geographical barriers, a shortage of specialized medical personnel, and the economic burdens associated with seeking treatment. The result is a system where timely access to appropriate care becomes a privilege rather than a universal right.

Consider a large metropolitan area grappling with precisely these issues. Within this sprawling urban landscape existed a network of public health clinics serving a diverse and often vulnerable population. These clinics were the primary point of care for many, offering essential services ranging from primary care and chronic disease management to emergency consultations for non-life-threatening conditions. However, the clinics were perpetually under siege. Daily, hundreds of individuals would present themselves, each with a unique set of symptoms and needs. The intake process, largely manual, involved administrative staff meticulously recording patient information, understanding the presenting complaint, and attempting to gauge the urgency of each case. This triage process, while handled with dedication by experienced nurses and healthcare assistants, was inherently subjective and prone to variability. It relied heavily on individual judgment, which, under immense pressure, could lead to inconsistencies. Patients with less obvious but potentially serious conditions might be overlooked, while those with minor issues could inadvertently occupy valuable clinician time, further exacerbating waitlists for those with more pressing needs. The system's limitations were starkly visible: patients languished in waiting areas for hours, some expressing growing anxiety, others succumbing to frustration. The clinical teams, meanwhile, found themselves constantly battling the tide, struggling to provide the focused, high-quality care that each patient deserved. The efficiency of the entire system was compromised, leading to burnout among staff and a palpable decline in patient satisfaction. This was the environment that underscored the urgent need for an innovative approach to manage patient flow and prioritize care effectively.

It was within this demanding milieu that the concept for "Project Genesis" began to take shape. The overarching goal was ambitious yet essential: to fundamentally re-engineer the patient intake and routing process within public healthcare settings, thereby improving healthcare access and the quality of care delivered. The project recognized that the initial point of contact—the triage—was a critical juncture. Optimizing this stage, it was theorized, could create a ripple effect of positive change throughout the entire patient journey. The vision for Project Genesis was to develop and implement an intelligent triage tool, powered by artificial intelligence, that could assist healthcare professionals in rapidly and accurately assessing patient needs, directing them to the most appropriate level of care, and thereby alleviating the systemic pressures. The project was not conceived as a replacement for human clinicians, but rather as a sophisticated augmentative tool, designed to empower them with more information, streamline workflows, and enable more equitable distribution of resources. The core problems it aimed to address were multi-pronged: reducing the excessive wait times that characterized the existing system, ensuring that patients were seen by the right specialist or clinician without unnecessary delays, and ultimately enhancing the overall efficiency and effectiveness of public healthcare delivery.

The genesis of Project Genesis was rooted in the tangible and often heart-wrenching realities faced by both patients and providers in under-resourced public healthcare facilities. The problem was not a lack of dedication or expertise, but a systemic deficit in the ability to manage overwhelming demand with finite resources. Anecdotal evidence abounded of patients presenting with early symptoms of serious conditions, only to be sent home with advice to return if symptoms worsened, by which time the disease had progressed significantly. Conversely, individuals with minor complaints, such as a common cold or a routine prescription refill, could tie up clinical resources that were desperately needed for more acute cases. This inefficiency bred a culture of constant pressure. Nurses and doctors often described feeling like they were constantly in crisis mode, unable to dedicate sufficient time to individual patient assessments, leading to a gnawing sense of professional dissatisfaction and the ever-present fear of missing something critical.

The logistical hurdles were also immense. The manual intake process involved a significant amount of administrative overhead. Gathering patient histories, verifying insurance information, and documenting chief complaints consumed valuable staff time that could otherwise be spent in direct patient care. Furthermore, the subjective nature of traditional triage meant that the urgency assigned to a case could vary depending on the individual performing the assessment and the context of the moment. While experienced professionals developed a keen intuition, the sheer volume of patients and the stress of the environment could still lead to variations in judgment. This lack of standardization contributed to perceptions of unfairness among patients, who often felt their needs were not being adequately prioritized.

The project team, comprising public health officials, clinicians, data scientists, and ethicists, embarked on an intensive discovery phase. This involved extensive site visits to various clinics, in-depth interviews with healthcare staff at all levels—from receptionists and nurses to physicians and administrators—and detailed observations of patient flow from arrival to departure. They meticulously documented the existing workflows, identified key pain points, and quantified the extent of the problem. Data was collected on average wait times, the types of conditions presenting at the clinics, the allocation of clinical resources, and patient satisfaction scores. The findings were sobering: average wait times for non-emergency consultations often exceeded three hours, with some patients waiting upwards of six hours. A significant percentage of patient visits, estimated to be around 30%, were for conditions that could potentially have been managed with self-care advice or by a less specialized healthcare provider, yet they occupied the time of physicians. This data painting a clear picture of a system under immense strain, where resources were not being optimally deployed.

The core mission of Project Genesis crystallized around several key objectives. Foremost among these was the
**reduction of patient wait times**. By intelligently pre-assessing and routing patients, the system aimed to significantly decrease the time individuals spent waiting for initial consultations and subsequent treatments. Secondly, the project sought to **improve the accuracy and consistency of triage**. Leveraging AI, the tool was designed to provide a more objective and data-driven assessment of patient acuity, reducing the variability inherent in manual triage and ensuring that the most critically ill patients received immediate attention. A third crucial

objective was to **optimize the allocation of clinical resources**. By directing patients to the most appropriate level of care—whether that be a physician, a nurse practitioner, a specialized clinic, or even a telehealth consultation—the AI tool could help ensure that clinician time was utilized most effectively, preserving it for those who truly required it. Furthermore, the project aimed to **enhance patient satisfaction** by providing a more transparent, efficient, and equitable care experience. Finally, and of paramount importance, was the commitment to ensuring that the AI system would operate within a robust ethical framework, prioritizing **patient safety, data privacy, and algorithmic fairness**.

The envisioned AI triage tool was designed to be more than just a symptom checker. It was conceptualized as an integrated system that would work in concert with existing healthcare infrastructure and human expertise. The initial interaction would typically involve a patient (or their caregiver) using a user-friendly interface, accessible via a kiosk in the clinic, a dedicated mobile application, or even a web portal. This interface would guide the individual through a series of questions related to their symptoms, medical history, and current health status. These questions would be dynamic, adapting based on the patient's responses, much like an experienced clinician would probe further based on initial information. The AI would process this input, drawing upon a vast, curated knowledge base of medical conditions, treatment protocols, and risk factors.

The underlying technology would involve sophisticated natural language processing (NLP) capabilities to understand free-text descriptions of symptoms, alongside structured data input for allergies, pre-existing conditions, and medications. Machine learning models, trained on extensive datasets of anonymized patient records and clinical outcomes, would then analyze this comprehensive profile to generate an initial risk assessment and a recommended course of action. This could range from immediate referral to an emergency physician, to a scheduled appointment with a primary care provider, to advice on self-care and follow-up. The system would also be capable of identifying potential red flags that might warrant a second opinion or a more urgent consultation than initially perceived.

Crucially, the AI's output would not be a definitive diagnosis or a command to override clinical judgment. Instead, it would serve as a powerful decision-support tool for the human healthcare professionals. The AI would present its assessment, along with the reasoning behind it, to a triage nurse or physician. This would provide the clinician with a consolidated summary of the patient's situation, highlighting key symptoms, potential risks, and a suggested routing. The clinician would then review this information, apply their own professional expertise and contextual understanding of the patient, and make the final decision regarding the patient's care pathway. This human-in-the-loop approach was fundamental to the design, ensuring that technology augmented, rather than replaced, the indispensable human element in healthcare. The aim was to reduce cognitive load on clinicians, allowing them to focus on critical decision-making and direct patient interaction, rather than spending excessive time gathering and initial data.

The implementation of Project Genesis was envisioned to be a phased approach. The initial phase would focus on developing and rigorously testing the AI model in a controlled environment, using historical data and simulations. This would be followed by pilot testing in a limited number of clinics, where the tool would be used in parallel with the existing manual triage process. This parallel operation would allow for direct comparison of outcomes, identification of any discrepancies, and refinement of the AI algorithms and user interfaces. Extensive feedback would be collected from both healthcare staff and patients during these pilot phases. The subsequent phases would involve scaling the deployment to a broader network of clinics, accompanied by comprehensive training programs for all staff who would interact with the system. Ongoing monitoring and evaluation would be a continuous process, ensuring that the AI tool remained effective, safe, and aligned with the evolving needs of the healthcare system and its users.

The success of Project Genesis hinged not only on the technological sophistication of the AI but also on its seamless integration into the complex ecosystem of public healthcare. This involved careful consideration of interoperability with existing electronic health record (EHR) systems, ensuring that patient data flowed smoothly and securely between different platforms. It also necessitated a clear understanding of the administrative and operational changes required to support the new workflow, including potential adjustments to staffing roles and responsibilities. The project team recognized that technology alone was insufficient; it required a holistic approach that addressed people, processes, and technology in equal measure. This comprehensive strategy was the bedrock upon which the ambitious goals of Project Genesis were built, aiming to transform healthcare access for countless individuals within the community.

The journey from conceptualizing an AI-powered triage system to its tangible implementation within the intricate landscape of public healthcare is profoundly shaped by the people it aims to serve and the professionals who dedicate their lives to its operation. Project Genesis, from its inception, was built on the understanding that technological innovation, especially in a sector as sensitive and vital as healthcare, cannot succeed in a vacuum. It necessitates a deep, ongoing engagement with all parties affected by its introduction. This led to a deliberate and systematic approach to stakeholder identification and a commitment to a co-design process, ensuring that the AI triage tool would not only be technically sound but also contextually relevant, ethically grounded, and genuinely beneficial to the community.

The initial step in this crucial phase was comprehensive stakeholder mapping. This was not a superficial exercise; it involved an in-depth analysis of all individuals and groups who had a vested interest in, or would be impacted by, the deployment of the AI triage system. This broad spectrum included, most critically, the
**patients** themselves – the ultimate beneficiaries of improved access and care. This category encompassed individuals from diverse socioeconomic backgrounds, varying levels of digital literacy, and a wide array of health conditions. Understanding their needs, concerns, and expectations was paramount. Following closely were the **healthcare providers**, a diverse group ranging from frontline triage nurses and administrative receptionists to general practitioners, specialists, and clinic managers. Each of these roles interacted with the patient intake process differently, possessing unique insights into its operational challenges and potential improvements. Beyond the direct users of the system, the map extended to **community leaders and representatives**. These individuals, often trusted voices within their neighborhoods, could offer invaluable perspectives on the cultural nuances, trust factors, and broader community health priorities that needed to be considered. Finally, **administrative and IT staff** within the healthcare facilities were essential stakeholders. Their understanding of existing infrastructure, data management protocols, and day-to-day operational workflows was critical for successful integration and sustainability.

With a clear understanding of who constituted the stakeholder ecosystem, the project team transitioned to the vital process of engagement. This was not about simply informing stakeholders about the project, but actively involving them in its development. The cornerstone of this engagement strategy was a series of meticulously planned

**co-design workshops**. These were not passive lectures but interactive sessions designed to foster open dialogue, collaborative problem-solving, and shared ownership. The workshops were structured to address specific aspects of the AI triage tool, from the initial patient interface to the information presented to clinicians.

The first set of workshops focused on the patient-facing elements of the system. Participants, including patients with varying degrees of technological familiarity and healthcare professionals who regularly interacted with patients during intake, were presented with early prototypes of the user interface. These prototypes were intentionally kept somewhat abstract, allowing for maximum creative input. Questions were posed that encouraged participants to think critically: "How would you prefer to describe your symptoms?", "What kind of reassurances would you need while waiting?", "What information would be most helpful to you at this stage?", and "What concerns would you have about using a digital tool for your health needs?". Patients shared their anxieties about technology, their preference for human interaction in moments of vulnerability, and their desire for clear, simple language. They also highlighted the importance of accessibility, advocating for larger font sizes, multiple language options, and the possibility of assistance from clinic staff for those less comfortable with digital interfaces. Healthcare providers, on the other hand, underscored the need for the system to be quick and intuitive, recognizing that patients arriving at the clinic were often in distress or discomfort. They also emphasized the importance of the system capturing essential demographic and insurance information without being overly burdensome, as administrative staff often juggled multiple tasks simultaneously.

Another critical area explored during co-design was the communication protocol between the AI system, the patient, and the clinical staff. This involved delving into how the AI's assessment and recommended pathway should be communicated. Should the AI provide a direct recommendation, or should it present a series of possibilities for the clinician to consider? How should the system inform patients about their expected wait times, and how should updates be delivered? Patients expressed a strong desire for transparency, wanting to understand *why* they were being routed to a particular service or expected to wait a certain amount of time. They also stressed the need for the system to manage expectations realistically, avoiding promises of immediate attention that might not be met. Healthcare professionals, particularly nurses, voiced concerns about how the AI's output would be presented to them. They needed the information to be concise, accurate, and easily digestible, allowing them to quickly verify the AI's assessment and integrate it into their own clinical judgment. The concept of a "confidence score" for the AI's recommendations was discussed, providing clinicians with an indication of how certain the model was about its assessment. The importance of clear, actionable alerts for critically ill patients was also a recurring theme, ensuring that no serious condition would be missed due to reliance on the technology.

Feedback sessions were another integral component of the participatory design process. These were often held after initial prototype demonstrations or following brief periods of simulated use. The project team actively solicited honest and critical feedback, creating a safe space for stakeholders to voice their concerns and suggestions. For instance, during one feedback session with a group of long-term clinic users, many expressed apprehension about the potential for the AI to be impersonal or to miss subtle, non-verbal cues that experienced staff often picked up on. This led to the incorporation of features that allowed patients to add free-text notes about their symptoms, providing an avenue for qualitative information that might otherwise be lost. It also reinforced the decision to maintain a human element at the triage desk, where a member of staff would review the AI's output and interact with the patient.

Community leaders played a pivotal role in ensuring that the project was culturally sensitive and aligned with community values. They helped identify potential barriers to adoption, such as historical distrust of technological interventions or specific cultural norms around health-seeking behaviors. For example, in one community, it was raised that direct questions about certain symptoms might be considered taboo. This insight prompted the development of more indirect questioning strategies within the AI's conversational flow and reinforced the need for a human intermediary who could navigate these sensitivities with grace. They also helped in disseminating information about the project to their constituents, building trust and encouraging participation. Their involvement ensured that the project was not perceived as an imposition but as a collaborative effort to improve local healthcare services.

The insights gleaned from these co-design workshops and feedback sessions directly shaped the AI triage tool's functionalities and user interface. For instance, the initial concept for the patient interface might have been text-heavy, but patient feedback led to the adoption of more visual elements, icons, and progress indicators. The language used in the AI's prompts was refined to be empathetic and easy to understand, avoiding clinical jargon. The flow of questions was adjusted based on how patients naturally described their ailments, rather than adhering strictly to a rigid medical ontology. The system's capacity to integrate with existing patient portals was also a direct result of feedback from administrative staff and IT professionals, who highlighted the importance of seamless data exchange to avoid duplication of effort and ensure data integrity.

Furthermore, the co-design process was instrumental in defining the communication protocols between the AI and the human clinical staff. The project team worked closely with triage nurses and physicians to ensure that the AI's output was presented in a way that was most useful for their decision-making. This included designing dashboards that highlighted key symptoms, patient-reported severity, potential risk factors, and the AI's recommended next steps, along with a confidence score. The system was also designed to flag urgent cases with clear visual and auditory alerts, ensuring that immediate attention was drawn to critical situations. The collaborative design of these interfaces meant that clinicians felt ownership over the tool and were more likely to trust and utilize its capabilities. They provided valuable input on how to best integrate the AI's recommendations into their existing clinical workflows, ensuring that the technology augmented rather than disrupted their practice.

The ethical considerations were woven into every stage of the co-design process. Discussions about data privacy and security were held with patients and community representatives, ensuring that their concerns were addressed and that the system was designed with robust safeguards. For example, patients were given clear information about how their data would be used and were assured of its anonymization for training purposes. The principle of algorithmic fairness was also a recurring theme. Stakeholders raised concerns about potential biases in the AI, particularly for marginalized communities. This led to a commitment to continuously monitor the AI's performance across different demographic groups and to implement corrective measures as needed. The co-design process provided a platform to proactively identify and mitigate these potential biases, rather than addressing them as an afterthought. The project team recognized that building trust required not just a functional tool, but one that was perceived as equitable and respectful of individual autonomy and privacy. This iterative engagement with stakeholders—listening, incorporating feedback, and demonstrating tangible changes based on their input—was not merely a step in the development process; it was the very foundation upon which the successful and ethical deployment of the AI triage system was built. It transformed the project from a technology initiative into a community-driven effort to improve healthcare access.

The successful implementation of any AI system, particularly one designed for public healthcare, hinges not just on its technical efficacy but, crucially, on its equitable impact. Project Genesis recognized from its outset that developing an AI triage tool that serves all segments of the population fairly was as important as its ability to accurately assess urgency. This commitment necessitated a rigorous and intentional approach to defining and validating equity metrics, moving beyond a theoretical understanding of fairness to practical, measurable outcomes. This subsection delves into the intricate process of establishing, implementing, and monitoring these metrics, underscoring the challenges and triumphs encountered in ensuring the AI triage system upheld the principle of equitable care for every individual.

Defining what constitutes "equity" in the context of an AI triage system requires careful consideration of the specific harms that could arise from algorithmic bias in a healthcare setting. The goal was not simply to avoid overt discrimination, but to proactively identify and mitigate subtler forms of disadvantage that could disproportionately affect vulnerable populations. This led to the selection of several key fairness criteria, tailored to the unique operational domain of healthcare access and triage. One primary focus was on **equitable referral rates across demographic groups**. This metric aimed to ensure that the AI system did not systematically under-refer or over-refer patients from particular racial, ethnic, socioeconomic, or gender groups to appropriate levels of care. For instance, if the AI tended to route patients from lower-income neighborhoods to less urgent pathways, even when their reported symptoms were similar to those from more affluent areas, this would indicate a critical equity failure.

Another critical area of concern was **minimizing bias in the assessment of urgency**. The AI's core function is to quickly and accurately gauge the severity of a patient's condition to direct them to the most appropriate service, whether it be immediate emergency care, a same-day appointment, or a routine follow-up. Bias here could manifest as the AI consistently downplaying the urgency of symptoms reported by certain groups, or conversely, overstating the urgency for others. This could be influenced by the training data, where historical biases in healthcare provision might have led to certain symptoms being treated with less seriousness when reported by specific demographics. Therefore, a key metric involved analyzing the AI's assigned urgency scores across different demographic strata, looking for statistically significant deviations that could not be explained by genuine clinical differences.

Beyond these broad categories, Project Genesis also focused on metrics related to **access and usability for diverse populations**. While not strictly algorithmic bias, ensuring that the interface and the process itself were accessible to individuals with varying levels of digital literacy, language proficiencies, and physical abilities was fundamental to equitable service delivery. Metrics here might include the success rate of patients completing the triage process without assistance, the time taken to complete triage by different demographic groups, and the rate at which patients from non-English speaking backgrounds successfully navigated the system. The project team recognized that an AI tool, however fair in its core assessment, would fail in its equity mission if significant portions of the population found it too difficult or intimidating to use.

The selection of these fairness criteria was an iterative process, heavily informed by the co-design workshops and consultations with stakeholders, particularly patient advocacy groups and community health workers. These groups brought invaluable insights into the lived experiences of communities often underserved by healthcare systems. They highlighted how subtle language choices, the order of questions, or even the visual design of the interface could create barriers for certain individuals. For example, a question phrased in overly clinical terms might be easily understood by a healthcare professional but confusing to a patient with limited health literacy, potentially leading to inaccurate symptom reporting and, consequently, an inequitable triage decision. This feedback directly influenced the refinement of the metrics, ensuring they encompassed not only the AI's output but also the entire patient journey from interaction to outcome.

A significant hurdle in developing and validating these equity metrics was the

**challenge of data collection and quality within the healthcare context**. To measure equitable referral rates or bias in urgency assessment, granular demographic data is essential. However, collecting this data in a public healthcare setting can be sensitive. Patient privacy laws, ethical considerations regarding data collection, and the potential for data incompleteness or inaccuracy all posed substantial challenges. Furthermore, historical data in healthcare systems often reflects existing societal inequities, meaning that simply training an AI on past data without careful adjustments could inadvertently perpetuate those same biases.

To address these challenges, Project Genesis implemented a multi-pronged strategy. Firstly, the project team worked closely with the healthcare facilities to ensure that demographic data was collected in a standardized, ethical, and privacy-preserving manner. This involved clear communication with patients about *why* this data was being collected – to ensure the AI served everyone fairly – and obtaining explicit consent. Where direct demographic data was unavailable or incomplete, the project explored the use of proxies, such as geographical data correlated with known demographic distributions, while acknowledging the limitations and potential for inaccuracy this introduced.

Secondly, the team focused on data augmentation and synthetic data generation to supplement real-world data, particularly for underrepresented groups. This allowed for a more robust testing of the AI's performance across a wider spectrum of scenarios without relying solely on potentially scarce real-world data for these groups. However, the use of synthetic data also required rigorous validation to ensure it accurately reflected real-world complexities and did not introduce novel biases. Expert review by clinicians and community representatives was crucial in this validation process.

Thirdly, and perhaps most importantly, Project Genesis established a continuous monitoring framework. Equity metrics were not viewed as a one-time validation exercise but as an ongoing commitment. This involved setting up dashboards that tracked key fairness indicators in near real-time once the AI triage system was deployed. These dashboards provided an aggregated view of referral patterns, urgency assessments, and system usage across different demographic segments. Automated alerts were configured to flag any statistically significant deviations from the expected equitable distribution, prompting immediate investigation by the project team and clinical governance committees.

The validation process itself was multi-layered. It began with offline testing using curated datasets that deliberately included scenarios designed to stress-test the AI for potential biases. This involved creating "counterfactuals" – pairs of patient profiles that were identical in all clinically relevant aspects except for a single demographic attribute, and then observing if the AI's triage recommendation differed. For example, two patient profiles reporting the exact same symptoms, medical history, and vital signs would be created, with the only difference being race or gender. Any divergence in the AI's output for these identical-but-differently-attributed profiles would be flagged as a critical bias.

Following offline validation, the system underwent phased rollouts, starting with limited pilot programs in specific clinics. During these pilot phases, the equity metrics were closely monitored by a dedicated team that included ethicists, data scientists, and representatives from patient advocacy groups. This allowed for real-world validation and the identification of any emergent biases that might not have been captured in offline testing. Feedback mechanisms were established to allow both patients and healthcare staff to report any perceived inequities or issues with the system.

One of the critical lessons learned during this phase was the interconnectedness of different fairness criteria. For example, an AI that was technically unbiased in its urgency assessment might still be inequitable if its interface was difficult for a specific demographic to use, leading to fewer patients from that group successfully completing the triage and thus appearing to have lower referral rates. This reinforced the need for a holistic approach to equity, considering the AI's performance not just in isolation but within the broader context of the healthcare delivery system and user experience.

The development and validation of equity metrics for the AI triage system was an ongoing, dynamic process that demanded constant vigilance and adaptation. It involved defining what equity truly meant in this specific context, wrestling with the complexities of data in healthcare, and establishing robust systems for monitoring and validation. This meticulous work was not merely a compliance exercise; it was a fundamental aspect of ensuring that Project Genesis delivered on its promise to create a healthcare system that was not only more efficient but also fundamentally more just and accessible for all members of the community it served. The metrics developed served as the guardians of fairness, providing concrete evidence that the AI triage tool was moving closer to its ideal of providing equitable care, and serving as a compass to guide future improvements.

The transition from a meticulously tested prototype to a fully operational AI triage system within a public healthcare setting is a complex undertaking, fraught with operational challenges and requiring a deliberate, phased approach. Project Genesis understood that the ultimate success of the AI triage tool would be measured not just by its technical accuracy or its adherence to equity metrics in a controlled environment, but by its seamless integration into the daily workflows of healthcare professionals and its genuine utility for patients navigating the often-stressful journey of seeking medical care. This phase, therefore, focused on the practical realities of deployment: getting the system into the hands of those who needed it, ensuring they knew how to use it effectively, and establishing robust mechanisms for ongoing oversight and refinement.

The strategy for deployment was rooted in a principle of calculated risk mitigation and iterative learning. A "big bang" rollout, where the system is deployed across all facilities simultaneously, was deemed too perilous.

Such an approach would have amplified the potential negative consequences of any unforeseen bugs, usability issues, or systemic integration problems. Instead, Project Genesis opted for a phased rollout, commencing with a limited number of pilot sites. These initial sites were carefully selected to represent a diverse range of operational environments within the public healthcare network – including a high-volume urban emergency department, a community health clinic serving a rural population, and a specialized urgent care center. This deliberate selection ensured that the AI triage tool would be exposed to a variety of patient demographics, clinical presentations, and existing IT infrastructures from the outset.

For each pilot site, a comprehensive deployment plan was developed in close collaboration with the local healthcare administration and clinical leadership. This plan encompassed technical integration, which involved mapping the AI triage system's data inputs and outputs to existing Electronic Health Record (EHR) systems, ensuring a smooth flow of information and avoiding data duplication or loss. Critical to this integration was establishing secure communication protocols to protect patient data, adhering to all relevant privacy regulations. The technical team worked tirelessly to ensure that the AI system was not an isolated entity but a connected component of the broader healthcare IT ecosystem. This often involved custom middleware development and close coordination with EHR vendors to ensure compatibility and interoperability. The goal was to make the AI triage tool feel like a natural extension of the existing clinical workflow, rather than an add-on that created additional administrative burden.

Beyond the technical architecture, the human element of deployment was given paramount importance. Effective training was recognized as a cornerstone of successful adoption. A multifaceted training program was designed, catering to the distinct needs of different user groups. For frontline clinical staff – nurses, paramedics, and administrative personnel who would directly interact with the AI system – the training focused on practical usage. This included hands-on sessions demonstrating how to initiate a triage, interpret the AI's recommendations, and override them when clinical judgment dictated. Special emphasis was placed on understanding the AI's limitations and the circumstances under which human intervention or skepticism was most crucial. Training materials were developed in multiple formats, including interactive online modules, printed user manuals, and in-person workshops, ensuring accessibility for staff with varying learning preferences and schedules. The training also included role-playing scenarios designed to simulate real-time patient interactions, allowing staff to practice their skills in a low-stakes environment.

For clinicians – physicians, specialist nurses, and other medical professionals who would receive the AI-generated triage recommendations and act upon them – the training focused on understanding the AI's underlying logic and its confidence levels. This was essential for building trust and enabling informed decision-making. Sessions were conducted by clinical informaticians and data scientists who could explain the principles behind the AI's assessments, its strengths, and its known areas of potential weakness. The importance of critically evaluating the AI's output, rather than blindly accepting it, was a recurring theme. This fostered a culture of informed skepticism and empowered clinicians to leverage the AI as a supportive tool rather than a replacement for their expertise.

A crucial component of the deployment was the establishment of robust user support mechanisms. Recognizing that questions and issues would inevitably arise once the system was live, a dedicated support team was assembled. This team comprised individuals with a blend of technical expertise, clinical knowledge, and an understanding of the AI triage tool's specific functionalities. They operated through multiple channels, including a dedicated helpdesk hotline, an email support system, and an internal messaging platform, ensuring that users could reach out for assistance through their preferred method. The support team was trained to not only troubleshoot technical glitches but also to address questions related to the AI's recommendations, its interpretation, and its impact on workflow. Furthermore, "super-users" were identified and trained at each pilot site – individuals who became go-to resources for their colleagues, providing on-the-spot assistance and escalating more complex issues to the central support team. This decentralized support model proved invaluable in providing timely and context-specific help, minimizing disruption to clinical operations.

As the AI triage tool began to operate in these pilot sites, the focus shifted to intensive monitoring. This was not merely about tracking uptime and technical performance, but about a holistic assessment of the system's impact, encompassing operational efficiency, user satisfaction, and, critically, the equity outcomes that had been so carefully defined during the development phase. A comprehensive monitoring framework was established, drawing upon data from multiple sources.

System performance was tracked through a suite of automated dashboards. These dashboards provided near real-time visibility into key operational metrics: the volume of triage requests processed, the average time taken for triage, the success rate of AI-driven recommendations (i.e., how often the AI's suggested pathway aligned with the eventual clinical decision), and the rate of system downtime. These metrics were crucial for identifying any technical bottlenecks or performance degradations that might impact the speed and reliability of the triage process. For instance, if the system consistently experienced slowdowns during peak hours, this would trigger an investigation into server capacity or network latency.

User satisfaction was gauged through a combination of direct feedback mechanisms and indirect indicators. Regular surveys were administered to both clinical staff and patients who interacted with the system. These surveys were designed to capture qualitative insights into their experiences, identifying areas of frustration or delight. For staff, questions focused on the ease of use, the perceived value of the AI in their workflow, and their confidence in its recommendations. For patients, surveys probed their understanding of the process, their comfort level with the technology, and their overall satisfaction with the triage experience. Beyond surveys, indirect indicators such as the frequency of support requests related to specific functionalities, the rate at which staff overrode AI recommendations, and the qualitative feedback received through the dedicated support channels provided further insights into user sentiment.

However, the most critical aspect of the monitoring framework was the continuous tracking of the pre-defined equity metrics. This involved diligently collecting and analyzing data across various demographic groups, as outlined in the previous chapter. Referral patterns were scrutinized to ensure that patients from different racial, ethnic, socioeconomic, and gender backgrounds were being directed to appropriate care pathways at equitable rates. Any significant deviations from expected distributions, after accounting for genuine clinical differences, were flagged for immediate investigation. Similarly, the AI's assessment of urgency was continuously monitored for any systematic biases. This involved comparing the AI-assigned urgency scores for patients with similar clinical presentations but different demographic attributes. Any statistically significant disparities would trigger a deep dive into the specific data inputs and algorithmic pathways that might be contributing to the inequity.

Furthermore, the monitoring framework extended to assessing the system's accessibility and usability for diverse populations. Data was collected on completion rates of the triage process by individuals with varying levels of digital literacy and language proficiency. The frequency with which patients required assistance to navigate the system was also tracked, disaggregated by demographic factors. This ensured that the AI triage tool was not inadvertently creating new barriers to care for already vulnerable groups.

The data generated by this comprehensive monitoring framework was not merely for archival purposes. It formed the bedrock of an iterative improvement process, a commitment to continuous refinement of the AI triage tool. Real-time feedback and performance data were fed back into the development cycle, leading to a series of targeted adjustments and enhancements.

One common area for iteration was the natural language processing (NLP) component of the AI. Based on user feedback and analysis of misinterpretations, the NLP models were retrained and fine-tuned to better understand colloquialisms, regional accents, and less formal descriptions of symptoms. For example, if patients frequently used phrases like "feeling under the weather" or "my tummy hurts," and the AI struggled to accurately categorize these descriptions, the NLP models were updated with more diverse linguistic examples. This involved enriching the training data with a wider range of conversational inputs, ensuring the AI could interpret a broader spectrum of patient language.

Algorithmic adjustments were also a frequent outcome of the monitoring process. If the data revealed subtle biases in urgency assessment for a particular demographic, the team would meticulously examine the relevant algorithms. This might involve recalibrating weighting factors for certain symptoms, re-evaluating the influence of specific medical history variables, or even re-training parts of the model on more balanced datasets. For instance, if the AI was found to consistently downplay the severity of chest pain in female patients compared to male patients with identical descriptions, the model's sensitivity to cardiac symptoms for women would be enhanced. This often involved collaborating with clinical experts to ensure that algorithmic changes were grounded in sound medical understanding and did not inadvertently introduce new clinical inaccuracies.

User interface (UI) and user experience (UX) improvements were also a significant part of the iterative cycle. Feedback from both patients and staff often highlighted areas where the interface could be simplified, made more intuitive, or rendered more accessible. This might involve redesigning the flow of questions, clarifying the wording of prompts, improving the visual hierarchy of information, or enhancing the accessibility features for users with visual or motor impairments. For example, if surveys indicated that older patients found the font size too small or the buttons too close together, adjustments would be made to improve readability and ease of interaction. The goal was to make the system as user-friendly and accessible as possible for the widest possible range of users.

Furthermore, the monitoring data often revealed opportunities to expand the AI's capabilities. As the system gained traction and clinicians became more accustomed to its use, requests emerged for it to handle a wider range of triage scenarios or to provide more nuanced guidance. This led to the gradual expansion of the AI's diagnostic and referral capabilities, always subject to rigorous re-validation of its performance and, crucially, its equity impact. For example, if initial deployment focused on common acute conditions, subsequent iterations might incorporate the AI's ability to triage patients presenting with chronic disease exacerbations or specific types of injuries, always ensuring that these expanded capabilities did not introduce new biases.

The iterative process was not solely driven by data analytics; it was deeply intertwined with ongoing engagement with stakeholders. Regular feedback sessions were held with clinical teams, patient advocacy groups, and community representatives. These sessions provided a vital forum for discussing the monitoring findings, gathering qualitative insights that data alone could not provide, and collaboratively brainstorming solutions. The co-design principles that had guided the development phase were extended into the post-deployment period, ensuring that the evolution of the AI triage tool remained responsive to the needs and concerns of its users and the communities it served. This continuous dialogue fostered a sense of shared ownership and responsibility for the system's success, building trust and ensuring that the AI remained a tool that truly served the public good.

The journey from initial concept to widespread implementation and continuous improvement is a testament to the dynamic nature of AI in public healthcare. The deployment phase of Project Genesis's AI triage tool was characterized by a strategic rollout, comprehensive training, dedicated user support, and a relentless commitment to monitoring and iteration. By embracing a phased approach, prioritizing human factors alongside technical integration, and establishing a robust framework for continuous feedback and refinement, Project Genesis aimed to ensure that the AI triage system not only met its clinical objectives but also lived up to its ethical imperative: to enhance healthcare access and deliver equitable care to all. The ongoing process of learning and adaptation, fueled by real-world data and stakeholder input, was fundamental to realizing this vision.

The deployment of an AI triage system in a public healthcare setting, as exemplified by Project Genesis, offers a rich tapestry of lessons learned, revealing both the profound potential and the inherent complexities of integrating advanced technology into deeply human systems. The journey from a carefully crafted prototype to a functional, integrated tool highlighted that while technical efficacy is a prerequisite, it is merely the first step in a much longer and more nuanced process. One of the most salient takeaways is the undeniable importance of context-specific adaptation. The idealized performance of an AI in a controlled research environment often diverges significantly from its performance in the dynamic, often unpredictable, real-world clinical setting. Factors such as varied patient populations, diverse levels of digital literacy among users, and the sheer unpredictability of human health all conspire to create scenarios that even the most sophisticated algorithms might not have fully anticipated. This underscores the necessity for extensive, site-specific pilot testing and a willingness to iterate based on granular, on-the-ground observations, rather than assuming a one-size-fits-all solution.

Furthermore, the project underscored that "AI readiness" is not solely a technical or financial consideration, but a profound organizational and cultural one. The success of the triage system was inextricably linked to the willingness and ability of healthcare professionals to embrace and adapt to a new way of working. This required more than just superficial training; it demanded a deep understanding of the AI's capabilities and limitations, fostering a sense of partnership between human expertise and artificial intelligence. Building this trust was a continuous process, nurtured through transparent communication, opportunities for feedback, and the consistent demonstration of the AI's value. The experience revealed that resistance, or even cautious skepticism, from frontline staff is not necessarily a sign of failure, but an opportunity to engage in dialogue, address concerns, and collaboratively refine the system. Ignoring or dismissing such feedback would have been a critical misstep, potentially undermining the entire initiative.

The ethical imperative to ensure equitable outcomes remained a central, guiding principle throughout the deployment. While the initial development phase focused on embedding fairness into the algorithm, the real-world application presented ongoing challenges in verifying and maintaining that equity. The continuous monitoring of demographic disparities in referral patterns and urgency assessments became a critical feedback loop. It highlighted that algorithmic bias is not a static problem, but one that can emerge or shift as the system interacts with a live population and as external factors evolve. For instance, a subtle shift in the types of conditions presenting at a particular clinic, or an increase in patients from a newly arrived demographic group with different health-seeking behaviors, could potentially manifest as new disparities if not proactively identified and addressed. This necessitates a proactive, rather than reactive, approach to ethical AI, with robust auditing mechanisms and a commitment to ongoing fairness evaluations.

The practicalities of data integration also emerged as a significant, often underestimated, challenge. While technical interoperability is a cornerstone of modern healthcare IT, the real-world implementation revealed the messy realities of disparate legacy systems, varying data standards, and the sheer volume of information that needs to be processed. Ensuring that the AI triage system could seamlessly and securely exchange data with existing Electronic Health Records (EHRs) and other clinical information systems required significant engineering effort and constant vigilance. The risk of data siloes, data duplication, or incomplete information transfer could not only compromise the AI's performance but also introduce new clinical risks. This experience emphasizes the need for a holistic approach to health IT infrastructure, where interoperability is not an afterthought but a foundational requirement for any new technology deployment, especially those that promise to influence clinical decision-making.

The role of human oversight and the principle of "human-in-the-loop" were repeatedly validated. The AI triage system was designed not to replace clinical judgment but to augment it, and this principle proved crucial for both safety and user acceptance. There were instances where the AI's recommendation, while statistically sound based on the provided data, did not fully capture the nuances of a patient's presentation or their unique circumstances. The ability of clinicians to critically review the AI's output, consider factors not captured by the algorithm (such as patient history, social determinants of health, or subtle clinical cues), and ultimately make the final decision was paramount. This highlighted the need for user interfaces that clearly indicate the AI's confidence levels, present its reasoning transparently, and facilitate straightforward overrides when necessary. The system's value was amplified when it served as an intelligent assistant, empowering clinicians with timely information and suggested pathways, rather than as an autonomous decision-maker.

The financial and resource implications of deploying and maintaining such a system also warrant careful consideration. While the long-term promise of AI in healthcare often centers on efficiency gains and improved outcomes, the initial investment in technology, training, infrastructure upgrades, and ongoing maintenance can be substantial. Public healthcare systems, often operating under tight budgetary constraints, need to carefully weigh these costs against the anticipated benefits. Demonstrating a clear return on investment, not just in financial terms but also in terms of improved patient access, reduced wait times, and enhanced clinical capacity, is essential for securing sustained support and funding. The Project Genesis experience suggests that a phased, iterative approach to deployment can help manage these costs, allowing for adjustments and optimization as the system matures and its value proposition becomes clearer.

Looking beyond the immediate lessons, the implications of Project Genesis for the future of AI in public healthcare are far-reaching. Firstly, it reinforces the understanding that AI is not a panacea but a tool that must be wielded with care, foresight, and a deep commitment to humanistic values. The successful integration of AI requires a multidisciplinary approach, bringing together clinicians, data scientists, ethicists, policymakers, and crucially, the patients and communities these systems are intended to serve.

The experience also points towards the need for more robust regulatory frameworks and ethical guidelines specifically tailored to AI in healthcare. While existing medical device regulations provide a starting point, the dynamic and adaptive nature of AI presents unique challenges. Clear standards for validation, ongoing performance monitoring, transparency, and accountability are essential to build public trust and ensure that AI technologies are developed and deployed responsibly. This includes addressing critical questions about data ownership, algorithmic transparency, and the legal implications of AI-assisted clinical decisions.

Furthermore, the project highlights the transformative potential of AI to democratize access to healthcare, particularly in underserved communities. By providing a standardized, accessible, and intelligent triage mechanism, AI can help alleviate pressure on overloaded healthcare facilities, reduce wait times for those seeking care, and ensure that patients are directed to the most appropriate level of care more efficiently. However, this potential can only be fully realized if the design and deployment of these systems prioritize equity and accessibility for all, proactively addressing potential digital divides and language barriers.

The ongoing evolution of AI technologies, including advancements in explainable AI (XAI), will likely offer even greater opportunities for building trust and facilitating human oversight. As AI models become more transparent in their decision-making processes, clinicians will be better equipped to understand and validate their recommendations, further strengthening the human-AI partnership. The pursuit of "responsible AI" is therefore not a static destination but an ongoing journey, requiring continuous innovation, critical evaluation, and a steadfast commitment to ethical principles.

The lessons learned from Project Genesis serve as a vital roadmap for other public healthcare systems contemplating the adoption of AI triage or similar technologies. It is a testament to the fact that while the path is complex, marked by technical hurdles, ethical dilemmas, and the need for profound organizational change, the ultimate reward - improved patient care, enhanced efficiency, and more equitable access to health services - is well worth the endeavor. The future of AI in public healthcare hinges on our ability to learn from these real-world experiences, to adapt our approaches, and to ensure that technology always serves humanity's most fundamental needs. This case study demonstrates that with careful planning, continuous engagement, and an unwavering ethical compass, AI can indeed become a powerful force for good in the pursuit of better health for all. The integration was not merely a technical implementation but a profound organizational learning experience, continuously shaping and reshaping the technology and the human systems that interact with it. The project's success was not defined by a singular moment of deployment, but by its sustained commitment to adaptation, learning, and an unwavering focus on serving the diverse needs of the public.

# Chapter 14: Adapting Templates for Participatory Practices

The successful integration of participatory practices into public sector AI projects is significantly accelerated and strengthened by the strategic use of reusable frameworks. These frameworks act as foundational blueprints, offering structured approaches to common challenges encountered in community engagement, co-design, and ethical AI development. Instead of reinventing the wheel for every new initiative, public sector teams can leverage these pre-defined templates to ensure that critical steps are not overlooked, thereby saving valuable time and resources. This is particularly crucial in the fast-paced environment of AI development, where agility and efficiency are paramount, yet the depth of stakeholder involvement and ethical consideration must remain unwavering.

Consider the process of stakeholder mapping, a fundamental step in understanding who needs to be involved in an AI project and what their interests and potential impacts might be. A reusable stakeholder mapping framework typically includes categories for identifying different user groups (e.g., end-users, administrators, impacted communities, domain experts), their levels of influence and interest, their current and potential relationship with the AI system, and key considerations for engaging them. Such a framework might prompt questions like: "Which community groups are most likely to be affected by this AI's decisions, and how can we ensure their voices are heard from the outset?" or "What existing local knowledge do domain experts possess that could enrich the AI's training data or inform its ethical guardrails?" By providing a structured checklist and guiding questions, a template streamlines this often complex and sensitive process. It ensures that the identification of stakeholders is systematic, comprehensive, and sensitive to power dynamics, rather than being an ad-hoc exercise that might miss crucial perspectives. Without such a framework, a team might default to engaging only the most readily accessible or vocal groups, inadvertently marginalizing those whose input is most vital for equitable AI deployment.

Another area where reusable frameworks prove invaluable is in the design of participatory workshops. Whether the goal is to gather user requirements, solicit feedback on AI prototypes, or co-design ethical guidelines, a well-structured workshop template can be a game-changer. These templates often outline different phases of a workshop, such as introductions and ice-breakers to foster a welcoming environment, information-sharing sessions to contextualize the AI project, brainstorming or ideation activities, and structured feedback mechanisms. For instance, a template for a co-design workshop focused on an AI-powered public service delivery system might include specific activities designed to elicit user needs related to accessibility, transparency, and accountability. It might suggest using journey mapping exercises to understand how citizens currently interact with the service and where an AI could realistically enhance or disrupt that experience. It could also incorporate methods for collaboratively defining the "red lines" – the ethical boundaries that the AI must not cross – through scenario-based discussions. Providing these templates equips facilitators with proven methods that have been refined through practice, ensuring that workshops are productive, inclusive, and yield actionable insights, rather than devolving into unstructured discussions that leave participants feeling unheard and the project team without clear direction.

The development of ethical AI principles and governance structures also benefits immensely from adaptable frameworks. Rather than starting from scratch, public sector organizations can utilize templates that outline key ethical considerations, such as fairness, accountability, transparency, privacy, and safety. These frameworks can guide teams in developing specific policies and procedures tailored to their AI projects. For example, a template for an AI ethics review board might include sections on mandatory documentation required from project teams, criteria for assessing potential harms, procedures for stakeholder consultation, and escalation pathways for ethical concerns. A framework for developing an AI transparency statement might prompt teams to consider what information needs to be disclosed to the public about an AI system's purpose, its data sources, its performance metrics, and how it can be contested or appealed. These pre-defined structures ensure that ethical considerations are not an afterthought but are systematically integrated into the AI lifecycle, fostering a culture of responsible innovation. The value lies in providing a robust starting point that can be customized, ensuring that no critical ethical dimension is overlooked in the pursuit of functionality and efficiency.

Furthermore, the iterative nature of AI development, especially when guided by participatory practices, necessitates flexible yet structured feedback loops. Reusable frameworks for collecting and synthesizing user feedback can be incredibly effective. These might include standardized survey instruments, interview protocols, or methods for analyzing qualitative data from workshops and focus groups. A template for a user feedback form for an AI-powered public transit navigation app, for instance, could include specific questions about perceived accuracy, ease of use, reliability, and any instances where the AI's recommendations led to unexpected or negative outcomes. It could also include open-ended prompts for suggestions on improvements. By using consistent feedback mechanisms across different stages of development or across different projects, organizations can build a rich dataset of user experiences over time. This allows for the identification of recurring issues, the tracking of improvements, and a deeper understanding of user needs and preferences, thereby enabling more informed and responsive AI system evolution.

The adoption of reusable frameworks also promotes consistency and scalability across an organization. When multiple teams are working on different AI initiatives, having common templates for stakeholder engagement, ethical reviews, or user testing ensures a degree of standardization. This makes it easier to share best practices, compare outcomes, and onboard new team members. For a large public sector entity, this organizational learning and capacity building is invaluable. It moves beyond individual project successes to creating a systemic capability for responsible and inclusive AI deployment. For example, a central repository of these frameworks, accompanied by guidance on their appropriate application, can serve as a vital resource for project managers, data scientists, and community liaisons across the organization. This not only enhances efficiency but also reinforces a shared commitment to the principles of participatory AI.

Moreover, these frameworks can be crucial in navigating the procurement and oversight of AI systems developed by external vendors. By incorporating requirements for participatory engagement and ethical considerations into standardized templates for requests for proposals (RFPs) or service level agreements (SLAs), public sector bodies can ensure that external partners are aligned with their values and objectives. A template RFP might include specific clauses requiring vendors to detail their approach to stakeholder consultation, data privacy, algorithmic transparency, and mechanisms for user feedback and grievance redressal. This proactive approach helps to embed participatory principles from the very inception of a project, preventing potential issues down the line.

The practical benefit of these templates lies in their ability to reduce cognitive load and overcome inertia. Public sector employees, often juggling multiple priorities and facing complex challenges, may find it daunting to conceptualize and implement participatory practices from scratch. A well-designed template provides a clear path forward, breaking down complex processes into manageable steps. It offers concrete examples of what successful engagement looks like and provides tools to achieve it. This lowers the barrier to entry for adopting these vital practices. For instance, a template for conducting an impact assessment of an AI system on vulnerable populations might include a checklist of potential risks, a guide for consulting with affected communities, and a framework for documenting mitigation strategies. This structured approach ensures that such critical assessments are conducted thoroughly and systematically.

Ultimately, the value of reusable frameworks for participatory practices in public sector AI initiatives lies in their dual capacity to enhance efficiency and deepen effectiveness. They provide the structure needed to navigate the complexities of AI development responsibly, ensuring that technology serves the public good. By offering adaptable, proven tools, these frameworks empower public sector teams to engage communities meaningfully, embed ethical considerations from the outset, and build AI systems that are not only technically sound but also equitable, trustworthy, and responsive to the diverse needs of the people they are designed to serve. They represent a tangible commitment to operationalizing the principles of responsible AI within the practical realities of public service delivery.

Understanding and engaging with all affected parties is a cornerstone of equitable and effective public sector AI deployment. This necessitates a structured approach to identifying, analyzing, and planning engagement with these stakeholders. While the specific groups and their concerns will vary significantly depending on the AI initiative—be it a predictive policing tool, an automated benefits eligibility system, or a public health diagnostic aid—the underlying process of mapping and analysis can be guided by a consistent, adaptable framework. This section introduces a practical template designed to facilitate this critical step, ensuring that no essential perspective is overlooked. The goal is to move beyond perfunctory consultation towards genuine co-creation and informed decision-making, grounded in a deep understanding of the diverse needs, capacities, and potential impacts on all individuals and communities.

The Stakeholder Mapping and Analysis Template serves as a comprehensive tool to systematically document and understand the landscape of individuals, groups, and organizations that have an interest in, or are impacted by, a public sector AI project. It encourages a proactive rather than reactive approach, allowing project teams to anticipate potential challenges, identify opportunities for collaboration, and tailor engagement strategies effectively. This template is not a rigid, one-size-fits-all document but a flexible structure that should be populated and adapted based on the unique context of each AI initiative. Its primary utility lies in providing a structured way to ask critical questions, ensuring a thorough and inclusive assessment.

At its core, the template begins with the identification of stakeholder groups. This initial step requires broad thinking, pushing teams to consider all potential actors. It's crucial to move beyond the obvious and include those who might be indirectly affected or whose voices are historically marginalized. For an AI system aimed at optimizing traffic flow in a city, for example, stakeholders might include not only city planners, traffic engineers, and technology providers but also residents of affected neighborhoods, local businesses, public transport users, cyclists, pedestrians, emergency services, and even environmental advocacy groups concerned about emissions. The template provides distinct columns for listing these groups, encouraging specificity. For instance, instead of simply listing "residents," one might break it down into "residents in the immediate vicinity of new traffic infrastructure," "commuters using major arterial roads," and "residents reliant on public transport." This granular approach is vital for designing targeted and relevant engagement.

Following the identification of stakeholder groups, the next critical element is to assess their interests and concerns. What do these groups hope to gain from the AI system, or what do they fear might happen? For the traffic flow AI, residents might be interested in reduced noise pollution and improved pedestrian safety, while businesses might focus on increased customer accessibility. Conversely, some residents might fear increased traffic volume on side streets or concerns about surveillance. Understanding these diverse, and sometimes conflicting, interests is paramount. The template facilitates this by including a section for detailing each stakeholder group's primary interests, their potential benefits from the AI, and their potential risks or negative impacts. This encourages a nuanced understanding beyond simple pro or con categorizations. It prompts questions such as: "What are the core needs this AI system is intended to address for this group?" and "What are their specific anxieties or potential harms that need to be acknowledged and mitigated?"

Equally important is the assessment of stakeholder influence and power. Not all stakeholders have equal capacity to impact a project's trajectory, whether through formal decision-making authority, public opinion, or resource control. This aspect of the analysis helps prioritize engagement efforts and identify potential allies or adversaries. For the traffic AI, a municipal council has significant formal power, while a vocal community group might wield considerable influence through media attention and public mobilization. The template includes a section for rating each stakeholder group's level of influence (e.g., High, Medium, Low) and their level of interest in the project (also High, Medium, Low). This "Influence/Interest Matrix" is a classic tool for strategic stakeholder management. It helps teams decide where to focus their energy: high-influence, high-interest groups often require close management and collaboration, while low-influence, low-interest groups might only need to be kept informed. However, it's also crucial to consider how to elevate the voices of low-influence, high-interest groups, particularly those who are most impacted, to ensure equity. This might involve specific outreach strategies or partnerships with intermediaries.

Once stakeholders are identified and their interests and influence are assessed, the focus shifts to engagement strategy. How will the project team communicate with each group, and what methods will be employed? This section of the template requires mapping communication channels and outlining engagement activities. For the traffic AI, communication might range from public town hall meetings for broad community input to targeted workshops with transportation advocacy groups, and regular briefings with elected officials. The template prompts for the identification of preferred communication methods for each group, their current relationship with the public sector organization, and the desired future relationship. It also calls for specific engagement activities planned for each stakeholder group, such as surveys, interviews, focus groups, participatory design sessions, or advisory committees. This ensures that engagement is not a one-off event but a continuous, tailored process. For example, a group with high interest and high influence might be invited to a co-design workshop, while a group with low interest but high potential impact might receive regular email updates and be offered opportunities to provide feedback through an online portal.

Furthermore, the template emphasizes the importance of identifying potential barriers to engagement and developing strategies to overcome them. These barriers can be numerous and varied, ranging from lack of awareness and distrust of government to practical issues like language differences, time constraints, accessibility challenges, and digital divides. For an AI project impacting vulnerable populations, such as a system for determining eligibility for social services, barriers might include limited internet access, lack of childcare during meeting times, or fear of reprisal for speaking out. The template includes a dedicated section to document these potential barriers for each stakeholder group and to brainstorm concrete mitigation strategies. This proactive identification allows teams to design inclusive engagement processes from the start, rather than being caught off guard by participation gaps. For instance, if language is a barrier, the strategy might involve providing translation services, multilingual materials, and hiring community liaisons who speak the relevant languages. If meeting times are an issue, flexible scheduling options or asynchronous feedback mechanisms might be employed.

A crucial aspect of this template is its emphasis on documenting assumptions and potential biases. Public sector AI projects, like all technology implementations, are susceptible to underlying assumptions and biases held by the development team or embedded within the data. The stakeholder analysis process can help uncover these by listening to the perspectives of those who might be differently situated. The template encourages teams to reflect on their own assumptions about stakeholder needs and behaviors, and to critically examine whether the AI system's design or intended use might inadvertently perpetuate existing inequalities or create new ones. This section serves as a prompt for critical self-reflection: "What are we assuming about how this AI will be used?" "What data are we using, and what biases might it contain?" "How might this AI disproportionately affect certain groups, and have we sought their input on this?"

To illustrate the practical application of this template, consider a hypothetical public sector AI initiative to develop a personalized learning platform for K-12 students.

**Stakeholder Group:** Students (various age groups, diverse backgrounds)
**Interests:** Engaging and effective learning tools, personalized support, fair assessment, privacy of their data.
**Risks/Negative Impacts:** Increased screen time, potential for algorithmic bias in content delivery or assessment, data privacy breaches, feeling monitored or pressured.
**Influence:** Low to Medium (direct users, but often not decision-makers).
**Interest:** High.
**Communication Channels:** School-based channels, parent-teacher associations, student-led forums (if established), age-appropriate surveys.
**Engagement Activities:** Co-design workshops with student focus groups (age-tailored), usability testing of prototypes, feedback surveys, student advisory panel.
**Barriers:** Difficulty articulating needs (younger students), time constraints due to curriculum, potential discomfort speaking critically to authority figures.
**Mitigation:** Use of gamified feedback tools, creation of safe spaces for feedback, involvement of trusted educators as facilitators, simplified language.

**Stakeholder Group:** Teachers and Educators
**Interests:** Tools that support teaching goals, reduce administrative burden, improve student outcomes, ease of integration into existing curricula, professional development.
**Risks/Negative Impacts:** Increased workload if the platform is complex, potential for de-skilling if AI replaces too much instructional judgment, concerns about data accuracy and reliability, fear of job displacement.
**Influence:** High (gatekeepers to student access, curriculum experts).
**Interest:** High.
**Communication Channels:** School district communications, professional development sessions, teacher union representatives, direct email.
**Engagement Activities:** Pilot programs, training sessions, feedback forums, curriculum integration workshops, advisory committees.
**Barriers:** Time constraints, resistance to new technology, varying levels of digital literacy, concerns about pedagogical soundness.
**Mitigation:** Comprehensive and ongoing training, clear demonstration of benefits (e.g., time-saving features), co-development of integration strategies, involvement of teacher leaders in design.

**Stakeholder Group:** Parents and Guardians
**Interests:** Ensuring their children receive quality education, understanding how the platform works, monitoring their child's progress, data privacy and security, equitable access for all students.
**Risks/Negative Impacts:** Concerns about screen time balance, lack of transparency about algorithmic decisions affecting their child, potential for increased educational inequality if access or quality varies.
**Influence:** Medium to High (advocates for their children, can influence school board decisions).
**Interest:** High.

**Communication Channels:** School newsletters, parent-teacher meetings, school websites, parent association meetings.
**Engagement Activities:** Information sessions, Q&A panels with project team and educators, surveys, feedback forms.
**Barriers:** Busy schedules, technical jargon, lack of understanding of AI concepts, digital divide.
**Mitigation:** Clear and accessible communication materials (e.g., FAQs, infographics), offering information sessions at various times, providing resources in multiple languages, ensuring parents can access information without high technical proficiency.

**Stakeholder Group:** Technology Developers and Vendors
**Interests:** Successful project delivery, technical feasibility, adherence to requirements, potential for future work, ethical and responsible AI development.
**Risks/Negative Impacts:** Unclear requirements, scope creep, inadequate data, ethical constraints that challenge technical solutions, reputational damage if AI fails or causes harm.
**Influence:** High (technical expertise, implementation capacity).
**Interest:** High.
**Communication Channels:** Project management platforms, regular technical meetings, contractually defined communication protocols.
**Engagement Activities:** Joint development sessions, technical reviews, code reviews (if applicable), testing and validation cycles.
**Barriers:** Misunderstandings regarding public sector processes or policy goals, differing priorities (technical vs. social outcomes).
**Mitigation:** Clear contract language, dedicated liaisons, regular interdisciplinary meetings, clearly defined ethical guidelines and performance metrics.

**Stakeholder Group:** District Administrators and Policymakers

**Interests:** Improving overall educational outcomes, ensuring equity across schools, managing budget effectively, compliance with regulations, positive public perception.
**Risks/Negative Impacts:** Project failure, budget overruns, negative press, potential for litigation due to bias or privacy issues, failing to meet educational mandates.
**Influence:** Very High (funding, policy direction, ultimate decision-making authority).
**Interest:** High.
**Communication Channels:** Formal reports, briefings, presentations, executive summaries.
**Engagement Activities:** Steering committee meetings, regular progress reports, strategic planning sessions.
**Barriers:** Limited time, focus on strategic overview rather than granular details, differing political priorities.
**Mitigation:** Concise and data-driven reporting, clear articulation of benefits and risks, aligning project goals with broader district strategic plans, transparency in decision-making processes.

This detailed mapping for the personalized learning platform example highlights how the template encourages a comprehensive view. It's not just about identifying who is involved, but understanding *how* they are involved, what they care about, and how to best work with them. The act of filling out such a template should be an iterative process, revisited and updated as the project evolves and new information or stakeholders emerge.

The "Communication Channels" and "Engagement Activities" sections are particularly crucial for operationalizing the analysis. Merely identifying stakeholders is insufficient; a clear plan for interaction is needed. This involves selecting appropriate methods that are accessible and meaningful to each group. For instance, a digital-first approach might exclude elderly residents or those in rural areas with poor connectivity. Therefore, the plan must incorporate a mix of online and offline strategies, formal and informal interactions. This adaptability is key: a framework used for a large-scale city-wide AI implementation will require different engagement tactics than one for a specialized departmental tool. The template encourages teams to think through this: Should we use a webinar? A series of in-person workshops? A dedicated online forum? Direct mail? Each method has its own strengths, weaknesses, and associated costs.

Moreover, the inclusion of "Potential Barriers to Engagement" and "Mitigation Strategies" moves the template from a passive analytical tool to an active planning instrument. It forces project teams to confront the realities of participation, acknowledging that meaningful engagement is not always easy. By identifying potential obstacles early—such as cultural misunderstandings, lack of trust in government, or practical accessibility issues—teams can proactively design inclusive processes. This might involve allocating budget for translation services, childcare at meetings, accessible venues, or hiring community liaisons who have established trust within specific populations. Failing to address these barriers can lead to skewed participation, where only the most vocal, privileged, or easily accessible groups are represented, undermining the very principles of participatory AI.

The "Assumptions and Biases" section is perhaps the most ethically critical component of the template. Public sector AI projects are often framed around notions of efficiency, objectivity, and fairness. However, these systems are built by humans using data that reflects societal structures, including historical inequalities and biases. By prompting teams to articulate their underlying assumptions and to consider how their AI might interact with existing societal biases, this template encourages a more critical and responsible approach. For example, an AI system designed to predict recidivism might be based on historical policing data that disproportionately targets certain communities. A thorough stakeholder analysis, including input from those communities, could reveal these biases and inform modifications to the data, the model, or the system's deployment to mitigate harm. This section ensures that ethical considerations are not an afterthought but are embedded in the initial planning stages.

Finally, the template should be viewed as a living document. Stakeholder landscapes are dynamic. New groups may emerge, existing stakeholders' interests may shift, and new barriers to engagement may arise as a project progresses. Therefore, the stakeholder mapping and analysis should be revisited and updated regularly throughout the AI project lifecycle. This continuous engagement ensures that the project remains responsive to evolving needs and concerns, fostering sustained trust and collaboration. By providing a structured yet flexible approach, this stakeholder mapping and analysis template empowers public sector organizations to navigate the complex social and ethical dimensions of AI development, ensuring that technology serves the public good in a truly inclusive and equitable manner. It is a foundational step in building AI systems that are not only technically sound but also socially responsible and trusted by the communities they are intended to serve.

The preceding sections have underscored the critical importance of deeply understanding and actively engaging with all stakeholders when deploying artificial intelligence in the public sector. This engagement is not a mere procedural step but a foundational element for building trust, ensuring equity, and ultimately, developing AI systems that genuinely serve the public good. Having meticulously mapped and analyzed these stakeholders, the next logical and perhaps most potent step is to bring them together in a structured, collaborative environment. This is where the concept of co-design truly takes flight, moving beyond consultation to a shared process of creation and refinement.

To translate the insights gleaned from stakeholder analysis into actionable progress, a well-structured co-design workshop is indispensable. This subsection provides a practical framework for planning and facilitating such workshops, ensuring they are productive, inclusive, and yield tangible outcomes. The goal is to empower all participants, regardless of their technical background or position within society, to contribute meaningfully to the design and development of public sector AI. This template is designed to be flexible, adaptable to diverse project needs, participant groups, and specific stages of AI development, from initial concept ideation to the refinement of deployed systems.

At its heart, a co-design workshop aims to bridge the gap between the lived experiences of those affected by AI and the technical expertise of its creators. It is a space where assumptions can be challenged, biases can be surfaced and addressed, and shared understanding can be built. The process itself should embody the principles of equity and inclusion that the AI system itself is striving to uphold. This means carefully considering who is invited, how they are invited, and what support is provided to ensure their full and equal participation.

**Planning Your Co-Design Workshop: A Structured Approach**

Effective co-design does not happen by accident; it requires deliberate planning. This planning process begins with clearly defining the workshop's objectives. What specific questions do you aim to answer? What decisions need to be made? What outputs are expected? These objectives should be concrete and measurable, guiding the selection of participants, activities, and the overall structure of the workshop. For instance, an objective might be: "To collaboratively identify and prioritize the key ethical considerations for an AI-powered citizen feedback platform," or "To co-develop user interface prototypes for an AI-assisted public service portal, focusing on accessibility for diverse user groups."

**Participant Selection: The Foundation of Inclusivity**

Drawing directly from the stakeholder analysis, the participant roster is paramount. Aim for a representative mix that includes individuals with direct lived experience, community representatives, domain experts, and technical team members. It is crucial to go beyond the usual suspects and actively seek out participants from marginalized or underrepresented communities who might be disproportionately impacted by the AI. Consider how to extend invitations: a formal letter might not reach everyone effectively. Employing community liaisons, utilizing trusted local organizations, and providing clear, jargon-free information about the workshop's purpose and its benefits for participants are essential. Offering compensation for participants' time and expertise, especially for those from less privileged backgrounds, is not only an ethical imperative but also a practical necessity to ensure equitable participation. This might include stipends, covering travel expenses, or providing childcare.

**Workshop Structure and Agenda: Guiding the Collaborative Journey**

A typical co-design workshop, ranging from half a day to two full days, benefits from a balanced structure that blends focused activities with opportunities for informal interaction and reflection. Here is a sample agenda that can be adapted:

**Sample Co-Design Workshop Agenda (1-Day Format)**

**9:00 AM - 9:30 AM: Welcome and Setting the Stage**
**Objective:** To create a welcoming atmosphere, establish common ground, and introduce the workshop's purpose and expected outcomes.
**Activities:**
**Icebreaker:** A brief, inclusive activity designed to help participants get to know each other and feel more comfortable. Examples include a "two truths and a lie" related to their experiences with public services, or a quick round of sharing what they hope to gain from the day.
**Welcome Remarks:** A senior leader or project sponsor who emphasizes the value of co-design and the commitment to incorporating participant input.
**Introduction of Project and Goals:** A concise overview of the AI initiative, its purpose, and the specific focus of this workshop. Crucially, this should be explained in plain language, avoiding technical jargon.
**Co-design Principles and Ground Rules:** Clearly articulate the principles of respectful dialogue, active listening, valuing diverse perspectives, and a commitment to constructive collaboration. Co-creating these ground rules with the participants can enhance buy-in.

**9:30 AM - 10:30 AM: Understanding the Landscape: Current Experiences and Challenges**
**Objective:** To surface participants' current experiences, pain points, and aspirations related to the public service or domain the AI aims to address.

**Activities:**
**Personal Storytelling:** Participants share brief personal anecdotes or observations related to the problem the AI is designed to solve. This can be done in small groups or as a whole, depending on group size and comfort levels. For example, a workshop on an AI for housing assistance might ask participants to share their journey navigating the housing application process.
**Empathy Mapping or Journey Mapping:** In small, mixed groups (e.g., community members, developers), participants collaboratively map out the steps, thoughts, feelings, and pain points associated with a specific user journey or interaction with a public service. This visually captures the user experience from different perspectives.

**10:30 AM - 10:45 AM: Coffee Break**

**10:45 AM - 12:00 PM: Exploring the AI Opportunity: Concepts and Possibilities**
**Objective:** To introduce AI concepts relevant to the project in an accessible way and brainstorm potential applications and benefits.
**Activities:**
**Demystifying AI (Briefly):** A non-technical explanation of the specific AI capabilities being considered, focusing on what *they can do and* how *they might help, rather than the complex* how-to. Use relatable analogies.
**Brainstorming AI Solutions:** Participants, in mixed groups, brainstorm ways AI could address the challenges identified earlier. This could involve "How Might We..." questions. For instance, "How might we use AI to ensure equitable distribution of public resources?" or "How might we leverage AI to make accessing public services more intuitive for everyone?"
**Affinity Mapping:** Ideas generated are grouped into themes or categories to identify patterns and emerging concepts.

**12:00 PM - 1:00 PM: Lunch Break and Informal Networking**

**1:00 PM - 2:30 PM: Deep Dive into Specific Design Challenges (Activity Rotations)**
**Objective:** To collaboratively work on concrete design elements or ethical dilemmas. This section often involves rotating through different stations or activities.
**Possible Activity Rotations:**
**Scenario Building:** Participants work in groups to develop detailed scenarios of how an AI system might be used or misused. They then discuss the potential impacts and desired outcomes for each scenario.
**Prototyping (Low-Fidelity):** Using paper, sketches, or simple digital tools, participants create basic mock-ups of interfaces or workflows. This focuses on functionality and user flow, not aesthetics. For an AI system, this might involve sketching out how a user would interact with an AI chatbot or how an AI-generated recommendation would be presented.
**Ethical Dilemma Framing:** Present participants with specific ethical quandaries related to the AI (e.g., data privacy vs. service personalization, bias in algorithms). Groups discuss the trade-offs, potential harms, and desired safeguards. They might develop "ethical design principles" for the AI.

**2:30 PM - 3:00 PM: Feedback and Prioritization**
**Objective:** To gather feedback on the co-developed ideas or prototypes and prioritize the most promising concepts or solutions.
**Activities:**
**Gallery Walk:** Participants review the outputs from the different activity rotations (e.g., sketches, scenario descriptions, ethical principles). They

provide feedback using dot voting, sticky notes, or brief written comments.

**Prioritization Exercise:** Using methods like dot voting or a simple ranking system, participants identify the top 2-3 concepts or solutions that the team should focus on next.

**3:00 PM - 3:15 PM: Coffee Break**

**3:15 PM - 4:00 PM: Synthesizing and Next Steps**

**Objective:** To summarize key takeaways, agree on immediate next steps, and outline the path forward.

**Activities:**

**Key Takeaways Synthesis:** Facilitators summarize the main themes, insights, and decisions from the day. This should be a collaborative process, with participants invited to add or correct.

**Action Planning:** Clearly define what happens next. This includes who is responsible for what, by when. This might involve further research, refining prototypes, scheduling follow-up meetings, or preparing reports.

**Commitment and Closing:** Reiterate the value of the participants' contributions and express gratitude. Clearly communicate how their input will be used and how they will be kept informed about progress.

**Facilitation: The Art of Guiding Collaboration**

Effective facilitation is the linchpin of a successful co-design workshop. A skilled facilitator creates and maintains a safe, inclusive, and productive environment. This requires a delicate balance of guiding the process, managing group dynamics, and ensuring that all voices are heard.

**Key Facilitation Strategies:**

1.
**Neutrality and Objectivity:** Facilitators should remain neutral, avoiding taking sides or imposing their own opinions. Their role is to enable the group to reach its own conclusions. They must be adept at spotting and gently addressing biases, both in participants and within the project itself.
2. **Active Listening and Observation:** Pay close attention to both verbal and non-verbal cues. Observe who is speaking, who is silent, and the overall energy of the room. This allows facilitators to intervene constructively when needed, perhaps by prompting quieter participants or redirecting unproductive discussions.
3. **Clear Communication and Direction:** State instructions for activities clearly and concisely. Check for understanding. Provide ample time for each activity but be prepared to adjust the schedule if a particular discussion is proving highly fruitful or is getting bogged down.
4. **Managing Group Dynamics:** Skilled facilitators can navigate disagreements constructively. They can encourage dissent by framing it as a valuable opportunity for deeper understanding, rather than a conflict to be avoided. Techniques like round-robin sharing, the "talking stick" method, or small breakout groups can ensure more equitable participation. They must also be prepared to manage dominant personalities and create space for more introverted or hesitant participants.
5. **Visual Facilitation:** Utilize whiteboards, flip charts, sticky notes, and digital collaboration tools to capture ideas and discussions visually. This makes the collective thinking process transparent and accessible to all. It also helps to externalize abstract concepts, making them easier to grapple with. For instance, drawing a journey map as participants describe it makes the experience tangible.

6. **Time Management:** Keep the workshop on track by respecting the allocated times for each segment. However, flexibility is key; a critical discussion might warrant a few extra minutes, while a less productive segment might need to be gently shortened.
7. **Empathy and Psychological Safety:** Create an environment where participants feel safe to express their genuine thoughts, concerns, and even frustrations without fear of judgment or reprisal. This is particularly crucial when discussing sensitive topics or potential harms of AI. Acknowledge emotions and validate participants' experiences.
8. **Bridging Technical and Non-Technical Language:** Facilitators often act as translators, helping to bridge the communication gap between technical experts and community members. They must be able to rephrase complex technical jargon into understandable terms and, conversely, help translate lived experiences into language that technical teams can understand and act upon.
9. **Documenting and Synthesizing:** The facilitator, or a dedicated scribe, must meticulously document the proceedings, capturing key ideas, decisions, action items, and any unresolved questions. This documentation is vital for follow-up and for demonstrating that participants' contributions have been heard and recorded.

### Adapting the Framework: Flexibility in Practice

The provided agenda and facilitation tips are a starting point, not a rigid prescription. The specific needs of the AI initiative and its stakeholders will dictate necessary adaptations.

**Project Stage:** For early-stage concept development, the focus might be more on brainstorming and visioning. For later stages, particularly around testing and refinement, the workshop might involve evaluating prototypes or testing specific features.

**Participant Demographics:** Workshops involving young people will require different activities and facilitation styles than those with senior citizens or busy professionals. Consider attention spans, preferred communication methods, and accessibility needs. For instance, a workshop for seniors might benefit from larger print materials, fewer transitions, and more verbal explanations.

**Virtual vs. In-Person:** While in-person workshops often foster stronger connections, virtual formats may be necessary for geographical dispersion or resource constraints. Virtual co-design requires different tools (e.g., Miro, Mural, Jamboard) and specific facilitation techniques to maintain engagement and prevent digital fatigue. Ensuring everyone has reliable internet access and is comfortable with the technology is paramount.

**Duration and Frequency:** A single intensive workshop might be suitable for some projects, while others may benefit from a series of shorter, more frequent sessions that allow for iterative development and deeper relationship building.

## Beyond the Workshop: Sustaining Engagement

The co-design workshop is a powerful catalyst, but it is not an endpoint. The commitment to collaborative practice must extend beyond the single event. This involves:

**Transparent Reporting:** Sharing the workshop outcomes and how the input will be integrated into the project with all participants and relevant stakeholders.

**Ongoing Communication:** Keeping participants informed about project progress, challenges, and future opportunities for involvement.
**Feedback Loops:** Establishing clear channels for participants to provide ongoing feedback as the AI system evolves.
**Iterative Co-design:** Incorporating feedback from the workshop into revised designs or prototypes, and then bringing participants back to co-evaluate these iterations. This demonstrates a genuine commitment to a continuous, collaborative process.

By embracing the principles of co-design and employing structured planning and skillful facilitation, public sector organizations can move towards developing AI systems that are not only technically sound but also deeply aligned with the needs, values, and aspirations of the communities they serve. This shift from top-down development to a truly participatory approach is essential for building public trust and ensuring that AI serves as a force for equitable progress. The investment in thoughtful co-design pays dividends in the form of more robust, relevant, and ultimately, more ethical public sector AI. It fosters a sense of shared ownership and ensures that the technologies deployed are truly for the people, by the people, and with the people.

The preceding sections have illuminated the transformative power of co-design in shaping public sector AI initiatives, moving us from abstract principles to concrete collaborative practices. We've explored how to foster inclusive environments, design engaging agendas, and facilitate meaningful dialogue that bridges technical expertise with lived experience. Now, as we transition from the 'how' of co-design to the 'what' of equitable AI deployment, a critical next step emerges: systematically measuring and verifying that our AI systems uphold the very principles of fairness and equity we strive for. This involves translating our ethical commitments and co-designed aspirations into quantifiable metrics, a process that demands careful planning and thoughtful execution. To support this vital endeavor, we introduce the

Equity Metrics Design Worksheet, a practical tool designed to guide teams through the complex but essential task of operationalizing fairness in AI.

**Operationalizing Equity: The Equity Metrics Design Worksheet**

The journey from identifying potential biases during co-design to actively mitigating them in a deployed AI system requires a robust framework for measurement. This is where the Equity Metrics Design Worksheet comes into play. It serves as a structured guide, prompting teams to think critically about how to define, measure, and monitor fairness and equity within their specific AI contexts. This worksheet is not merely a data collection form; it's a strategic planning tool designed to embed equity considerations at every stage of the AI lifecycle, from initial development to ongoing operation.

The worksheet encourages a holistic approach by asking teams to consider multiple dimensions of fairness, recognizing that a single metric rarely captures the full picture of equitable impact. It prompts reflection on who might be disproportionately affected by the AI system, how those impacts manifest, and what constitutes a fair outcome for each group. By systematically addressing these questions, teams can move beyond aspirational statements about fairness to demonstrable, evidence-based practices.

The process begins with a clear understanding of the AI system's intended purpose and its potential beneficiaries, as well as those who might be negatively impacted. This understanding, ideally informed by the co-design workshops, forms the bedrock upon which equity metrics are built. Without this foundational knowledge, metrics can become arbitrary or misaligned with the actual ethical challenges.

**Section 1: System Context and Ethical Objectives**

Before diving into specific metrics, it's crucial to ground the exercise in the reality of the AI system and its intended equitable outcomes. This initial section of the worksheet serves to capture this essential context.

**AI System Description:** A brief, clear description of the AI system. What problem does it solve? What are its core functionalities? For instance, is it an AI for prioritizing social housing applications, an AI-powered chatbot for citizen inquiries, or an AI for optimizing public transport routes? The description should be accessible, avoiding overly technical jargon.
**Intended Beneficiaries and Potential Impacted Groups:** Who is this AI system designed to serve? More importantly, who are the specific populations or demographic groups that might be most affected by its decisions or outputs? This requires a nuanced understanding, informed by stakeholder analysis and co-design insights. For example, an AI for loan application processing might serve all citizens seeking loans but could disproportionately impact low-income individuals, recent immigrants, or individuals with non-traditional credit histories.
**Ethical Principles and Equity Goals:** What are the overarching ethical principles guiding the development and deployment of this AI? What specific equity goals are we trying to achieve? This is where the aspirations discussed in co-design sessions are articulated formally. Examples might include:

"To ensure equitable access to public services, irrespective of socioeconomic status."
"To prevent algorithmic discrimination against protected demographic groups."
"To promote fairness in resource allocation."
"To ensure transparency and explainability for all users, especially vulnerable populations."
"To maintain user privacy and data security for all citizens."

This section is not just about documentation; it's about alignment. It ensures that the subsequent metric selection is directly tied to the stated mission and ethical commitments of the AI initiative. It's a checkpoint to confirm that the team is focusing its measurement efforts on the most relevant ethical challenges.

## Section 2: Identifying Potential Biases and Harms

With the system context established, the next step is to proactively identify where bias might creep in and what harms could result. This section encourages a speculative, critical examination of the AI system's potential failure modes from an equity perspective.

**Sources of Potential Bias:** For each identified demographic group that could be impacted, teams should brainstorm potential sources of bias. These could include:
**Data Bias:** Are there historical biases present in the data used to train the model? (e.g., underrepresentation of certain groups, biased labeling). For an AI system processing job applications, if historical hiring data

reflects past discriminatory practices, the AI might perpetuate those biases.
**Algorithmic Bias:** Does the algorithm itself introduce or amplify biases? (e.g., feature selection that disproportionately affects certain groups, proxy variables that correlate with protected attributes). If an AI for credit scoring uses zip codes as a feature, and certain zip codes are strongly correlated with racial or ethnic groups due to historical segregation, this could introduce algorithmic bias.
**Deployment Bias:** Does the way the AI is implemented or used in practice introduce bias? (e.g., differential access to the AI, biased human interpretation of AI outputs). If an AI-powered medical diagnostic tool is only made available in well-resourced hospitals, it could lead to disparities in healthcare access.
**Measurement Bias:** Are the evaluation metrics themselves biased, or do they fail to capture important differences across groups? If an AI for predicting recidivism is evaluated solely on overall accuracy, it might mask higher false positive rates for certain racial groups.

**Potential Harms:** For each identified source of bias, what are the potential negative consequences for the impacted groups? These harms can be direct or indirect, individual or systemic. Examples include:
**Unfair Allocation:** Denial of services, benefits, or opportunities (e.g., denial of loans, housing, or essential services).
**Discriminatory Outcomes:** Being subjected to unwarranted surveillance, profiling, or harsher penalties.
**Reinforcement of Stereotypes:** Perpetuating harmful societal stereotypes through biased outputs or recommendations.
**Erosion of Trust:** Damaging public trust in government services and technology.
**Exclusion:** Preventing certain groups from accessing essential information or resources.

**Disproportionate Burden:** Imposing additional effort or hardship on specific communities.

This proactive identification of biases and harms is crucial. It shifts the focus from simply reacting to problems after they arise to building equity into the system from the outset. It ensures that the subsequent selection of metrics is targeted towards preventing or mitigating these specific, identified risks.

**Section 3: Selecting Fairness Metrics**

This is often the most technically challenging section, requiring careful consideration of different definitions of fairness and their applicability to the specific AI system. The worksheet prompts teams to select metrics that best align with their identified ethical goals and potential harms. It's essential to acknowledge that different fairness metrics can sometimes be in tension with each other, and the choice involves trade-offs.

For each identified group and potential harm, teams should consider appropriate fairness metrics. It's beneficial to group metrics by the "type" of fairness they aim to measure.

**Group Fairness Metrics:** These metrics assess whether the AI system's outcomes are comparable across different demographic groups.
**Demographic Parity (or Statistical Parity):** This metric requires that the proportion of individuals receiving a positive outcome is the same across all protected groups. For example, if an AI is used to determine eligibility for a job training program, demographic parity would mean

that the program acceptance rate is equal for men and women, or for different racial groups.

Formulaic Representation: $P(\hat{Y}=1 \mid G=g) = P(\hat{Y}=1 \mid G=h)$ for all groups g, h, where $\hat{Y}$ is the predicted outcome and G is the protected group attribute.

When to Consider: Useful when the goal is to ensure equal representation or access, and when the base rates of the outcome are not expected to differ significantly across groups for legitimate reasons.

Potential Downsides: Can lead to inefficient outcomes if base rates differ. May not be appropriate if the AI is predicting a risk that legitimately varies by group.

**Equalized Odds:** This metric requires that the true positive rates (sensitivity) and false positive rates are equal across different groups. This is a more stringent condition than demographic parity.

Formulaic Representation: $P(\hat{Y}=1 \mid Y=1, G=g) = P(\hat{Y}=1 \mid Y=1, G=h)$ AND $P(\hat{Y}=1 \mid Y=0, G=g) = P(\hat{Y}=1 \mid Y=0, G=h)$ for all groups g, h, where Y is the true outcome.

When to Consider: Appropriate when accurately identifying positive cases (true positives) and avoiding incorrectly classifying negative cases as positive (false positives) are equally important across groups. For instance, in a fraud detection system, you want the rate of correctly identifying fraud to be the same across all customer segments, and the rate of flagging legitimate transactions as fraudulent to also be the same.

Potential Downsides: Can be difficult to satisfy simultaneously, especially if base rates differ.

**Equal Opportunity:** This metric focuses on ensuring that the true positive rates are equal across groups, while allowing false positive rates to differ.

Formulaic Representation: $P(\hat{Y}=1 \mid Y=1, G=g) = P(\hat{Y}=1 \mid Y=1, G=h)$ for all groups g, h.

When to Consider: *Applicable when the primary concern is ensuring that individuals who* should receive a positive outcome are identified equally across groups. For example, in an AI used for identifying candidates for a

scholarship, equal opportunity would ensure that qualified candidates from all backgrounds have an equal chance of being flagged.
Potential Downsides: Does not address potential disparities in false positive rates, which could still lead to unfairness for some groups.

**Individual Fairness Metrics:** These metrics focus on ensuring that similar individuals are treated similarly by the AI system, regardless of their group membership. This often requires defining a similarity metric between individuals.
**Counterfactual Fairness:** This metric aims to ensure that an individual's outcome would remain the same even if their protected attribute (e.g., race, gender) were changed, assuming all other relevant factors remain constant. This is a conceptually powerful but often difficult-to-implement metric, as it requires understanding causal relationships and having access to counterfactual data.
When to Consider: When the goal is to eliminate the influence of protected attributes on decisions, ensuring that decisions are based on merit or relevant factors only.
Potential Downsides: Requires strong assumptions about causal models and can be challenging to measure in practice without controlled experiments or sophisticated causal inference techniques.

**Other Relevant Metrics:**
**Predictive Rate Parity:** The positive predictive value (PPV) should be equal across groups. This means that among individuals predicted to have a positive outcome, the proportion who actually have that outcome is the same across groups.
When to Consider: Useful when the cost of false positives is high, and it's important that a positive prediction means the same thing regardless of group membership.

**Calibration:** The predicted probability of an outcome should reflect the true probability of that outcome within each group. For example, if the AI predicts a 70% chance of a positive outcome for a certain group, then approximately 70% of individuals in that group for whom the AI predicted a 70% chance should actually experience the positive outcome.
When to Consider: Important for decision-making processes where probabilities are used directly, or when users need to trust the accuracy of the predicted probabilities.

**Worksheet Prompt for Section 3:**

For each identified impacted group and potential harm, specify:

1.
**Primary Fairness Metric(s):** Which specific metric(s) (e.g., Demographic Parity, Equalized Odds, Equal Opportunity) best addresses this harm and aligns with our equity goals? Justify the choice.
2. **Secondary Fairness Metric(s):** Are there other metrics that should be monitored to provide a more complete picture of fairness, even if they are not the primary focus?
3. **Definition of "Positive Outcome":** Clearly define what constitutes a "positive outcome" for the purpose of calculating these metrics in this specific context. (e.g., "receiving the service," "being approved for the loan," "being flagged as high risk").
4. **Protected Attribute(s):** What are the demographic or group attributes (e.g., race, gender, age, disability status, socioeconomic status) for which fairness will be measured? Ensure these are defined clearly and are relevant to the identified potential harms.

**Section 4: Defining Target Values and Thresholds**

Selecting a metric is only the first step; the next is to define what constitutes an acceptable level of fairness. This involves setting target values or acceptable ranges for the chosen metrics. This is a crucial step that translates abstract fairness into concrete operational targets.

**Target Value/Range:** For each selected primary fairness metric, what is the desired target value or acceptable range? This is where the co-designed aspirations meet practical realities.
For Demographic Parity: Target might be a difference of no more than 5% between groups. Or, ideally, a difference of 0%.
For Equalized Odds/Equal Opportunity: Target might be that the difference in true positive rates and false positive rates between any two groups is no more than 3%.
For Calibration: Target might be that predicted probabilities are within a +/- 5% margin of actual probabilities for each group.

**Justification for Target Values:** Why were these specific target values chosen? This justification should be informed by:
**Legal and Regulatory Requirements:** Are there any legal mandates or guidelines that dictate minimum fairness standards?
**Societal Norms and Expectations:** What level of fairness is generally considered acceptable by the public or stakeholders?
**Severity of Potential Harms:** If the harms of unfairness are severe, the target values for fairness should be stricter.
**Data Availability and Limitations:** Sometimes, achieving perfect parity is not feasible with current data or models. The targets should be ambitious yet realistic, with a plan to improve over time.

**Co-design Input:** What did community members and stakeholders deem acceptable or desirable in terms of fairness outcomes?

**Minimum Acceptable Threshold:** What is the absolute minimum acceptable performance before the system is deemed unfair and requires intervention? This threshold is critical for triggering alerts or decision-making processes when fairness degrades. This should be a more conservative value than the target value.

**Worksheet Prompt for Section 4:**

For each primary fairness metric identified in Section 3:

1.
**Target Value/Range:** Specify the target value or acceptable range.
2. **Minimum Acceptable Threshold:** Specify the minimum threshold for acceptable fairness.
3. **Justification:** Provide a clear rationale for these targets and thresholds, referencing legal, societal, harm severity, data, and co-design considerations.

**Section 5: Data Collection and Monitoring Plan**

Defining metrics and targets is meaningless without a plan for how to collect the necessary data and monitor performance over time. This section focuses on the practicalities of implementation.

**Data Requirements:** What specific data is needed to calculate the chosen fairness metrics? This includes:
**Outcome Variable:** The actual outcome of the AI's decision or prediction.
**Predicted Outcome:** The output of the AI system.
**Protected Attribute(s):** The demographic or group identifiers for which fairness is being measured.
**Other Relevant Variables:** Any other data points necessary for calculating specific metrics or for stratification (e.g., relevant covariates).

**Data Collection Method:** How will this data be collected?
Will it be logged automatically by the system?
Will it require manual data entry or annotation?
What is the process for ensuring the accuracy and reliability of collected data?
How will privacy be maintained during data collection and storage? (e.g., anonymization, aggregation, differential privacy techniques).

**Monitoring Frequency:** How often will the fairness metrics be calculated and reviewed?
Real-time? Daily? Weekly? Monthly? Quarterly?
The frequency should depend on the criticality of the AI system and the speed at which outcomes can change. For high-stakes applications, more frequent monitoring is essential.

**Reporting Mechanism:** Who will receive the fairness reports? In what format?
Internal development teams? Ethics committees? Regulatory bodies? Public-facing dashboards?

Reports should clearly present the metrics, trends over time, and comparisons against targets and thresholds.

**Alerting and Intervention Procedures:** What happens when a fairness metric falls below the minimum acceptable threshold?
Who is responsible for investigating the issue?
What are the pre-defined steps for intervention (e.g., pausing the system, retraining the model, human review of decisions)?
How will remediation efforts be tracked and verified?

**Data Disaggregation:** Ensure that data is collected and analyzed in a disaggregated manner for each relevant protected attribute and intersectional groups (e.g., not just race, but also Black women, Hispanic men, etc.). This is crucial for uncovering nuanced disparities.

**Worksheet Prompt for Section 5:**

1.
**Required Data Fields:** List all data fields needed to calculate each metric.
2. **Data Collection Process:** Describe the method for collecting and storing this data, including privacy safeguards.
3. **Monitoring Cadence:** Specify the frequency of metric calculation and review.
4. **Reporting Structure:** Outline who receives reports, what they contain, and how often.
5. **Alert and Intervention Protocol:** Detail the steps to be taken when fairness thresholds are breached.

**Section 6: Iteration and Continuous Improvement**

Equity in AI is not a one-time achievement; it's an ongoing process. This final section emphasizes the need for continuous learning and adaptation.

**Feedback Loops:** How will feedback from monitoring and audits be incorporated back into the AI development lifecycle?
This includes not only technical adjustments but also potential revisions to the ethical goals or fairness metrics themselves, based on new insights or evolving societal understanding.

**Regular Review and Updates:** How often will the entire equity metrics framework (including metrics, targets, and monitoring plans) be reviewed and updated?
This review should ideally involve stakeholders and co-design participants to ensure the framework remains relevant and effective.

**Auditing and External Verification:** Consider the role of internal or external audits to verify the fairness of the AI system and the effectiveness of the implemented metrics.

**Learning and Documentation:** How will the lessons learned from monitoring, audits, and remediation efforts be documented and shared to inform future AI development within the organization?

**Worksheet Prompt for Section 6:**

1.
**Feedback Integration:** Describe the process for incorporating monitoring feedback into AI development and strategy.
2. **Framework Review Schedule:** Specify the cadence for reviewing and updating the equity metrics framework.
3. **Auditing Strategy:** Outline plans for internal or external fairness audits.
4. **Knowledge Management:** Describe how lessons learned will be documented and disseminated.

By systematically working through the Equity Metrics Design Worksheet, public sector organizations can move beyond good intentions towards measurable, demonstrable equity in their AI systems. This structured approach ensures that fairness is not an afterthought but a core component of responsible AI development and deployment, fostering trust and ensuring that technology truly serves the public good. It provides a tangible bridge between the ideals of co-design and the practicalities of building equitable AI solutions. This detailed process, when completed collaboratively and with genuine commitment, can transform abstract fairness principles into actionable operational guidelines, paving the way for AI that is not only effective but also just.

The preceding sections have illuminated the transformative power of co-design in shaping public sector AI initiatives, moving us from abstract principles to concrete collaborative practices. We've explored how to foster inclusive environments, design engaging agendas, and facilitate meaningful dialogue that bridges technical expertise with lived experience. Now, as we transition from the 'how' of co-design to the 'what' of equitable AI deployment, a critical next step emerges: systematically measuring and verifying that our AI systems uphold the very principles of fairness and equity we strive for. This involves translating our ethical commitments and co-designed aspirations into quantifiable metrics, a process that demands careful planning and thoughtful execution. To support this vital endeavor, we introduce the Equity Metrics Design Worksheet, a practical tool designed to guide teams through the complex but essential task of operationalizing fairness in AI.

## Operationalizing Equity: The Equity Metrics Design Worksheet

The journey from identifying potential biases during co-design to actively mitigating them in a deployed AI system requires a robust framework for measurement. This is where the Equity Metrics Design Worksheet comes into play. It serves as a structured guide, prompting teams to think critically about how to define, measure, and monitor fairness and equity within their specific AI contexts. This worksheet is not merely a data collection form; it's a strategic planning tool designed to embed equity considerations at every stage of the AI lifecycle, from initial development to ongoing operation.

The worksheet encourages a holistic approach by asking teams to consider multiple dimensions of fairness, recognizing that a single metric rarely captures the full picture of equitable impact. It prompts reflection on who might be disproportionately affected by the AI system, how those impacts manifest, and what constitutes a fair outcome for each group. By systematically addressing these questions, teams can move beyond aspirational statements about fairness to demonstrable, evidence-based practices.

The process begins with a clear understanding of the AI system's intended purpose and its potential beneficiaries, as well as those who might be negatively impacted. This understanding, ideally informed by the co-design workshops, forms the bedrock upon which equity metrics are built. Without this foundational knowledge, metrics can become arbitrary or misaligned with the actual ethical challenges.

**Section 1: System Context and Ethical Objectives**

Before diving into specific metrics, it's crucial to ground the exercise in the reality of the AI system and its intended equitable outcomes. This initial section of the worksheet serves to capture this essential context.

**AI System Description:** A brief, clear description of the AI system. What problem does it solve? What are its core functionalities? For instance, is it an AI for prioritizing social housing applications, an AI-powered chatbot for citizen inquiries, or an AI for optimizing public transport routes? The description should be accessible, avoiding overly technical jargon.
**Intended Beneficiaries and Potential Impacted Groups:** Who is this AI system designed to serve? More importantly, who are the specific populations or demographic groups that might be most affected by its decisions or outputs? This requires a nuanced understanding, informed by stakeholder analysis and co-design insights. For example, an AI for loan application processing might serve all citizens seeking loans but could disproportionately impact low-income individuals, recent immigrants, or individuals with non-traditional credit histories.
**Ethical Principles and Equity Goals:** What are the overarching ethical principles guiding the development and deployment of this AI? What specific equity goals are we trying to achieve? This is where the aspirations discussed in co-design sessions are articulated formally. Examples might include:

"To ensure equitable access to public services, irrespective of socioeconomic status."
"To prevent algorithmic discrimination against protected demographic groups."
"To promote fairness in resource allocation."
"To ensure transparency and explainability for all users, especially vulnerable populations."
"To maintain user privacy and data security for all citizens."

This section is not just about documentation; it's about alignment. It ensures that the subsequent metric selection is directly tied to the stated mission and ethical commitments of the AI initiative. It's a checkpoint to confirm that the team is focusing its measurement efforts on the most relevant ethical challenges.

**Section 2: Identifying Potential Biases and Harms**

With the system context established, the next step is to proactively identify where bias might creep in and what harms could result. This section encourages a speculative, critical examination of the AI system's potential failure modes from an equity perspective.

**Sources of Potential Bias:** For each identified demographic group that could be impacted, teams should brainstorm potential sources of bias. These could include:
**Data Bias:** Are there historical biases present in the data used to train the model? (e.g., underrepresentation of certain groups, biased labeling). For an AI system processing job applications, if historical hiring data

reflects past discriminatory practices, the AI might perpetuate those biases.
**Algorithmic Bias:** Does the algorithm itself introduce or amplify biases? (e.g., feature selection that disproportionately affects certain groups, proxy variables that correlate with protected attributes). If an AI for credit scoring uses zip codes as a feature, and certain zip codes are strongly correlated with racial or ethnic groups due to historical segregation, this could introduce algorithmic bias.
**Deployment Bias:** Does the way the AI is implemented or used in practice introduce bias? (e.g., differential access to the AI, biased human interpretation of AI outputs). If an AI-powered medical diagnostic tool is only made available in well-resourced hospitals, it could lead to disparities in healthcare access.
**Measurement Bias:** Are the evaluation metrics themselves biased, or do they fail to capture important differences across groups? If an AI for predicting recidivism is evaluated solely on overall accuracy, it might mask higher false positive rates for certain racial groups.

**Potential Harms:** For each identified source of bias, what are the potential negative consequences for the impacted groups? These harms can be direct or indirect, individual or systemic. Examples include:
**Unfair Allocation:** Denial of services, benefits, or opportunities (e.g., denial of loans, housing, or essential services).
**Discriminatory Outcomes:** Being subjected to unwarranted surveillance, profiling, or harsher penalties.
**Reinforcement of Stereotypes:** Perpetuating harmful societal stereotypes through biased outputs or recommendations.
**Erosion of Trust:** Damaging public trust in government services and technology.
**Exclusion:** Preventing certain groups from accessing essential information or resources.

**Disproportionate Burden:** Imposing additional effort or hardship on specific communities.

This proactive identification of biases and harms is crucial. It shifts the focus from simply reacting to problems after they arise to building equity into the system from the outset. It ensures that the subsequent selection of metrics is targeted towards preventing or mitigating these specific, identified risks.

**Section 3: Selecting Fairness Metrics**

This is often the most technically challenging section, requiring careful consideration of different definitions of fairness and their applicability to the specific AI system. The worksheet prompts teams to select metrics that best align with their identified ethical goals and potential harms. It's essential to acknowledge that different fairness metrics can sometimes be in tension with each other, and the choice involves trade-offs.

For each identified group and potential harm, teams should consider appropriate fairness metrics. It's beneficial to group metrics by the "type" of fairness they aim to measure.

**Group Fairness Metrics:** These metrics assess whether the AI system's outcomes are comparable across different demographic groups.
**Demographic Parity (or Statistical Parity):** This metric requires that the proportion of individuals receiving a positive outcome is the same across all protected groups. For example, if an AI is used to determine eligibility for a job training program, demographic parity would mean

that the program acceptance rate is equal for men and women, or for different racial groups.

Formulaic Representation: $P(\hat{Y}=1 \mid G=g) = P(\hat{Y}=1 \mid G=h)$ for all groups g, h, where $\hat{Y}$ is the predicted outcome and G is the protected group attribute.

When to Consider: Useful when the goal is to ensure equal representation or access, and when the base rates of the outcome are not expected to differ significantly across groups for legitimate reasons.

Potential Downsides: Can lead to inefficient outcomes if base rates differ. May not be appropriate if the AI is predicting a risk that legitimately varies by group.

**Equalized Odds:** This metric requires that the true positive rates (sensitivity) and false positive rates are equal across different groups. This is a more stringent condition than demographic parity.

Formulaic Representation: $P(\hat{Y}=1 \mid Y=1, G=g) = P(\hat{Y}=1 \mid Y=1, G=h)$ AND $P(\hat{Y}=1 \mid Y=0, G=g) = P(\hat{Y}=1 \mid Y=0, G=h)$ for all groups g, h, where Y is the true outcome.

When to Consider: Appropriate when accurately identifying positive cases (true positives) and avoiding incorrectly classifying negative cases as positive (false positives) are equally important across groups. For instance, in a fraud detection system, you want the rate of correctly identifying fraud to be the same across all customer segments, and the rate of flagging legitimate transactions as fraudulent to also be the same.

Potential Downsides: Can be difficult to satisfy simultaneously, especially if base rates differ.

**Equal Opportunity:** This metric focuses on ensuring that the true positive rates are equal across groups, while allowing false positive rates to differ.

Formulaic Representation: $P(\hat{Y}=1 \mid Y=1, G=g) = P(\hat{Y}=1 \mid Y=1, G=h)$ for all groups g, h.

When to Consider: *Applicable when the primary concern is ensuring that individuals who* should receive a positive outcome are identified equally across groups. For example, in an AI used for identifying candidates for a

scholarship, equal opportunity would ensure that qualified candidates from all backgrounds have an equal chance of being flagged.
Potential Downsides: Does not address potential disparities in false positive rates, which could still lead to unfairness for some groups.

**Individual Fairness Metrics:** These metrics focus on ensuring that similar individuals are treated similarly by the AI system, regardless of their group membership. This often requires defining a similarity metric between individuals.
**Counterfactual Fairness:** This metric aims to ensure that an individual's outcome would remain the same even if their protected attribute (e.g., race, gender) were changed, assuming all other relevant factors remain constant. This is a conceptually powerful but often difficult-to-implement metric, as it requires understanding causal relationships and having access to counterfactual data.
When to Consider: When the goal is to eliminate the influence of protected attributes on decisions, ensuring that decisions are based on merit or relevant factors only.
Potential Downsides: Requires strong assumptions about causal models and can be challenging to measure in practice without controlled experiments or sophisticated causal inference techniques.

**Other Relevant Metrics:**
**Predictive Rate Parity:** The positive predictive value (PPV) should be equal across groups. This means that among individuals predicted to have a positive outcome, the proportion who actually have that outcome is the same across groups.
When to Consider: Useful when the cost of false positives is high, and it's important that a positive prediction means the same thing regardless of group membership.

**Calibration:** The predicted probability of an outcome should reflect the true probability of that outcome within each group. For example, if the AI predicts a 70% chance of a positive outcome for a certain group, then approximately 70% of individuals in that group for whom the AI predicted a 70% chance should actually experience the positive outcome.
When to Consider: Important for decision-making processes where probabilities are used directly, or when users need to trust the accuracy of the predicted probabilities.

**Worksheet Prompt for Section 3:**

For each identified impacted group and potential harm, specify:

1.
**Primary Fairness Metric(s):** Which specific metric(s) (e.g., Demographic Parity, Equalized Odds, Equal Opportunity) best addresses this harm and aligns with our equity goals? Justify the choice.
2. **Secondary Fairness Metric(s):** Are there other metrics that should be monitored to provide a more complete picture of fairness, even if they are not the primary focus?
3. **Definition of "Positive Outcome":** Clearly define what constitutes a "positive outcome" for the purpose of calculating these metrics in this specific context. (e.g., "receiving the service," "being approved for the loan," "being flagged as high risk").
4. **Protected Attribute(s):** What are the demographic or group attributes (e.g., race, gender, age, disability status, socioeconomic status) for which fairness will be measured? Ensure these are defined clearly and are relevant to the identified potential harms.

**Section 4: Defining Target Values and Thresholds**

Selecting a metric is only the first step; the next is to define what constitutes an acceptable level of fairness. This involves setting target values or acceptable ranges for the chosen metrics. This is a crucial step that translates abstract fairness into concrete operational targets.

**Target Value/Range:** For each selected primary fairness metric, what is the desired target value or acceptable range? This is where the co-designed aspirations meet practical realities.
For Demographic Parity: Target might be a difference of no more than 5% between groups. Or, ideally, a difference of 0%.
For Equalized Odds/Equal Opportunity: Target might be that the difference in true positive rates and false positive rates between any two groups is no more than 3%.
For Calibration: Target might be that predicted probabilities are within a +/- 5% margin of actual probabilities for each group.

**Justification for Target Values:** Why were these specific target values chosen? This justification should be informed by:
**Legal and Regulatory Requirements:** Are there any legal mandates or guidelines that dictate minimum fairness standards?
**Societal Norms and Expectations:** What level of fairness is generally considered acceptable by the public or stakeholders?
**Severity of Potential Harms:** If the harms of unfairness are severe, the target values for fairness should be stricter.
**Data Availability and Limitations:** Sometimes, achieving perfect parity is not feasible with current data or models. The targets should be ambitious yet realistic, with a plan to improve over time.

**Co-design Input:** What did community members and stakeholders deem acceptable or desirable in terms of fairness outcomes?

**Minimum Acceptable Threshold:** What is the absolute minimum acceptable performance before the system is deemed unfair and requires intervention? This threshold is critical for triggering alerts or decision-making processes when fairness degrades. This should be a more conservative value than the target value.

**Worksheet Prompt for Section 4:**

For each primary fairness metric identified in Section 3:

1.
**Target Value/Range:** Specify the target value or acceptable range.
2. **Minimum Acceptable Threshold:** Specify the minimum threshold for acceptable fairness.
3. **Justification:** Provide a clear rationale for these targets and thresholds, referencing legal, societal, harm severity, data, and co-design considerations.

**Section 5: Data Collection and Monitoring Plan**

Defining metrics and targets is meaningless without a plan for how to collect the necessary data and monitor performance over time. This section focuses on the practicalities of implementation.

**Data Requirements:** What specific data is needed to calculate the chosen fairness metrics? This includes:

**Outcome Variable:** The actual outcome of the AI's decision or prediction.

**Predicted Outcome:** The output of the AI system.

**Protected Attribute(s):** The demographic or group identifiers for which fairness is being measured.

**Other Relevant Variables:** Any other data points necessary for calculating specific metrics or for stratification (e.g., relevant covariates).

**Data Collection Method:** How will this data be collected?

Will it be logged automatically by the system?

Will it require manual data entry or annotation?

What is the process for ensuring the accuracy and reliability of collected data?

How will privacy be maintained during data collection and storage? (e.g., anonymization, aggregation, differential privacy techniques).

**Monitoring Frequency:** How often will the fairness metrics be calculated and reviewed?

Real-time? Daily? Weekly? Monthly? Quarterly?

The frequency should depend on the criticality of the AI system and the speed at which outcomes can change. For high-stakes applications, more frequent monitoring is essential.

**Reporting Mechanism:** Who will receive the fairness reports? In what format?

Internal development teams? Ethics committees? Regulatory bodies? Public-facing dashboards?

Reports should clearly present the metrics, trends over time, and comparisons against targets and thresholds.

**Alerting and Intervention Procedures:** What happens when a fairness metric falls below the minimum acceptable threshold?
Who is responsible for investigating the issue?
What are the pre-defined steps for intervention (e.g., pausing the system, retraining the model, human review of decisions)?
How will remediation efforts be tracked and verified?

**Data Disaggregation:** Ensure that data is collected and analyzed in a disaggregated manner for each relevant protected attribute and intersectional groups (e.g., not just race, but also Black women, Hispanic men, etc.). This is crucial for uncovering nuanced disparities.

**Worksheet Prompt for Section 5:**

1.
**Required Data Fields:** List all data fields needed to calculate each metric.
2. **Data Collection Process:** Describe the method for collecting and storing this data, including privacy safeguards.
3. **Monitoring Cadence:** Specify the frequency of metric calculation and review.
4. **Reporting Structure:** Outline who receives reports, what they contain, and how often.
5. **Alert and Intervention Protocol:** Detail the steps to be taken when fairness thresholds are breached.

**Section 6: Iteration and Continuous Improvement**

Equity in AI is not a one-time achievement; it's an ongoing process. This final section emphasizes the need for continuous learning and adaptation.

**Feedback Loops:** How will feedback from monitoring and audits be incorporated back into the AI development lifecycle?
This includes not only technical adjustments but also potential revisions to the ethical goals or fairness metrics themselves, based on new insights or evolving societal understanding.

**Regular Review and Updates:** How often will the entire equity metrics framework (including metrics, targets, and monitoring plans) be reviewed and updated?
This review should ideally involve stakeholders and co-design participants to ensure the framework remains relevant and effective.

**Auditing and External Verification:** Consider the role of internal or external audits to verify the fairness of the AI system and the effectiveness of the implemented metrics.

**Learning and Documentation:** How will the lessons learned from monitoring, audits, and remediation efforts be documented and shared to inform future AI development within the organization?

**Worksheet Prompt for Section 6:**

1.
**Feedback Integration:** Describe the process for incorporating monitoring feedback into AI development and strategy.
2. **Framework Review Schedule:** Specify the cadence for reviewing and updating the equity metrics framework.
3. **Auditing Strategy:** Outline plans for internal or external fairness audits.
4. **Knowledge Management:** Describe how lessons learned will be documented and disseminated.

By systematically working through the Equity Metrics Design Worksheet, public sector organizations can move beyond good intentions towards measurable, demonstrable equity in their AI systems. This structured approach ensures that fairness is not an afterthought but a core component of responsible AI development and deployment, fostering trust and ensuring that technology truly serves the public good. It provides a tangible bridge between the ideals of co-design and the practicalities of building equitable AI solutions. This detailed process, when completed collaboratively and with genuine commitment, can transform abstract fairness principles into actionable operational guidelines, paving the way for AI that is not only effective but also just.

### Community Feedback Collection and Analysis Tool

Beyond the technical metrics designed to quantify algorithmic fairness, the ongoing engagement with the community provides an invaluable qualitative and contextual layer of understanding. While the Equity Metrics Design Worksheet focuses on system performance against

predefined fairness goals, the community feedback mechanism ensures that the
*perception* and *lived experience* of fairness are continuously assessed and integrated. This is not about re-auditing the algorithm's outputs in a quantitative sense, but rather about understanding how the system's impacts are experienced by those it serves, identifying unforeseen consequences, and gathering insights for iterative improvement that go beyond numerical targets. A well-structured tool for collecting and analyzing this feedback is essential for maintaining a truly participatory approach throughout the AI lifecycle.

### Purpose and Design Principles of the Tool

The primary purpose of this tool is to systematically capture, organize, and analyze the diverse feedback received from community engagement activities. This ensures that valuable insights are not lost, that recurring themes are identified, and that the impact of community input on project decisions is transparently tracked. The design principles guiding such a tool should prioritize accessibility, clarity, verifiability, and actionability.

**Accessibility:** The tool should be usable by various team members, including project managers, ethicists, developers, and community liaisons. Its structure should be intuitive, allowing for easy data entry and retrieval.
**Clarity:** Feedback categorization and theme identification should be clear and consistent, reducing ambiguity. Definitions for sentiment analysis or impact assessment should be well-documented.
**Verifiability:** It should be possible to trace specific feedback items back to their source (e.g., a particular workshop, survey, or individual

consultation) and to see how they have been addressed or why they were not addressed.
**Actionability:** The ultimate goal is to inform decisions. The tool should facilitate the translation of feedback into concrete actions, policy adjustments, or design modifications.

## Components of the Community Feedback Tool

A comprehensive tool for community feedback collection and analysis can be conceptualized with several interlocking components, often best managed within a shared digital platform or a well-organized set of documents.

### 1. Feedback Source and Context Capture

Every piece of feedback, whether solicited or unsolicited, needs to be logged with its origin clearly identified. This provides crucial context for interpretation and verification.

**Feedback ID:** A unique identifier for each submission.
**Source Type:** Categorizes where the feedback came from (e.g., Community Workshop, Online Survey, Public Forum, Direct Email/Call, Social Media, Stakeholder Meeting, Co-design Session).
**Date Received:** Timestamp of when the feedback was submitted.
**Engagement Event/Activity (if applicable):** Specific name or date of the event where feedback was gathered.
**Demographic Information (Optional and Anonymized):** If collected and permissible, relevant demographic data about the feedback provider

(e.g., age range, geographic area, primary language, self-identified group) can help identify patterns related to specific community segments. This must be handled with extreme care regarding privacy and consent.
**Project/AI System Involved:** Which specific AI initiative or project phase this feedback pertains to.
**Original Feedback Text:** The verbatim feedback as provided by the community member.

**2. Categorization and Thematic Analysis Module**

This module is central to making sense of the volume of feedback. It involves classifying feedback into predefined or emergent categories and identifying overarching themes.

**Predefined Categories:** Based on the AI system's objectives, potential impacts, and ethical considerations identified during co-design. Examples might include:
**System Functionality/Performance:** Comments on how the AI works or performs.
**Fairness and Bias Concerns:** Feedback directly related to potential or perceived discrimination.
**Transparency and Explainability:** Questions or concerns about understanding AI decisions.
**Data Privacy and Security:** Feedback on how personal data is handled.
**Accessibility and Usability:** Comments on ease of access and use for diverse populations.
**Service Delivery Impact:** How the AI affects the delivery or quality of public services.
**Trust and Reliability:** Perceptions of whether the system can be trusted.

**Positive Feedback/Appreciation:** Comments highlighting successful aspects or benefits.
**Suggestions for Improvement:** Ideas for enhancing the system or its implementation.
**Emergent Themes:** A process for identifying new themes that arise organically from the feedback, which may not have been anticipated. This is particularly important for novel AI applications or complex socio-technical systems.
**Tagging System:** Implementing tags that can cross-reference multiple categories or themes. For instance, a comment about "AI denying loans to people in my neighborhood" could be tagged under "Fairness and Bias Concerns," "Service Delivery Impact," and potentially a neighborhood identifier if anonymized data is available.
**Sentiment Analysis (Automated or Manual):** Assessing the emotional tone of the feedback. This can range from positive, neutral, to negative. For public sector AI, a negative sentiment often warrants deeper investigation. Advanced tools can offer nuanced sentiment scoring (e.g., frustration, confusion, anger, satisfaction, hope).

### 3. Impact Assessment and Action Tracking

This component bridges the gap between receiving feedback and acting upon it. It ensures accountability and demonstrates that community input is valued and integrated.

**Impact Level:** Assigning a level of impact or importance to the feedback. This could be a simple scale (e.g., Low, Medium, High) or more detailed:
**Informational:** Provides context but requires no immediate action on the system itself.

**Minor Adjustment:** Suggests a small tweak to UI, documentation, or process.
**Significant Rework:** Indicates a need for substantial changes to algorithms, features, or data handling.
**Policy/Process Change:** Requires modification of organizational policies or operational procedures.
**Escalation:** Feedback that needs to be raised to higher levels of management, oversight bodies, or regulators.
**Action Taken:** A clear description of the response or action taken based on the feedback. Examples:
"No action required; feedback noted."
"User interface updated to improve clarity on X."
"Bias mitigation strategy for Y feature is being reviewed."
"Further investigation into fairness metric Z is underway."
"Decision made to pause deployment pending community consultation."
"Training materials revised based on user confusion."
**Status:** Tracking the progress of the action (e.g., Open, In Progress, Resolved, Deferred, Rejected).
**Responsible Party:** Identifying the individual or team responsible for addressing the feedback.
**Date of Resolution/Update:** Timestamp for when the action was completed or an update was made.
**Link to Decision Documentation:** A reference to any formal decision-making records, meeting minutes, or design documents that explain how the feedback influenced a decision. This provides auditable proof of responsiveness.

**4. Reporting and Visualization Dashboard**

To make the insights from collected feedback accessible and actionable for the project team and stakeholders, a reporting mechanism is essential.

**Summary Statistics:** Overall counts of feedback received, broken down by source, category, and sentiment.
**Trend Analysis:** Visualizations (e.g., line graphs) showing how feedback volume, sentiment, or specific concerns change over time, especially in relation to project milestones or deployments.
**Top Themes/Categories:** Identification of the most frequently raised issues or concerns. This can be presented as word clouds or bar charts.
**Sentiment Distribution:** Pie charts or bar graphs showing the proportion of positive, neutral, and negative feedback.
**Action Status Overview:** A dashboard view of the progress on addressing feedback, highlighting any overdue items or critical issues.
**Impact Over Time:** Visualization of how feedback has led to concrete changes or decisions throughout the project lifecycle.
**Heatmaps of Concerns:** If geographical or demographic data is available (and anonymized appropriately), heatmaps can show where specific concerns are most prevalent.

**Implementation Considerations**

Implementing such a tool requires careful planning and a commitment to its ongoing use.

**Platform Choice:** The tool can be implemented using various technologies, from dedicated feedback management software and CRM systems to robust spreadsheets (like Google Sheets or Excel) with defined structures and workflows, or project management platforms with custom fields. The key is consistency and shared access.

**Team Training:** All team members involved in collecting or analyzing feedback must be trained on how to use the tool, its categories, and the process for tagging and assessment.
**Regular Review Cadence:** Feedback analysis should not be a one-off activity. Regular team meetings (e.g., weekly or bi-weekly) should be dedicated to reviewing new feedback, discussing themes, and updating the action tracking.
**Feedback Loop Closure:** A crucial aspect is communicating back to the community about how their feedback has been used. This can be done through public reports, updates on project websites, or direct communication where appropriate. This closes the loop and reinforces the value of participation.
**Ethical Data Handling:** Strict adherence to privacy regulations (like GDPR) is paramount, especially when dealing with potentially sensitive demographic information or personal accounts of impact. Anonymization, aggregation, and informed consent are critical.

By systematically applying a tool for community feedback collection and analysis, public sector organizations can ensure that the voices of those most affected by AI systems are not only heard but are demonstrably integrated into the development and governance process. This complements the quantitative fairness metrics by providing the essential qualitative context, the human experience, and the continuous learning mechanism necessary for building truly equitable and trustworthy public sector AI. It moves beyond a one-time consultation to an embedded, ongoing dialogue, transforming the relationship between technology developers and the communities they serve into a genuine partnership. This dynamic engagement is fundamental to the long-term success and ethical integrity of any AI deployed in the public interest.

# Chapter 15: The Future of Responsible AI in Public Service

The trajectory of Artificial Intelligence within the public sector is not static; it is a dynamic landscape continually reshaped by technological advancements, evolving societal needs, and emergent policy considerations. As we stand on the precipice of further AI integration, several key trends are poised to redefine how public services are conceived, delivered, and experienced. Understanding these emerging frontiers is crucial for proactive planning, ensuring that innovation aligns with the core principles of equity, accountability, and public good.

One of the most significant technological waves cresting over the public sector is the rapid evolution and increasing accessibility of generative AI. These sophisticated models, capable of creating text, images, audio, and even code, hold immense potential for enhancing public engagement and communication. Imagine a future where government agencies can generate personalized responses to citizen inquiries across multiple channels, adapt complex policy documents into easily digestible summaries for diverse audiences, or even create tailored educational materials on public health initiatives. Generative AI could democratize access to information by breaking down language barriers and simplifying bureaucratic jargon, making government services more approachable and understandable for all. For instance, a local council might deploy a generative AI to draft initial versions of press releases, social media updates, or even answers to frequently asked questions on its website, freeing up human communication officers to focus on strategic messaging and complex stakeholder engagement. However, the deployment of such powerful tools necessitates a robust framework for ethical oversight. The potential for generating misinformation, perpetuating biases embedded in training data, or creating outputs that lack the nuanced understanding of human empathy demands careful consideration. Public sector organizations must establish clear guidelines for the responsible use of generative AI, including rigorous human review processes, transparency about AI involvement in communication, and mechanisms to detect and correct erroneous or biased outputs. The challenge lies in harnessing the efficiency and creativity of generative AI without compromising the integrity and trustworthiness of public communication.

Beyond communication, AI is emerging as a critical tool in tackling some of the most pressing global challenges, chief among them climate change. The intersection of AI and climate resilience is a rapidly expanding domain. AI algorithms can analyze vast datasets from weather patterns, satellite imagery, and environmental sensors to predict extreme weather events with greater accuracy, allowing for more timely and effective disaster preparedness and response. For example, AI can assist in identifying areas most vulnerable to flooding, wildfires, or heatwaves, enabling targeted interventions and infrastructure investments. Furthermore, AI can optimize resource allocation for climate adaptation strategies, such as managing water resources in drought-prone regions or optimizing energy grids to incorporate renewable sources more efficiently. Predictive models can forecast energy demand and supply, reducing waste and reliance on fossil fuels. AI-powered systems can also monitor deforestation, track carbon emissions from industrial sources, and help design more sustainable urban environments. The ethical considerations here are multifaceted. Ensuring equitable access to these predictive tools and the resources they help allocate is paramount, as vulnerable communities are often disproportionately affected by climate change. There's also a need for transparency in the models used for climate prediction, especially when these predictions inform critical infrastructure decisions or emergency protocols. The public must understand how AI is being used to protect them and how decisions are being made, fostering trust and enabling collective action.

The concept of personalized public service delivery, often discussed in the private sector, is gaining traction within government as AI capabilities mature. Historically, public services have often operated on a one-size-fits-all model, which can inadvertently disadvantage individuals with unique needs or circumstances. AI offers the potential to tailor services to individual requirements, improving efficiency and user satisfaction. This could manifest in numerous ways: an AI system could proactively identify citizens who might be eligible for social benefits they haven't applied for, based on anonymized data patterns. It could help create customized educational pathways for students struggling with specific subjects or assist in matching individuals with vocational training programs that align with their skills and local job market demands. In healthcare, AI could support personalized treatment plans or provide tailored health advice based on individual risk factors and preferences. The key to responsible personalization lies in careful data governance and a commitment to avoiding discriminatory practices. Systems must be designed to ensure that personalization does not lead to exclusion or the creation of tiered service systems that disadvantage certain groups. Robust anonymization techniques, clear consent mechanisms, and continuous auditing for bias are essential. The goal is to leverage AI to make services more inclusive and responsive, not to create new forms of digital redlining. For instance, an AI helping to allocate park maintenance resources could consider local population density, usage patterns, and community feedback to ensure equitable distribution, rather than simply relying on historical maintenance budgets which might reflect past inequities.

Another significant emerging trend is the increasing sophistication of AI in supporting public sector decision-making, moving beyond simple data analysis to more complex predictive and prescriptive capabilities. This includes AI models that can forecast demand for services, such as healthcare appointments, public transport usage, or emergency services responses, allowing for proactive resource allocation and staffing.

Predictive analytics can help identify individuals at risk of falling into poverty or requiring social support, enabling early intervention. In the justice system, AI is being explored for tasks like risk assessment for parole decisions or predicting potential crime hotspots, though these applications are fraught with significant ethical challenges related to fairness and bias that require extreme caution and rigorous oversight. The trend towards AI-assisted decision-making requires a profound shift in organizational culture, demanding that public servants develop new skills in interpreting AI outputs, understanding their limitations, and critically evaluating their recommendations. Moreover, transparency and explainability become paramount. If AI is informing decisions that significantly impact citizens' lives, the rationale behind those AI-driven recommendations must be understandable, auditable, and open to scrutiny. The development of "explainable AI" (XAI) techniques is therefore a critical enabler for this trend, aiming to shed light on the "black box" of complex algorithms. For example, if an AI recommends a particular allocation of limited housing resources, the justification for that recommendation, including the key factors considered and their weight, must be accessible to both decision-makers and potentially the affected individuals.

The rise of AI also necessitates a renewed focus on the infrastructure and skills required to support its widespread adoption. This includes investing in robust data management systems, secure cloud infrastructure, and ensuring interoperability between different government systems. However, perhaps more importantly, it requires a significant investment in human capital. Public sector employees will need to develop new competencies, ranging from data literacy and AI ethics awareness to the ability to work alongside AI systems effectively. This means rethinking education and training programs within government. Universities and professional development institutions will play a critical role in equipping the future public workforce with the necessary skills. Furthermore, a culture of continuous learning and adaptation will be essential, as AI technologies evolve at an unprecedented pace. Without a skilled and informed workforce, the potential benefits of AI will remain largely unrealized, and the risks of misuse or unintended consequences will be amplified.

Finally, the evolving landscape of AI adoption compels a forward-looking approach to governance and regulation. As AI systems become more pervasive, the need for clear ethical guidelines, robust regulatory frameworks, and effective oversight mechanisms becomes increasingly urgent. This involves not only addressing the immediate challenges of bias, transparency, and accountability but also anticipating future developments and their societal implications. Policymakers, ethicists, technologists, and the public must engage in ongoing dialogue to shape the future of AI in a way that aligns with public values and democratic principles. This might involve establishing dedicated AI ethics boards, developing sector-specific AI regulations, or fostering international collaboration to set global standards. The proactive identification and mitigation of emerging risks, coupled with a commitment to harnessing AI for public benefit, will be the hallmark of responsible innovation in the years to come. The journey is one of continuous adaptation, learning, and a steadfast dedication to ensuring that AI serves as a force for good in the public sphere.

The evolution of Artificial Intelligence within the public sector is characterized not only by technological advancements and new applications but also by a parallel and equally critical development: the continuous strengthening and adaptation of ethical AI governance frameworks. As governments increasingly leverage AI for public service delivery, decision-making, and operational efficiency, the imperative to establish robust, forward-looking governance structures becomes paramount. This involves a dynamic process of policy development, standard-setting, and regulatory adaptation, designed to ensure that AI deployment remains aligned with public values, societal well-being, and democratic principles. The very nature of AI, with its rapid pace of innovation and potential for unforeseen consequences, demands a governance approach that is not static but inherently adaptive and resilient.

At national and international levels, a discernible trend is the concerted effort to move beyond foundational ethical principles towards more concrete, actionable governance mechanisms. Early discussions often focused on broad ideals such as fairness, transparency, and accountability. While these remain foundational, the current phase of AI governance is increasingly concerned with translating these ideals into tangible policies, practical standards, and enforceable regulations. This shift reflects a growing understanding that abstract principles, without clear implementation pathways, can fall short in guiding the complex realities of AI development and deployment in public services. Governments are actively exploring various models, from the establishment of dedicated AI regulatory bodies to the integration of AI ethics considerations into existing sector-specific regulations. The challenge lies in creating frameworks that are sufficiently detailed to provide meaningful guidance and oversight, yet flexible enough to accommodate the rapid evolution of AI technologies and their applications. For instance, a nation might be developing a comprehensive AI Act, drawing inspiration from existing comprehensive data protection laws, to address issues like algorithmic bias, data privacy in AI systems, and the requirements for high-risk AI applications in critical public sectors like healthcare or criminal justice. This act would likely incorporate provisions for impact assessments, mandatory human oversight for certain AI-driven decisions, and mechanisms for redress for individuals adversely affected by AI.

The effectiveness of these governance frameworks is intrinsically linked to their ability to adapt to technological change. The AI landscape is characterized by continuous innovation, with new algorithms, models, and applications emerging at an accelerated pace. Governance structures that were conceived even a few years ago may already be outdated or insufficient to address emerging challenges. This necessitates a proactive and iterative approach to policy-making. Instead of attempting to regulate specific AI technologies, which can quickly become obsolete, governance frameworks are increasingly focusing on the underlying principles and processes that govern AI development and deployment. This includes establishing requirements for continuous risk assessment, ongoing monitoring and auditing of AI systems in operation, and mechanisms for updating regulations as new risks and capabilities emerge. For example, a government agency responsible for AI oversight might implement a policy that mandates regular re-certification of AI systems used in critical public functions, requiring developers to demonstrate that the systems still meet current ethical and performance standards, especially after significant updates or changes in operating environments. This continuous evaluation process ensures that governance remains relevant and effective in the face of ongoing technological shifts, rather than becoming a static, easily circumvented set of rules. The focus is shifting towards building adaptive capacity within the governance structures themselves.

A critical component of strengthening ethical AI governance is the imperative for multi-stakeholder collaboration. The development of effective and inclusive governance structures cannot be the sole purview of government or technology developers. A broad range of stakeholders, including civil society organizations, academic researchers, industry representatives, and the public at large, must be involved in shaping these frameworks. This collaborative approach ensures that governance structures are not only technically sound but also reflect a diverse array of societal values, concerns, and lived experiences. For instance, when developing guidelines for the use of AI in social welfare systems, it is crucial to involve individuals who rely on these services, as well as the frontline workers who administer them. Their insights can highlight potential unintended consequences, biases, or practical challenges that might be overlooked by policymakers or technologists alone. This engagement can take many forms, such as public consultations, advisory committees, and participatory design workshops. The goal is to foster a sense of shared ownership and responsibility for the ethical development and deployment of AI in the public sphere. International collaboration also plays a vital role, as AI technologies and their impacts transcend national borders. Sharing best practices, developing common standards, and coordinating regulatory approaches can prevent regulatory fragmentation and ensure a more consistent and ethical global AI ecosystem.

The practical implementation of adaptive governance also hinges on developing clear standards and benchmarks. While abstract principles are important, concrete standards provide measurable criteria against which AI systems and their development processes can be evaluated. These standards can cover a wide range of aspects, including data quality and integrity, algorithmic fairness metrics, transparency requirements, and security protocols. For instance, national standardization bodies, in collaboration with industry and academia, might develop specific technical standards for bias detection and mitigation in AI algorithms used for loan applications or hiring processes. These standards would outline acceptable levels of disparity across different demographic groups and specify methodologies for testing and reporting on these metrics. Similarly, standards for explainability could define the level of detail and format required for AI system explanations, ensuring they are understandable to the intended audience, whether that be a citizen seeking to understand a decision about their benefits or a regulator auditing a system for compliance. The development and adoption of such standards require a concerted effort to foster consensus and ensure that they are informed by both technical feasibility and ethical considerations.

Furthermore, the governance of AI in the public sector must grapple with the concept of accountability. As AI systems become more autonomous and their decision-making processes more complex, clearly defining who is responsible when something goes wrong becomes a significant challenge. This involves establishing clear lines of accountability for the design, development, deployment, and ongoing operation of AI systems. It means moving beyond diffuse responsibility and ensuring that individuals and organizations can be held answerable for the outcomes of AI. For example, a government department deploying an AI-powered traffic management system that inadvertently causes significant traffic disruptions due to a flaw in its predictive algorithms must have a clear chain of command and responsibility for addressing the issue, rectifying the system, and providing recourse to those affected. This might involve internal disciplinary procedures, external audits, or even legal liabilities depending on the severity of the impact. Establishing mechanisms for redress and appeal is therefore a crucial element of an accountable governance framework, ensuring that citizens have recourse when they believe an AI system has made an unfair or erroneous decision that affects them.

The ongoing refinement of these governance frameworks also necessitates a commitment to continuous learning and capacity building within public institutions. It is not enough to simply create policies and standards; public servants at all levels need to be equipped with the knowledge and skills to understand, implement, and uphold them. This involves investing in training programs that cover AI literacy, ethical AI principles, risk assessment methodologies, and the responsible use of AI tools. It also means fostering a culture that encourages critical thinking about AI and its potential impacts, and that supports employees in raising ethical concerns without fear of reprécussion. For example, a public sector training academy might develop a curriculum on "AI Ethics for Public Servants," covering topics such as identifying potential biases in datasets, understanding the limitations of AI models, and navigating the legal and ethical considerations of AI deployment. This ongoing investment in human capital ensures that the governance frameworks are not just documents on a shelf but are actively and effectively embedded in the daily operations of government.

Moreover, the evolving nature of AI governance requires a robust approach to enforcement and oversight. The most comprehensive ethical guidelines and regulations will have little impact if there are no effective mechanisms to ensure compliance. This involves establishing independent oversight bodies, conducting regular audits and assessments of AI systems, and imposing meaningful sanctions for non-compliance. The nature of these oversight mechanisms will likely vary depending on the context and the perceived risk associated with specific AI applications. For high-risk AI systems, such as those used in healthcare diagnostics or judicial proceedings, a more rigorous and intrusive form of oversight might be necessary, including pre-deployment assessments and ongoing monitoring by regulatory authorities. For lower-risk applications, a lighter touch approach might be more appropriate, relying on self-assessment and industry best practices, coupled with mechanisms for public reporting of issues. The key is to strike a balance that ensures accountability without stifling innovation or imposing undue burdens on public service delivery. The development of effective enforcement mechanisms is a critical step in building public trust in the use of AI by government, demonstrating a genuine commitment to responsible innovation.

Finally, the iterative and adaptive nature of AI governance means that it must be viewed as an ongoing journey rather than a destination. As AI technologies continue to advance and their societal impacts become more pronounced, the governance frameworks will need to be continually revisited, revised, and strengthened. This requires a long-term commitment from governments, a willingness to engage in continuous dialogue with stakeholders, and a proactive approach to anticipating and addressing emerging challenges. The goal is to create a governance ecosystem that is not only responsive to current needs but also resilient enough to guide the public sector through the uncertainties of the future of AI, ensuring that this powerful technology serves the public good in an equitable, transparent, and accountable manner. The ongoing process of strengthening ethical AI governance frameworks is, therefore, a fundamental prerequisite for realizing the full potential of AI in transforming public services for the better.

The integration of Artificial Intelligence into public service delivery and governance has often sparked debate, with concerns frequently raised about its potential to erode democratic principles. However, a more nuanced perspective reveals AI's significant capacity to act as a powerful catalyst for enhancing democratic participation and civic engagement. Rather than being an inherently anti-democratic force, AI, when thoughtfully designed and implemented, can democratize access to information, amplify citizen voices, and foster more inclusive and responsive governance. This section delves into the multifaceted ways in which AI can be harnessed to invigorate democratic processes, moving beyond theoretical possibilities to practical applications that empower citizens and strengthen the fabric of representative government.

One of the most immediate and impactful applications of AI in fostering democratic participation lies in its ability to democratize access to information. In an era often characterized by information overload and the proliferation of misinformation, AI can serve as a vital tool for filtering, organizing, and presenting complex governmental information in accessible formats. Government websites, legislative documents, public spending data, and policy proposals are often dense, technical, and difficult for the average citizen to navigate. AI-powered search engines and natural language processing (NLP) tools can significantly improve the discoverability of this information. Imagine a citizen seeking to understand a new piece of legislation. Instead of sifting through hundreds of pages of legal jargon, an AI assistant could provide a concise summary, highlight key provisions, and even answer specific questions about the bill's implications for their community. This is not about replacing the need for informed deliberation, but about lowering the barrier to entry for engaging with critical public information. Tools that can translate complex policy documents into plain language, identify relevant sections based on user queries, or even generate infographics and visualizations from raw data can empower individuals who might otherwise be deterred by the sheer complexity of governmental affairs. Furthermore, AI can be used to proactively disseminate relevant information. By analyzing public discourse and identifying emerging concerns, AI systems can help government agencies tailor and deliver information to specific demographic groups or geographic areas, ensuring that citizens are informed about decisions that affect them. This proactive approach transforms information access from a passive act of searching to an active and personalized engagement, thereby fostering a more informed citizenry, which is the bedrock of any healthy democracy.

Beyond mere information access, AI holds immense potential to facilitate and enhance public deliberation, a cornerstone of democratic decision-making. Traditional town hall meetings or public consultations, while valuable, are often limited by logistical constraints, geographical barriers, and the tendency for louder or more organized voices to dominate. AI can help overcome these limitations by creating new avenues for structured and inclusive dialogue. AI-powered platforms can be developed to moderate online discussions, summarize key arguments, identify areas of consensus and disagreement, and even detect the sentiment of participants. For instance, an AI system could be deployed to analyze thousands of comments submitted on a proposed policy, categorizing them by theme, identifying common concerns, and flagging outlier opinions. This allows policymakers to gain a more comprehensive and nuanced understanding of public sentiment than might be possible through manual review. Furthermore, AI can facilitate "deliberative polling" on a larger scale. By presenting participants with balanced information and structured prompts, AI tools can guide citizens through complex issues, encouraging them to consider different perspectives and engage in thoughtful discussion. Chatbots, for example, could be programmed to act as informed facilitators in online forums, posing clarifying questions, reminding participants of key facts, and encouraging respectful engagement. The goal is not to automate deliberation, but to augment human interaction with AI's capacity for processing large volumes of input and identifying patterns, thereby enabling richer and more representative public discourse. This can lead to policy outcomes that are more attuned to the needs and desires of the populace, strengthening the legitimacy of governmental decisions.

Empowering citizens to have a greater say in public policy is another critical area where AI can play a transformative role. This goes beyond providing feedback and delves into co-creation and participatory budgeting. AI can analyze citizen-generated data – such as reports of infrastructure issues, suggestions for local improvements, or priorities for public spending – and help translate this qualitative input into actionable insights. For example, an AI system could process thousands of geotagged photos and descriptions of potholes submitted by residents, automatically prioritizing repairs based on severity, traffic impact, and historical data. Similarly, in participatory budgeting processes, AI can help visualize the potential trade-offs of different spending proposals, allowing citizens to make more informed choices about how public funds are allocated. Imagine an AI tool that allows citizens to propose their own budget allocations, and then simulates the likely impact of these proposals on different public services. Such tools make complex budget decisions more tangible and understandable, thereby fostering a deeper sense of ownership and engagement. Furthermore, AI can be used to identify and engage underrepresented voices in policy discussions. By analyzing demographic data and participation patterns, AI systems can proactively reach out to communities that are traditionally marginalized or disengaged, ensuring their perspectives are included in policy development. This could involve using AI to identify preferred communication channels for specific communities and tailoring outreach efforts accordingly. The ultimate aim is to create a feedback loop where citizen input is not only collected but actively integrated into the policy-making cycle, making governance truly responsive to the people it serves.

However, realizing AI's potential to enhance democratic participation requires a commitment to responsible design and ethical implementation. The same AI tools that can democratize information could, if poorly designed, create new forms of digital exclusion. For instance, if AI-powered information systems are only accessible through high-speed internet or require advanced digital literacy, they risk further marginalizing those who lack these resources. Therefore, it is imperative that these tools are designed with inclusivity at their core, offering multiple access points, supporting various languages, and incorporating user-friendly interfaces that cater to diverse levels of technical proficiency. The principle of "digital inclusion" must be a guiding tenet in the development of any AI-driven civic engagement platform.

Transparency in how AI is used to facilitate participation is equally crucial. Citizens need to understand how their data is being collected, processed, and used to inform policy decisions. If an AI system is summarizing public comments, it should be clear how it is doing so, what criteria it uses to categorize feedback, and what its limitations are. Similarly, if AI is used to identify underrepresented groups, the methodology should be transparent and open to scrutiny to ensure it is not inadvertently reinforcing existing biases. This transparency builds trust and ensures that citizens can have confidence in the integrity of the participatory processes. Without it, AI-driven engagement could be perceived as a sophisticated form of public relations rather than genuine civic empowerment.

Accountability in the use of AI for democratic purposes is non-negotiable. When AI systems assist in summarizing public input or identifying areas of consensus, it is essential to establish clear lines of responsibility for the outcomes. Policymakers must remain ultimately accountable for the decisions made, even when informed by AI. The AI should be viewed as a tool to augment human judgment, not replace it. Mechanisms for human oversight and review must be embedded in any process where AI contributes to decision-making. This means ensuring that human officials can interrogate the AI's outputs, question its assumptions, and override its recommendations when necessary. For example, if an AI system flags a particular community concern as low priority based on its analysis, a human decision-maker should have the ability to review that assessment and consider contextual factors that the AI might have missed. Furthermore, mechanisms for citizen recourse are vital. If a citizen believes their input was misrepresented or overlooked due to an AI error, there must be a clear process for them to raise these concerns and seek a human review.

The potential for AI to exacerbate existing inequalities or introduce new forms of bias into democratic processes is a significant concern that must be proactively addressed. If AI systems are trained on biased historical data, they may perpetuate or even amplify those biases in their analysis of public sentiment or in their recommendations for policy. For example, an AI used to identify key themes in public comments might disproportionately prioritize concerns raised by demographic groups that have historically been more vocal, thereby overlooking the needs of quieter populations. Mitigating algorithmic bias requires a concerted effort to ensure that training data is representative and that AI models are rigorously tested for fairness across different demographic groups. This involves employing diverse development teams, conducting regular bias audits, and implementing fairness-aware machine learning techniques. The goal is to ensure that AI serves to bridge divides and amplify marginalized voices, rather than entrenching existing power structures.

Furthermore, the ethical deployment of AI in democratic participation demands a focus on fostering digital literacy and capacity building among citizens. For AI to truly empower citizens, they need to understand how these tools work, what their capabilities and limitations are, and how to engage with them effectively. Governments have a role to play in providing educational resources and training opportunities that demystify AI and equip citizens with the skills to participate meaningfully in AI-enhanced democratic processes. This could involve public workshops, online tutorials, and accessible explanations of how AI is being used in civic engagement initiatives. Without such capacity building, the benefits of AI-driven participation risk being limited to a digitally savvy elite.

The integration of AI into democratic processes also presents an opportunity to reimagine the very nature of civic engagement. It allows for more continuous and dynamic forms of participation, moving beyond periodic elections or occasional consultations. AI can facilitate ongoing dialogues between citizens and government, enabling a more fluid and responsive relationship. For instance, imagine an AI system that continuously monitors public sentiment on specific policy issues, providing real-time feedback to policymakers. This allows for more agile governance, where policies can be adjusted and refined in response to evolving public opinion. This also shifts the focus from a top-down model of governance to a more collaborative and co-creative approach, where citizens are active partners in shaping their communities.

In conclusion, the role of AI in enhancing democratic participation is not about replacing human deliberation or decision-making with machines. Instead, it is about leveraging AI's unique capabilities to overcome existing barriers to engagement, democratize access to information, amplify citizen voices, and foster more inclusive and responsive governance. By focusing on responsible design, transparency, accountability, and continuous capacity building, governments can harness AI to strengthen democratic institutions and ensure that the future of public service is one where every citizen has the opportunity to be informed, engaged, and heard. The potential is immense, but its realization hinges on a deliberate and ethical commitment to placing citizens at the center of technological innovation in governance.

The journey towards truly responsible AI in public service is not a solitary endeavor confined within national borders. Instead, it is a complex, multifaceted challenge that demands a collective, global response. As governments and public institutions worldwide grapple with the profound implications of integrating artificial intelligence into their operations, a critical need emerges for a robust and interconnected ecosystem of learning and collaboration. This is where the concept of a "global community of practice" for responsible AI in public services becomes not just beneficial, but imperative. Such a community would serve as a vibrant hub for knowledge exchange, a crucible for forging best practices, and a shared forum for dissecting and learning from the inevitable lessons—both positive and cautionary—that arise from deploying these transformative technologies.

The rationale for establishing such a global network is deeply rooted in the shared nature of the challenges and opportunities presented by AI. While the specific contexts of public service delivery may vary significantly from one nation to another, the fundamental ethical quandaries, the technical hurdles, and the societal impacts often share striking similarities. Consider the issue of algorithmic bias. A government in Scandinavia facing concerns about racial bias in an AI-powered child welfare risk assessment tool, or a ministry in Southeast Asia grappling with gender bias in an AI system used for loan application screening, are confronting essentially the same underlying problem. By fostering a global community of practice, these entities can move beyond isolated problem-solving. They can share anonymized case studies, discuss mitigation strategies that have proven effective, and collectively develop more sophisticated methods for detecting and rectifying bias. This collaborative approach accelerates the pace at which effective solutions are identified and disseminated, preventing each nation from having to "reinvent the wheel" at considerable cost in time and resources.

Beyond bias, consider the complexities of ensuring transparency and explainability in AI systems used for public decision-making. When an AI algorithm determines eligibility for social benefits, or flags individuals for closer scrutiny in law enforcement, the public's right to understand how these decisions are made is paramount. Different countries are experimenting with various technical and policy approaches to achieve this, from developing standardized explainability metrics to implementing robust audit trails and public registries of AI systems. A global community of practice would provide a platform for sharing these diverse approaches, allowing member nations to learn from each other's successes and failures. For instance, a country that has successfully implemented a framework for public AI audits could share its methodology, challenges encountered, and best practices with others who are in the nascent stages of developing similar initiatives. This cross-pollination of ideas and practical experience is invaluable, as it provides a broader toolkit of potential solutions and a clearer understanding of the implementation pathways that are most likely to be effective and build public trust.

Furthermore, the development of ethical guidelines and regulatory frameworks for AI in public service is an ongoing and evolving process. Many governments are in the process of crafting legislation, policy directives, and ethical codes to govern the responsible use of AI. These efforts are often complex, involving intricate trade-offs between innovation, public safety, privacy, and equity. A global community of practice would serve as a crucial forum for policymakers and legal experts to share their drafts, discuss emerging legal precedents, and collectively identify common principles that can underpin robust and adaptable regulations. Imagine a scenario where several nations are developing legislation around the use of AI in healthcare. By convening, they could share their approaches to data privacy in AI-driven medical diagnostics, discuss ethical considerations for AI-assisted surgical robots, and collaborate on developing international standards for the validation and deployment of such technologies. This collaborative approach can lead to more coherent and harmonized international regulations, which is particularly important in an increasingly interconnected world where AI technologies often transcend national boundaries.

The practical benefits of such a community extend to the realm of capacity building and talent development. Implementing responsible AI in public service requires a skilled workforce capable of understanding both the technical intricacies of AI and its ethical and societal implications. Many public sector organizations are facing a deficit of such expertise. A global community of practice can facilitate the sharing of training materials, curriculum development frameworks, and even joint training initiatives. Experts from countries with more advanced AI public sector programs could offer workshops or mentorship to their counterparts in nations with developing programs. This not only helps to build the necessary technical and ethical capacity within individual public administrations but also fosters a shared understanding of what constitutes "responsible AI" in practice, moving beyond abstract principles to concrete operationalization.

Moreover, a global community of practice can act as a powerful advocacy group, amplifying the collective voice of public sector AI practitioners and policymakers on the international stage. As AI technologies continue to evolve rapidly, there is a growing need for cohesive representation in global dialogues concerning AI governance, standards, and ethics. By presenting a united front, members of this community can better influence the development of international norms and standards, ensuring that the perspectives and needs of public service are adequately considered. This is crucial for ensuring that the development of AI technologies is aligned with public interest and societal well-being, rather than solely driven by commercial imperatives.

The establishment of a global community of practice would likely involve a multi-pronged approach. It could begin with the creation of a dedicated online platform – a secure portal where members can share documents, participate in forums, and engage in virtual discussions. This platform could host repositories of best practices, case studies, ethical assessment frameworks, and policy templates. Regular virtual and, where feasible, in-person convenings would be essential for deeper engagement, allowing for more dynamic brainstorming, collaborative problem-solving, and the building of stronger personal relationships that underpin effective collaboration. These convenings could take the form of workshops focusing on specific challenges, conferences showcasing innovative applications, or expert roundtables to tackle emerging issues.

To ensure the effectiveness and sustainability of such a community, clear governance structures and operational principles would be necessary. This might involve establishing a steering committee composed of representatives from diverse geographical regions and sectors, defining membership criteria, and outlining protocols for knowledge sharing and intellectual property. A crucial aspect would be ensuring inclusivity, actively reaching out to and supporting the participation of countries and public institutions that may have fewer resources or less established AI programs. Mechanisms for translating key documents and providing language support could be vital in fostering genuine global participation.

The challenges in building and sustaining such a community are not insignificant. They include navigating different legal and cultural contexts, ensuring data privacy and security when sharing sensitive information, and securing ongoing funding and institutional support. However, the potential rewards far outweigh these obstacles. A thriving global community of practice for responsible AI in public services would be a powerful engine for progress, enabling nations to learn from each other, accelerate the adoption of ethical AI, and ultimately build more effective, equitable, and trustworthy public services for citizens around the world. It represents a commitment to a shared future where technological advancement serves the public good, guided by collective wisdom and a unified ethical compass. This collaborative spirit is not merely an idealistic aspiration; it is a pragmatic necessity in navigating the profound societal transformation that AI represents. By pooling our knowledge, our experiences, and our ethical commitments, we can collectively chart a course towards a future where AI in public service truly enhances human well-being and strengthens the foundations of democratic governance.

The very act of forming such a global community underscores a profound shift in how we approach technological innovation within public service. It acknowledges that the challenges posed by AI—from ensuring algorithmic fairness to maintaining public trust and safeguarding democratic values—are not isolated incidents confined to specific jurisdictions but are shared, interconnected concerns. Therefore, the solutions must also be shared and interconnected. This collaborative ethos moves beyond a model of individual nations independently developing their AI strategies and policies, often in silos, to a more integrated and synergistic approach. By pooling resources, expertise, and lessons learned, public institutions can significantly de-risk the adoption of AI, mitigate potential harms more effectively, and accelerate the realization of AI's benefits for citizens.

Consider the practical implications of sharing open-source tools and frameworks. Many countries are developing internal tools for AI risk assessment, bias detection, or data anonymization. Instead of each nation building its own from scratch, a global community could collaborate on developing and maintaining a suite of robust, well-documented, and rigorously tested open-source tools. These tools could then be adapted and deployed by public administrations worldwide, significantly reducing development costs and ensuring a baseline level of quality and ethical consideration. For example, a framework developed by one national government for conducting algorithmic impact assessments, detailing the steps, questions, and ethical considerations involved, could be shared, iterated upon by others, and eventually become a de facto international standard. This not only promotes efficiency but also fosters interoperability and shared understanding of what constitutes responsible AI deployment.

Furthermore, a global community of practice can facilitate joint research and development initiatives. By identifying common research priorities—such as developing more resilient AI systems against adversarial attacks, improving methods for AI explainability in complex public service domains like criminal justice or social welfare, or exploring novel ways to use AI for citizen empowerment in areas with low digital literacy—nations can collaborate on funded projects. This pooling of research capacity can lead to breakthroughs that might be unattainable for individual nations working alone. Such collaborations can also help to decolonize AI development, ensuring that solutions are not solely driven by the perspectives and priorities of a few dominant technological powers, but are inclusive of diverse global needs and contexts.

The establishment of such a community also has profound implications for procurement and standardization. As public sector organizations increasingly seek to procure AI systems, they often face challenges in defining clear requirements for ethical performance, transparency, and accountability. A global community of practice can play a vital role in developing shared procurement guidelines, technical standards, and certification mechanisms for AI systems intended for public service. This would not only streamline the procurement process for governments but also incentivize AI vendors to develop products that meet high ethical and performance standards, thereby driving a more responsible AI market. For instance, a consensus reached within the community on specific benchmarks for AI fairness in public-facing applications could inform the criteria used by governments when evaluating vendor proposals.

The continuous learning aspect of a community of practice is also critical. The field of AI is evolving at an unprecedented pace, with new techniques, applications, and ethical challenges emerging constantly. A dynamic global network can serve as an early warning system, flagging emerging risks and opportunities. Members can share insights from pilot projects, alert each other to unforeseen consequences of deployed systems, and collectively develop adaptive strategies. This creates a learning loop that is essential for navigating the inherent uncertainty and rapid change associated with AI. For instance, if a particular AI application in one country begins to show unintended negative societal impacts, the community can quickly disseminate this information, allowing other nations to preemptively address similar issues before they arise in their own contexts.

Finally, and perhaps most importantly, a global community of practice for responsible AI in public service is an affirmation of a shared commitment to human-centric governance in the age of intelligent machines. It signifies a recognition that technology, while powerful, must ultimately be subservient to human values, democratic principles, and the public good. By coming together, sharing our knowledge, and holding each other accountable, we can ensure that AI is not an instrument of control or exclusion, but a tool that empowers citizens, enhances public services, and strengthens the fabric of our societies for the benefit of all. This collective endeavor is not just about managing technology; it is about shaping a more just, equitable, and responsive future for public service on a global scale.

The preceding discussions have illuminated the complex landscape of responsible AI in public service, from foundational ethical principles and robust governance frameworks to the critical importance of global collaboration. We have explored the nuances of bias mitigation, the imperative of transparency, the challenges of data governance, and the transformative potential of AI when wielded with foresight and integrity. As we stand at this pivotal juncture, where the integration of artificial intelligence into the fabric of public life is no longer a distant possibility but a present reality, it is crucial to transition from understanding to action. This is not a moment for passive observation, but for active engagement. The profound impact of AI on our societies necessitates a dedicated and unwavering commitment from those at the forefront of public service – the engineers, the policy architects, the frontline practitioners, and the leaders who guide these vital institutions.

The responsibility for ensuring that AI serves the public good, with equity, dignity, and fairness at its core, rests squarely on your shoulders. You are the custodians of public trust, the stewards of public resources, and the architects of services that touch the lives of millions. The decisions made today, the systems developed tomorrow, and the ethical considerations prioritized in every step will define the relationship between citizens and their governments for decades to come. This is not merely a technical undertaking; it is a profound civic duty. It demands a deep understanding not only of the algorithms and data structures that power these intelligent systems but also of the human values, democratic principles, and societal aspirations they are intended to uphold.

Therefore, this final section is a direct call to action, an earnest plea for proactive leadership and a sustained commitment to responsible innovation. It is an invitation to embrace the mantle of ethical stewardship and to champion the transformative potential of AI in ways that uplift all citizens. The journey has been one of learning and preparation; now, it is time to move forward with conviction, guided by the principles we have explored.

One of the most critical aspects of this call to action is the unwavering dedication to continuous learning. The field of artificial intelligence is characterized by its relentless pace of innovation. New techniques emerge, algorithms are refined, and entirely novel applications are conceived with astonishing frequency. For public servants tasked with harnessing AI for societal benefit, standing still is not an option. It requires a commitment to lifelong learning, a proactive pursuit of knowledge, and a willingness to adapt. This means actively seeking out training opportunities, engaging with emerging research, attending workshops and conferences, and fostering a culture of curiosity within your teams. It means understanding not just the "how" of AI, but also the "why" and the "what if." Explore the latest developments in explainable AI, delve into emerging cybersecurity threats against AI systems, and stay abreast of evolving ethical guidelines and regulatory frameworks. This continuous engagement ensures that your understanding remains current, allowing you to anticipate challenges and leverage opportunities effectively, rather than being caught reacting to them.

Ethical leadership is equally paramount. It is not enough to simply implement AI; it must be done with a strong moral compass. Ethical leadership in this context means championing transparency and accountability from the outset. It means fostering an environment where concerns about bias, fairness, and privacy can be raised and addressed without fear of reprisal. It involves actively integrating ethical considerations into every stage of the AI lifecycle – from conception and design to deployment and ongoing monitoring. This requires courage: the courage to question assumptions, the courage to challenge potentially harmful applications, and the courage to advocate for the needs of the most vulnerable populations. Ethical leaders set the tone for their organizations, demonstrating through their actions that the pursuit of innovation is inextricably linked to the upholding of public trust and democratic values. This might involve establishing internal ethics review boards, developing clear ethical codes of conduct for AI development and deployment, and actively engaging citizens in discussions about how AI is being used to serve them.

Furthermore, the spirit of collaborative action, so vital for navigating the global complexities of AI, must be embraced at every level within public service. The challenges we face are too significant, and the stakes too high, to operate in silos. Public servants must actively seek out partnerships, both within their own institutions and across governmental boundaries. This means sharing best practices, openly discussing failures and lessons learned, and working together to develop common standards and solutions. The global community of practice for responsible AI in public service, as discussed previously, is a testament to this necessity. However, this spirit of collaboration must also permeate your day-to-day work. Engage with academic institutions, collaborate with civil society organizations, and foster dialogue with the private sector, all while maintaining a steadfast focus on the public interest. Joint research initiatives, shared procurement frameworks, and collaborative development of open-source AI tools can all accelerate progress and ensure that AI development is inclusive and equitable.

Consider the practical steps that can be taken immediately. For engineers and data scientists, this means embedding ethical considerations into the very design of AI systems. It involves rigorously testing for bias, developing robust explainability mechanisms, and prioritizing data privacy from the ground up. It means documenting your work thoroughly, creating clear audit trails, and being prepared to explain the functionality and limitations of your systems. For policymakers and program managers, it means demanding transparency and accountability from AI vendors, establishing clear guidelines for AI procurement and deployment, and ensuring that robust oversight mechanisms are in place. It involves actively seeking input from affected communities and ensuring that AI solutions are designed to enhance, rather than diminish, citizen autonomy and participation.

For leaders, the call to action is to foster an environment where responsible AI can flourish. This means allocating resources for ethical AI development, providing training and professional development opportunities for your staff, and championing a culture that prioritizes ethical considerations alongside technical innovation. It means leading by example, demonstrating a deep commitment to the principles of fairness, accountability, and transparency. It involves creating structures and processes that support responsible AI, such as dedicated ethics committees, robust risk assessment frameworks, and clear channels for public feedback and redress.

The transformative potential of AI in public service is immense. It can help us to deliver services more efficiently, to make more informed decisions, to understand complex societal challenges, and to empower citizens in new and profound ways. However, this potential can only be fully realized if we approach the integration of AI with caution, with intention, and with an unwavering commitment to our ethical obligations. The future of responsible AI in public service is not a predetermined outcome; it is a future that is being built, day by day, by the actions and decisions of public servants like you.

This is your moment to lead. Embrace the responsibility that comes with wielding such powerful technology. Commit to continuous learning, to ethical leadership, and to collaborative action. Let us work together to ensure that artificial intelligence becomes a force for good, serving humanity with equity, dignity, and fairness, and strengthening the foundations of public service for generations to come. The path forward requires vigilance, dedication, and a shared vision of a future where technology and public service are aligned in the pursuit of a better society for all. This is not just about building smarter systems; it is about building a more just, equitable, and responsive public realm. Let the work begin, with purpose and with unwavering integrity.

**AI Lifecycle**: The entire process of developing, deploying, and managing an AI system, from initial concept and data collection through to its eventual decommissioning.

**Algorithmic Bias**: Systematic and repeatable errors in a computer system that create unfair outcomes, such as privileging one arbitrary group of users over others.

**Explainable AI (XAI)**: A set of tools and techniques in artificial intelligence that allow human users to understand and trust the results and output created by machine learning algorithms.

**Fairness (in AI)**: The principle that AI systems should not discriminate against individuals or groups, and should treat all people equitably. This often involves addressing and mitigating algorithmic bias.

**Governance Framework**: A structured system of rules, policies, and procedures designed to guide decision-making and ensure accountability in the development and deployment of AI technologies.

**Public Trust**: The confidence that citizens have in their government and its institutions, particularly in the context of technology adoption and data usage.

**Transparency (in AI)**: The degree to which the inner workings of an AI system are understandable, allowing for scrutiny of its decision-making processes and potential biases.

www.ingramcontent.com/pod-product-compliance
Lightning Source LLC
LaVergne TN
LVHW052017200926
848481LV00005B/1512